Integrated Intelligent Vehicle Dynamics and Control

Integrated Intelligent Vehicle Dynamics and Control

Guest Editors

Wuwei Chen
Hongbo Wang

MDPI

Basel • Beijing • Wuhan • Barcelona • Belgrade • Novi Sad • Cluj • Manchester

Guest Editors
Wuwei Chen
School of Automotive and Transportation Engineering
Hefei University of Technology
Hefei
China

Hongbo Wang
School of Automotive and Transportation Engineering
Hefei University of Technology
Hefei
China

Editorial Office
MDPI AG
Grosspeteranlage 5
4052 Basel, Switzerland

This is a reprint of the Special Issue, published open access by the journal *Actuators* (ISSN 2076-0825), freely accessible at: https://www.mdpi.com/journal/actuators/special_issues/EDX672YRD4.

For citation purposes, cite each article independently as indicated on the article page online and as indicated below:

Lastname, A.A.; Lastname, B.B. Article Title. *Journal Name* **Year**, *Volume Number*, Page Range.

ISBN 978-3-7258-5067-9 (Hbk)
ISBN 978-3-7258-5068-6 (PDF)
https://doi.org/10.3390/books978-3-7258-5068-6

Contents

actuators

MDPI

Article

Influence Analysis of Design Parameters of Elastic Valve Plate and Structural Types of Guide Flow Disc on the Performance of Relief Valve

Junjie Chen [1,*], Peng Huang [1], Xinrong Xie [1], Changyu Guan [2] and Hao Liu [3]

1 Vehicle Engineering, School of Mechanical and Electrical Engineering, Jiangxi University of Science and Technology, Ganzhou 341000, China; 1873996172hp@gmail.com (P.H.); xxr11567@foxmail.com (X.X.)
2 Zhejiang Leap Technology Co., Ltd., Hangzhou 310051, China; guanchangyu163@163.com
3 Hunan Yunjian Group Co., Ltd., Changsha 410100, China; 18475366391@163.com
* Correspondence: cjj852456@163.com; Tel.: +86-188-1926-2324

Abstract: The elastic valve plate and guide flow disc are key components that influence parameters such as opening and pressure difference of pilot-relief valve, which are also the core components enabling continuous damping adjustment in valve-controlled continuously variable dampers. Based on deformation characteristics of elastic valve plate and various structural types of guide flow disc, this paper reveals the impact of structural types of guide flow disc and design parameters of elastic valve plate on the performance of pilot-relief valve and obtains the relationship curves between opening pressure, pressure difference and opening of relief valve versus structural types and the angle, width and the number of arc plates of elastic valve plate. It shows that the pressure difference of the relief valve reaches maximum with min angle, max width, most arc plates and irregular-shaped type, and the opening reaches maximum with max angle, min width, fewest plates and round hole type. By adjusting structural types of guide flow disc and design parameters of elastic valve plate, the pressure difference and opening of the relief valve can be precisely controlled, providing theoretical support for the precise design of pilot-relief valve and the optimization of damping characteristics.

Keywords: pilot-relief valve; elastic valve plate; key design parameters; guide flow disc; structural type

1. Introduction

Pilot-relief valve provides continuous adjustability of damper damping force through flow and pressure. The special poppet structure and appropriate spring stiffness provide excellent steady-state position retention and fast dynamic response. However, it is complicated to analyze the performance of the relief valve under the influence of the elastic valve plate and guide flow disc, especially when it involves the complex poppet structure and the change of the flow field, and it is exceptionally difficult to analytically calculate the liquid force through the momentum equation. Therefore, based on the deformation theory of the elastic valve plate, it is crucial to study the influence of key design parameters of the elastic valve plate on the relief valve opening and investigate the effect of the structural types of the guide flow disc on the pressure difference-opening relationship of the relief valve, to optimize the design of pilot-relief valve.

At present, the computational fluid dynamics method is the mainstream method for solving the dynamic stroke of complex valve bodies. Zhang [1], Song [2], and Beune [3]

established a transient flow model of a pressure valve based on the transient computational fluid dynamics method, investigated the internal flow characteristics of the pressure valve at different stages of the opening process, and revealed the influence of the key parameters of the relevant valve system on the liquid flow. Roberto [4] and others established a three-dimensional pilot-relief valve geometric model and carried out numerical simulation and experimental study on the pressure-flow characteristics of the relief valve with flow compensation. All the above scholars used a three-dimensional model for calculation, which is a complicated calculation process and consumes a long calculation time. The simplified two-dimensional model can save computational resources, which provides a reference for efficiently identifying the operating characteristics of the pilot-relief valve.

Yuan et al. [5,6] established a nonlinear column coordinate control equation by establishing equivalence between concentrated forces and localized distributed forces. Through finite element analysis, they investigated the transient characteristics of relief valve opening dynamics, providing a solution framework for the nonlinear thin plate equation. Zhang [7] proposed a differential quadrature method based on the Fourier expansion and numerically analyzed the nonlinear bending deformation of an axially symmetric thin plate analysis. Hedayati et al. [8] investigated the free vibration of embedded annular sector plates based on three-dimensional elasticity theory. Tornabene [9] analyzed the deformation of an annular thin plate of medium thickness using first-order shear deformation theory and discretized it using a generalized differential quadrature method. Xu [10] et al. used a hybrid method of large deflection and beam deflection to develop a shock absorber elastic valve plate mathematical model of deformation. To address the deformation characteristics of the radial and circumferential boundaries of the valve plate, Bao [11] and others established a nonlinear mathematical model based on the principle of arithmetic and analyzed the effects of the damping holes and the damping valve plate structural parameters on the dynamic characteristics of the relief valve. Yin et al. [12] investigated the influence of dynamic pressure feedback holes on valve transient characteristics through frequency-domain analysis. By optimizing the geometry of the guided valve plate, they significantly improved the stability of a hydraulically piloted safety valve system.

The problem of deformation at the radial and circumferential edges of thin plates under the special case settings of fixed and free ends has not been resolved in the above studies. In this regard, the bending deformation of a circular elastic valve has been analyzed by this group [13] using the variational approximation method, with the addition of torque. At the same time, the effect of the thickness of the elastic valve on their performance has been considered, providing proof of the deformation behavior of the circular elastic valve.

To summarize the above, both domestic and international research on the relief valve has primarily concentrated on the calculation and analysis of three-dimensional flow field models. However, there has been a notable lack of research on the mapping relationship between the design parameters of the elastic valve plate and the opening characteristics of the pilot-relief valve. Furthermore, the influence of the structural types of guide flow disc on the operating characteristics of the relief valve has been less considered, resulting in deviations in predicting the operating characteristics of the pilot relief valve. To address these shortcomings, the six-degree-of-freedom two-dimensional flow field model has been established in this paper. It is demonstrated that changes in the opening of the pilot valve directly affect the operating characteristics of the relief valve. Additionally, based on the deformation theory of elastic plate, the influence of key design parameters of the elastic valve plate on the pressure difference of the pilot-relief valve has been studied. The opening characteristics of the pilot-relief valve have been obtained under different structural types of guide flow disc, as well as the different arc plate angle, arc plate width, and the number of arc plates of elastic valve plate. The pressure difference-opening mathematical model has been developed to provide theoretical support for the optimal design of the pilot-relief valve.

2. Influence Analysis of Design Parameters of Elastic Valve Plate on the Operating Characteristics of the Pilot-Relief Valve

As illustrated in Figure 1a, the structure of the pilot-relief valve consists of a two-stage valve, with the upper part functioning as an electromagnetic pilot valve and the lower part serving as the relief valve. These two components work in unison. The operating characteristics of the pilot-relief valve are characterized by the opening and pressure difference of the relief valve, which illustrates their effects on the regulation of flow and pressure within the hydraulic system.

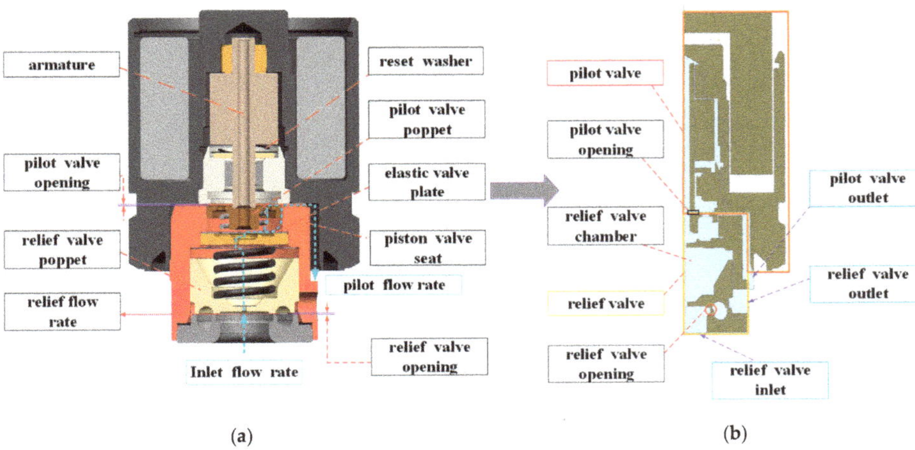

Figure 1. Structure schematic diagram of pilot-relief valve: (**a**) 3D Geometric Modeling; (**b**) 2D Geometric Model.

As shown in Figure 1b, since the pilot-relief valve under study is a symmetrical structure, the three-dimensional model is split into two-thirds of the three-dimensional model, the cross section of the two-thirds of the three-dimensional model is extracted, and finally the cross section is simplified into two-thirds of the two-dimensional model. The fluid domain is represented by the light blue region.

The direction of liquid flow is indicated by the light blue arrow in Figure 1a. When the pilot-relief valve is in the closed state, the pilot valve poppet is in close contact with the piston seat due to the pressure of the return spring, and the relief valve poppet is in the closed state due to the spring force of the series spring system composed of the elastic valve plate and the main spring, and the opening of the pilot valve and the relief valve are both zero. When the relief valve opening reaches a certain value, the solenoid valve is energized, causing the armature to move downward due to electromagnetic force. This movement separates the pilot valve poppet from the piston valve seat, creating the opening of the pilot valve. As the pilot increases, the pressure in the relief valve chamber decreases, which causes the relief valve poppet to move upward. When the pressure in the relief valve chamber falls to a certain level, the solenoid valve is de-energized, and the armature is restored to its initial position by the action of the reset washer. This completes the working stroke of the pilot-relief valve.

2.1. Deformation Model of the Elastic Valve Plate

As shown in Figure 2 [13], the four connecting regions of the elastic valve plate are designated as arc plates n (where $n = 1, 2, 3, 4$). A parallel relationship is established among the four segments of the arc plate, and the deformation of each segment is assumed to be independent of the others. Furthermore, the size, shape, and deformation characteristics of each segment are considered identical. Consequently, the deformation of each segment can be approximated as the deformation of the elastic valve plate.

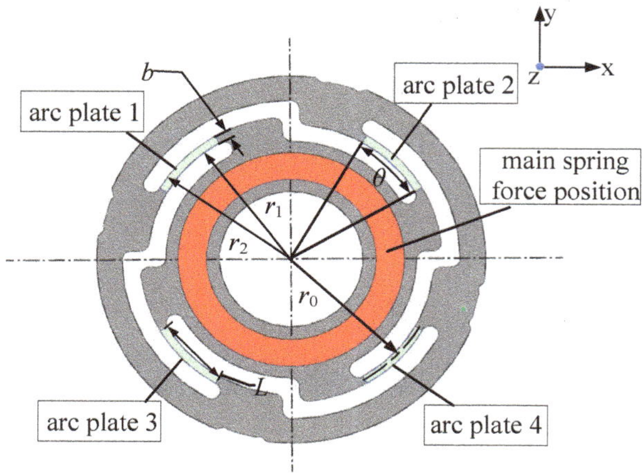

Figure 2. Structural diagram of the elastic valve plate.

Considering the torque generated by the external load at the free end of the arc plate, in order to more fully describe the deformation characteristics of the arc plate, introducing the torque T, and the deformation control differential equation of the arc plate is established as,

$$\frac{T}{r^2}\frac{\partial}{\partial r}\left(r^2\frac{\partial^3 w}{\partial r^3}\right) - \frac{1}{r}\frac{\partial}{\partial r}\left(\frac{1}{r}\frac{\partial w}{\partial r}\right) - K^2\omega^2 w = 0 \tag{1}$$

where, $K = \sqrt{24\rho h/Ebh^3}$, $\omega = \sqrt{c\pi^2/b^2\rho}$, K and ω are both dimensionless quantities, r is the distance from the inner diameter to the outer diameter of the arc plate, and its value range is $r \in [r_1, r_2]$, r_1 is the inner diameter of the arc plate, r_2 is the outer diameter of the arc plate; w is the deformation of the arc plate, ρ is the density of the elastic valve plate, h is the thickness of the valve plate, D is the flexural rigidity, E is the elastic modulus, b is the width of the arc plate, θ is the angle of the arc plate, and c is the shear modulus, Torque T can be represented as,

$$T = FL = 2Fr_0\sin\left(\frac{90\theta}{\pi}\right) \tag{2}$$

In the formula, F is the action force of the circular elastic valve plate, and its action position is shown in the red area of Figure 2, and the direction is perpendicular to the surface of the valve plate, r_0 is the mid-diameter of the arc plate, and L is the chord length at the position of the arc plate's mid-diameter.

At the same time, the boundary conditions are

$$\frac{\partial w}{\partial r}|_{r=0} = 0, w|_{r=0} = 0, T|_{r=0} = 0 \tag{3}$$

Associating Equations (1) and (2), the actual control equations for the deformation amount of the arc plate are obtained as follows,

$$2Fr^3r_0\sin\left(\frac{90\theta}{\pi}\right)\frac{\partial^4 w}{\partial r^4}+4Fr^2r_0\sin\left(\frac{90\theta}{\pi}\right)\frac{\partial^3 w}{\partial r^3}-r\frac{\partial^2 w}{\partial r^2}+\frac{\partial w}{\partial r}-24\frac{\rho\pi^2cr^3}{Eb^3h^2\rho}w=0 \quad (4)$$

2.2. Influence Analysis of Key Design Parameters of the Elastic Valve Plate on the Pressure Difference-Opening Characteristics of the Relief Valve

The key design parameters of the elastic valve plate mainly include arc plate angle (θ), arc plate width (b), and the number of arc plates (n), which affect the size of the stiffness of the elastic valve plate and further lead to the change of the stiffness of the spring system controlling the relief poppet, and ultimately affect the pressure difference, opening, and flow characteristics of the pilot-relief valve. Based on the elastic valve plate's deformation characteristics and structural features, this paper sets θ as $\pi/18$rad–$\pi/6$rad, b as 0.3–0.7 mm, and n as 2, 3, and 4 for the case to be studied.

The stiffness of the elastic valve plate was analyzed based on the deformation model described above. By applying different displacement values to the main spring force contact area, as shown in Figure 2, the elastic force generated by the elastic valve plate with different arc plate angle θ, arc plate width b, and the number of arc plates n is obtained, and then the corresponding stiffness values of the elastic valve plate are obtained, as shown in Table 1.

Table 1. Relationship between key design parameters and the stiffness of the elastic valve plate.

θ/(Rad)	Stiffness of Elastic Valve Plate/(N·mm^{-1})	b/(mm)	Stiffness of Elastic Valve Plate/(N·mm^{-1})	n	Stiffness of Elastic Valve Plate/(N·mm^{-1})
$\pi/18$	8.2	0.2	3.7	2	1.1
$\pi/12$	6.2	0.3	3.8		
$\pi/9$	5.1	0.4	4.1	3	2.2
$5\pi/36$	3.9	0.5	6.9	4	3.7
$\pi/6$	3.7	0.6	7.3		

The opening of the relief valve in the pilot-relief valve is primarily regulated by the movement of the relief poppet, which is also affected by the opening (d) of the pilot valve to a certain extent.

The opening d of the pilot valve range is set to 0–0.25 mm; set A = [d θ b n], and the opening-pressure difference characteristic of the relief valve is analyzed with the example as A = [0.25 $\pi/18$rad 0.3 4].

2.2.1. Pressure Difference Analysis of Relief Valve Based on Key Design Parameters of the Elastic Valve Plate

- Influence analysis of arc plate angle θ on pressure difference.

As shown in Figure 3, in just passing into the oil, due to the oil in the valve port there is a build-up phenomenon, increasing the slope of the curve. The pressure difference increased rapidly, and when the relief valve began to overflow, the valve system pressure difference gradually decreased, and the valve port pressure difference increased as the oil continued to flow. When the relief valve begins to open, the pressure difference with the valve system gradually decreases, while the pressure difference at the valve opening increases as oil continues to flow.

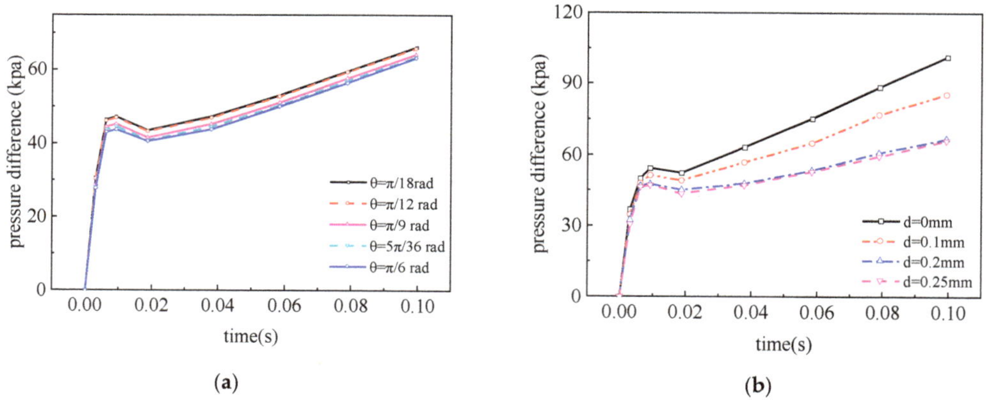

Figure 3. The influence of the arc plate angle on the pressure difference of the relief valve: (**a**) Pressure difference of the relief valve port for different θ at 0.25 mm pilot valve opening; (**b**) Pressure difference of the relief valve port for different pilot valve openings at $\theta = \pi/18$rad.

As seen in Figure 3a,b, when the pilot valve opening d remains constant, increasing the arc plate angle θ reduces the elastic valve plate's stiffness. This stiffness reduction propagates through the series-connected spring system that governs the relief valve poppet movement, ultimately lowering the inlet pressure at the relief valve opening. The diminished inlet pressure attenuates oil accumulation at the valve opening, thereby decreasing the opening resistance. These coupled effects lead to a progressive reduction in the pressure difference across the relief valve opening with increasing θ values.

- Influence analysis of arc plate width b on pressure difference.

As can be seen from Figure 4, when the oil is initially passed, the accumulation of oil at the valve opening causes an increase in the slope of the curve, leading to a rapid rise in the pressure difference. When the relief valve begins to open, the pressure difference gradually decreases, but it continues to increase as oil flows further into the valve.

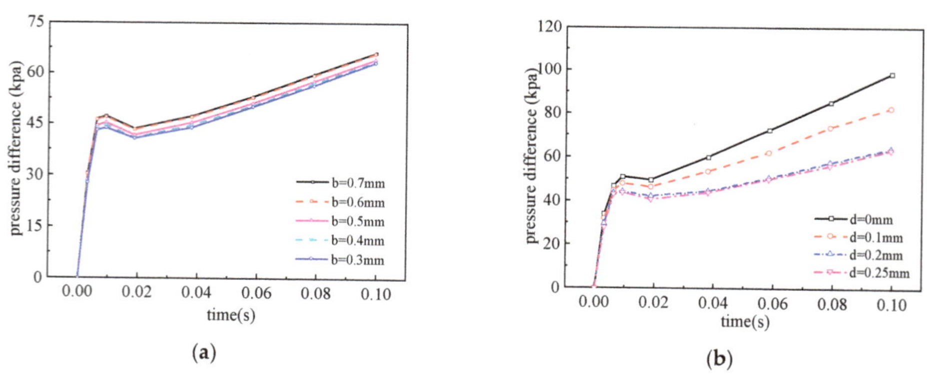

Figure 4. The influence of the arc plate width on the pressure difference of the relief valve: (**a**) Pressure difference of the relief valve port for different b at 0.25 mm pilot valve opening; (**b**) Pressure difference of the relief valve port for different pilot valve openings at b = 0.3 mm.

As seen in Figure 4a,b, when the opening d of the pilot valve is fixed, the stiffness of the elastic valve plate increases as the arc plate width b increases, which leads to an increase in the stiffness of the series spring system controlling the movement of the relief valve poppet. This in turn enhances the accumulation of oil at the valve opening, increasing the opening resistance of the relief valve. Therefore, as b increases, the pressure difference of the relief valve increases. When b is unchanged, the flow area increases with the increase in the opening d of the pilot valve, making it easier for oil to flow. The pilot flow rate increases and the relief flow rate decreases, which reduces the pressure in the relief valve chamber, causing a corresponding decrease in the pressure difference of the valve system.

- Influence analysis of the number of arc plates n on pressure difference.

From Figure 5, when oil is initially passed, the accumulation of oil at the valve opening causes localized high pressure, which increases the slope of the curve and leads to a rapid rise in the pressure difference. As the relief valve begins to open and release pressure, the pressure difference gradually decreases. However, as oil continues to flow in, the pressure difference at the valve opening increases once again.

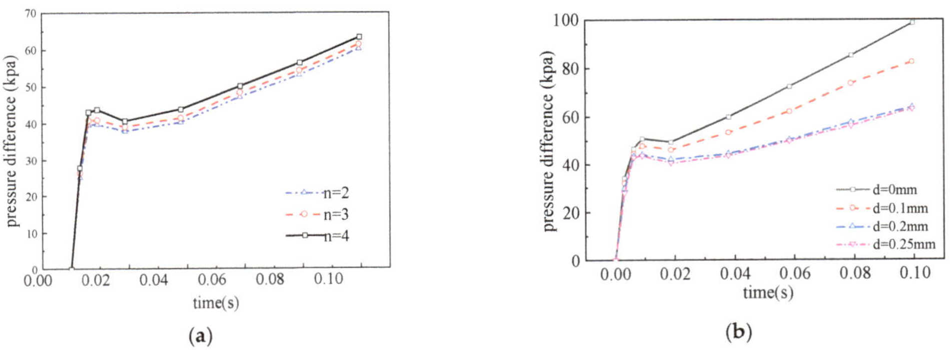

Figure 5. The influence of the number of the arc plate on the pressure difference of the relief valve: (**a**) Pressure difference of the relief valve port for different n at 0.25 mm pilot valve opening; (**b**) Pressure difference of the relief valve port for different pilot valve openings at $n = 4$.

From Figure 5a,b, maintaining a constant pilot valve opening d while increasing the number of arc plates n enhances the elastic valve plate's stiffness. This stiffness augmentation directly strengthens the series spring system controlling the relief valve poppet movement. The heightened system stiffness triggers intensified oil accumulation at the valve port, consequently amplifying the opening resistance. These cumulative effects systematically elevate the pressure difference across the relief valve with increasing n values.

When n is held constant, an increase in the opening d of the pilot valve leads to an increase in the flow area, facilitating the flow of oil. This results in an increase in the pilot flow rate and a decrease in the relief flow rate, which lowers the pressure within the relief valve chamber. Consequently, the pressure difference across the valve system decreases.

2.2.2. Opening Analysis of the Relief Valve Based on Key Design Parameters of the Elastic Valve Plate

- Influence analysis of arc plate angle θ on the opening;

As seen from Figure 6, at time 0 s, the relief valve is closed, with an opening of 0 mm. As oil flows into the system, the relief valve opens once the opening pressure is reached. At this point, the pressure difference across the valve port increases dramatically, causing the slope of the curve to steepen. The rate of valve opening also increases, exhibiting a nonlinear change. As the relief valve briefly relieves pressure, slope of the curve becomes less steep, and the rate of increase in valve opening slows. Since the relief valve is a velocity-type inlet, the oil flow rate increases proportionally with time, causing the curve's slope to stabilize. Subsequently, the rate of increase in valve opening continues, albeit at a slower pace.

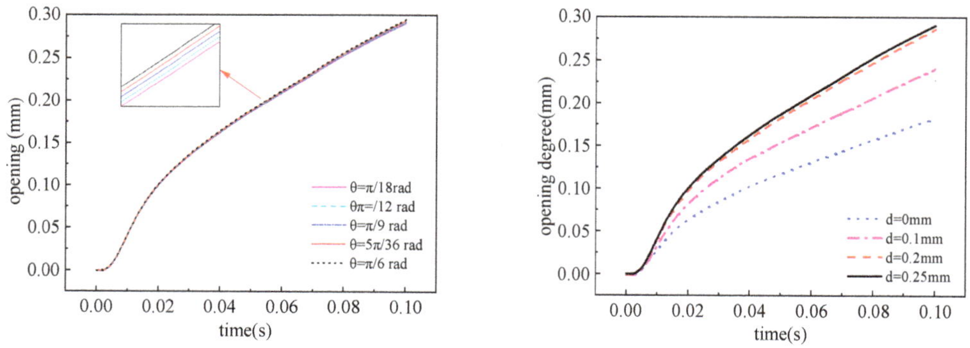

(**a**) (**b**)

Figure 6. The influence of the angle of the arc plate on the opening of the relief valve: (**a**) Opening of the relief valve port for different θ at 0.25 mm pilot valve opening; (**b**) Opening of the relief valve port for different pilot valve openings at $\theta = \pi/18$ rad.

From Figure 6a,b, it can be observed that, for a constant the opening d of the pilot valve, as the stiffness of the elastic valve plate decreases with an increase in the arc plate angle θ of the elastic valve plate, the stiffness of the tandem spring system controlling the movement of the relief valve poppet also decreases. As a result, the pressure at the valve port decreases, and the accumulation of oil at the valve port becomes less pronounced. This leads to a reduction in the valve system's pressure difference, causing the resistance to valve opening to diminish. Consequently, the opening of the relief valve increases as the arc angle θ of the elastic valve plate increases.

When θ is held constant, the increase in the opening d of the pilot valve increases the pilot flow, which lowers the pressure within the relief valve chamber. This reduction in pressure difference across the relief valve orifice weakens the resistance to valve opening, thus increasing the relief valve opening.

- Influence analysis of the arc plate width b on the opening;

As illustrated in Figure 7, the relief valve starts in a closed position. As oil flows in, the valve opens once the pressure reaches the valve-opening threshold. The rate at which the valve opens increases in a nonlinear manner. After a brief period of pressure relief, the slope of the opening curve begins to decrease, leading to a slower rate of increase in the valve opening. Since the relief valve is equipped with a velocity-type inlet, the flow of oil increases proportionally over time. At this stage, the slope of the curve stabilizes and the relief valve opening continues to expand.

As seen in Figure 7a,b, when the opening d of the pilot valve remains constant, the stiffness of the elastic valve plate increases as the arc plate width b of the arc plate increases. This enhancement in stiffness of the series spring system regulates the movement of the relief valve poppet. Consequently, the inlet pressure at the valve port increases, leading to an increase in the system's pressure difference. As a result, the resistance to valve opening increases, causing a decrease in the relief valve opening as b increases.

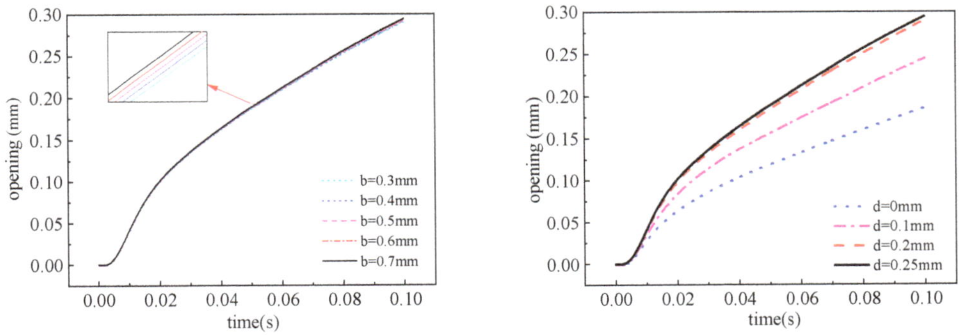

(**a**) (**b**)

Figure 7. The influence of the width of the arc plate on the opening of the relief valve: (**a**) Opening of the relief valve port for different b at 0.25 mm pilot valve opening; (**b**) Opening of the relief valve port for different pilot valve openings at b = 0.3 mm.

When b is held constant, the increase in the opening of the pilot valve results in a corresponding increase in the pilot flow. This lowers the pressure in the relief valve chamber, reducing the pressure difference across the relief valve. The decreased pressure difference weakens the resistance to the valve opening, thereby causing an increase in the relief valve opening.

- Influence analysis of arc plate number n on the opening;

As shown in Figure 8, at time 0 s, the relief valve is closed with an opening of 0 mm. As oil flows in, the relief valve opens once the valve-opening pressure is reached. At this point, oil rapidly enters the valve body, causing a sharp increase in the pressure difference at the valve port. This results in a steeper curve and a faster valve opening rate, exhibiting nonlinear behavior. As the relief valve briefly relieves pressure, the slope of the curve flattens, and the rate of increase in the valve opening slows. Given that the relief valve operates as a velocity-type inlet, the oil flow increases proportionally with time. Consequently, the slope of the curve stabilizes, and the valve opening continues to increase.

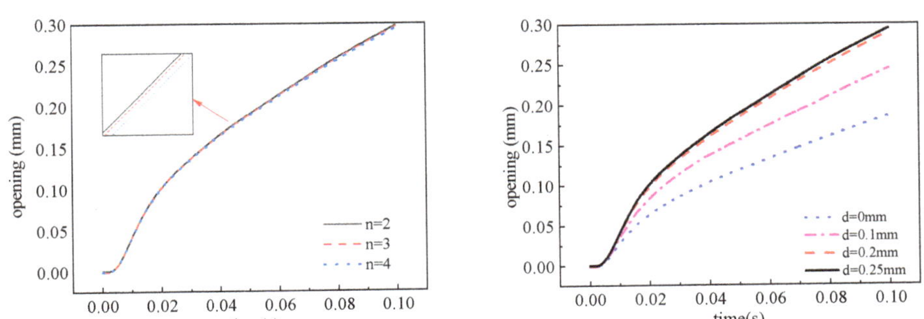

(**a**) (**b**)

Figure 8. The influence of the number of arc plates on the opening of the relief valve: (**a**) Opening of the relief valve port for different n at 0.25 mm pilot valve opening; (**b**) Opening of the relief valve port for different pilot valve openings at n = 4.

Figure 8a,b illustrate that when the opening d of the pilot valve remains constant, the stiffness of the elastic valve plate increases with the increase in the number of arc plates n, increasing the stiffness of the series spring system that controls the relief valve poppet. The oil buildup at the relief valve is strengthened, the relief valve opening resistance increases, and the opening of all the relief valves decreases with the increase in n.

When the value of n is unchanged, the pilot flow increases with the increase of the opening d of the pilot valve, and the pressure in the relief valve cavity decreases, resulting in a reduction of the pressure difference at the relief valve orifice. Consequently, the resistance to opening of the relief valve is weakened, so the relief valve opening increases accordingly.

3. Influence Analysis of the Structural Types of Guide Flow Disc on the Operating Characteristics of the Pilot-Relief Valve

3.1. Structural Types of the Guide Flow Disc

As a critical component of the valve, the guide flow disc determines the pressure difference, valve opening, flow characteristics, and overall system performance of the relief valve, thereby playing a pivotal role in the valve's control effectiveness. The primary function of the guide flow disc is to connect the pilot valve chamber to the relief valve chamber while exerting a throttling effect on the oil flowing through its throttling orifice. As depicted in Figure 9, the structural types of the guide flow disc can be categorized into five types based on the shape of the throttling orifice: irregular-shaped, bend-shaped, runway-shaped, arc-shaped, and round hole-shaped. The opening-pressure difference characteristics of the relief valve are analyzed with the opening d of the pilot valve of 0.25 mm as an example.

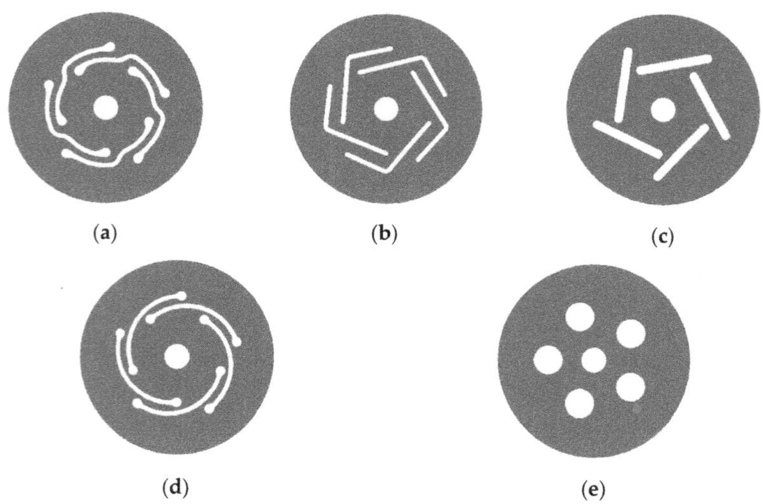

Figure 9. Structural types of the guide flow disc: (**a**) Irregular-shaped throttling orifice; (**b**) Bend-shaped throttling orifice; (**c**) Runway-shaped throttling orifice; (**d**) Arc-shaped throttling orifice; (**e**) Round hole-shaped throttling orifice.

3.2. Influence Analysis of Structural Types of the Guide Flow Disc on the Opening-Pressure Difference Characteristics of the Relief Valve

3.2.1. Inlet Pressure Analysis of Relief Valve Based on Structural Types of the Guide Flow Disc

The graph in Figure 10 illustrates the relationship between the inlet pressure of the relief valve and time. At 0.15 s, the curve reaches its peak, and the point corresponding to this peak is defined as the opening point of the valve, with the associated pressure referred to as the opening pressure.

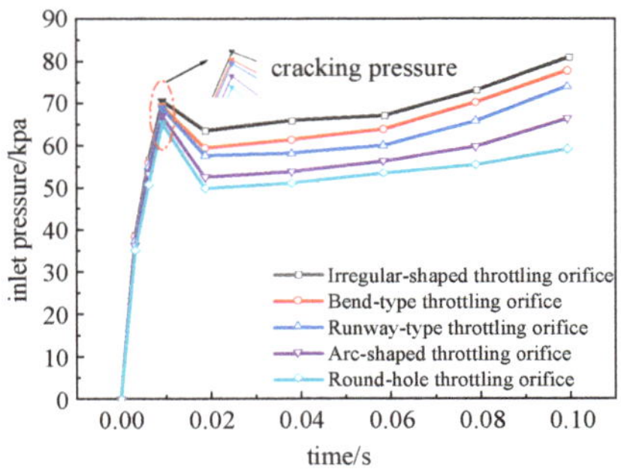

(**a**)

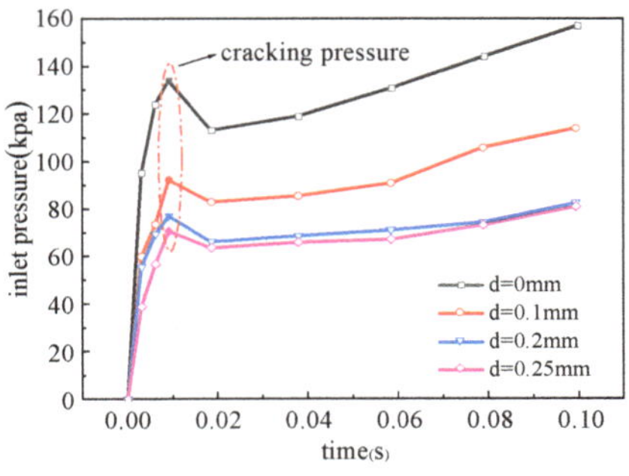

(**b**)

Figure 10. The inlet pressure of the relief valve under different structural types of the guide flow disc: (**a**) Inlet pressure of the relief valve port for different structural types of guide flow disc at 0.25 mm pilot valve opening; (**b**) Inlet pressure of the relief valve port for different pilot valve openings under irregular-shaped guide flow disc.

As shown in Figure 10, before the valve opening point, oil accumulates at the relief valve, generating localized high pressure, which causes a sharp rise in inlet pressure over time. At this stage, the relief valve remains closed, and fluid enters the valve body solely through the cylindrical valve port. Once the pressure curve reaches the valve opening point, the relief valve opens, and the inlet pressure at this point is defined as the opening pressure. Overflow begins, resulting in a subsequent decrease in inlet pressure. Following a brief period of pressure relief, the inlet pressure continues to rise again as the relief valve poppet slowly moves upward, and the continuous flow of fluid into the valve causes a further increase in inlet pressure.

As depicted in Figure 10a,b, when the opening of the pilot valve is maintained at a constant level, the throttling orifice area of the guide flow disc increases in the following order: irregular-shaped, bend-shaped, racetrack-shaped, arc-shaped, and round hole-shaped. As the orifice area increases, the flow rate of oil passing through the guide flow disc also rises, resulting in a reduction in the flow rate of oil entering the relief valve. Consequently, the pressure within the relief valve chamber decreases, which results in a reduction in both the inlet pressure and the opening pressure. After the valve opens, the inlet pressure continues to decrease.

When the type of guide flow disc is unchanged, the increase in the pilot valve opening leads to an increase in the flow area, thereby facilitating fluid flow. This results in an increase in the pilot flow and a decrease in the overflow flow, which in turn reduces the pressure within the relief valve chamber. Consequently, the opening pressure of the relief valve decreases as the opening of the pilot valve increases.

3.2.2. Pressure Difference Analysis of Relief Valve Based on Structural Types of the Guide Flow Disc

As shown in Figure 11, when oil is first passed through, the accumulation of fluid at the valve port causes the slope of the curve to increase, and the pressure difference rises rapidly. When the relief valve begins to relieve, the pressure difference gradually decreases, and as the fluid continues to flow in, the pressure difference increases accordingly.

From Figure 11a,b, it can be seen that, when the opening of the pilot valve remains constant, after the valve opening point, the inlet pressure of the relief valve decreases in the order of irregular-shaped throttling orifice, bent-shaped throttling orifice, runway-shaped throttling orifice, arc-shaped throttling orifice, and round hole-shaped throttling orifice. This hierarchical pressure decline correlates with diminished fluid accumulation at the valve port, leading to a corresponding decrease in the pressure difference generated by different structural types of guide flow disc.

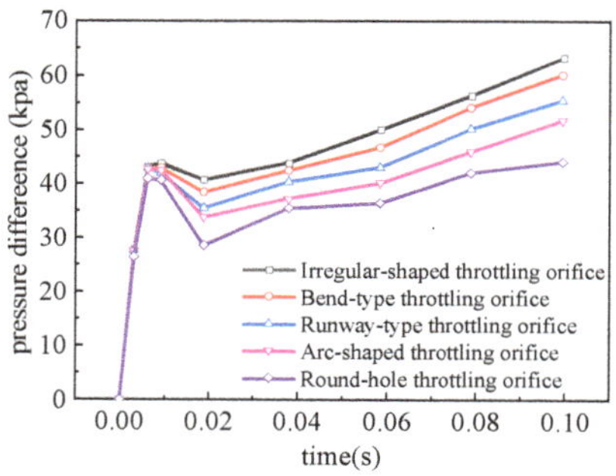

(a)

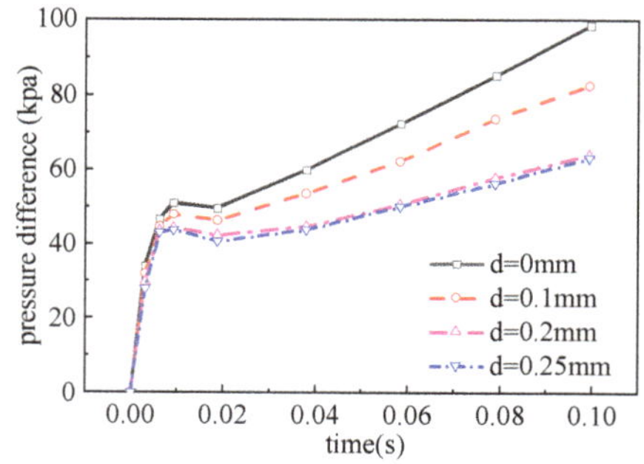

(b)

Figure 11. The pressure difference of the relief valve under different structural types of the guide flow disc: (**a**) Pressure difference of the relief valve port for different structural types of guide flow disc at 0.25 mm pilot valve opening; (**b**) Pressure difference of the relief valve port for different pilot valve openings under irregular-shaped guide flow disc.

Furthermore, when maintaining a fixed guide flow disc structure, the increase in the opening of the pilot valve leads to a larger flow area, making it easier for the fluid to flow. As a result, the pilot flow increases, which reduces the overflow flow, and consequently, the opening pressure of the relief valve decreases, causing a corresponding reduction in the pressure difference at the valve port.

3.2.3. Opening Analysis of Relief Valve Port Based on Structural Types of the Guide Flow Disc

As illustrated in Figure 12, at time 0 s, the relief valve is in a closed position with an opening of 0 mm. As fluid begins to flow, the valve opens once the opening pressure is reached. At this moment, the pressure difference across the relief valve rises sharply, causing the slope of the curve to steepen. The resulting nonlinear opening dynamics demonstrate progressively intensifying displacement rates until transient pressure release initiates slope attenuation. Since the valve port is the velocity-type inlet, the fluid flow increases at a constant rate over time. At this point, the slope of the curve stabilizes, and the opening of the relief valve continues to rise.

From Figure 12a,b with an unchanged opening d of the pilot valve, the relief valve opening increases in the following order: irregular-shaped throttling orifice, bent-shaped throttling orifice, runway-shaped throttling orifice, arc-shaped throttling orifice, and round hole-shaped throttling orifice. This phenomenon arises from the sequential mitigation of inlet fluid stagnation through optimized orifice geometries, leading to a corresponding reduction in pressure difference. As a result, the resistance to the opening of the relief valve decreases, and the opening increases gradually.

When the structure of the guide flow disc is fixed, the flow area increases as the opening of the pilot valve increases, increasing pilot flow and causing a corresponding decrease in overflow flow. This leads to a reduction in the pressure difference of the valve port, which weakens the resistance to the relief valve opening. As a result, the opening of the relief valve increases.

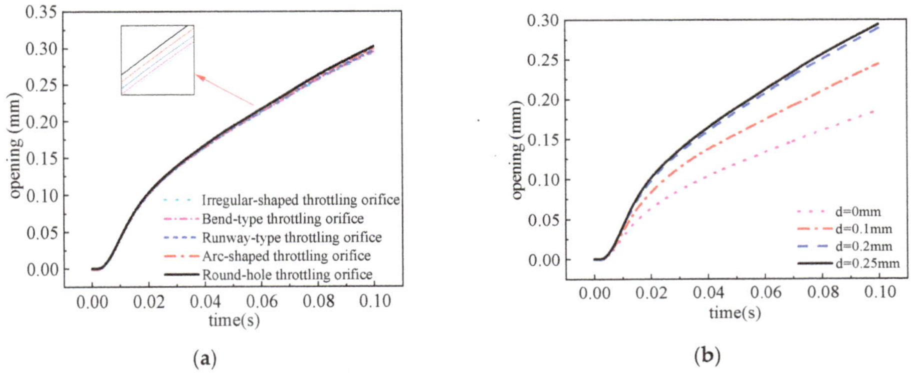

Figure 12. The opening of the relief valve under different structural types of the guide flow disc: (**a**) Opening of the relief valve port for different structural types of guide flow disc at 0.25 mm pilot valve opening; (**b**) Opening of the relief valve port for different pilot valve openings under irregular-shaped guide flow disc.

4. Conclusions

This paper presents a mathematical model for the deformation of the elastic valve plate. It analyzes the effects of structural types of guide flow disc, as well as the key design parameters of the elastic valve plate, on the inlet pressure, pressure difference, and opening characteristics of the relief valve under different openings of the pilot valve. The conclusions drawn from the analysis are as follows:

(1) Studies have shown that the opening variation of the pilot valve directly influences both the opening pressure and the opening characteristics of the relief valve. When the opening of the pilot valve increases, pilot flow rises, relief flow decreases, and pressure of the relief valve chamber drops. Meanwhile, the valve port's opening pressure decreases while the opening of the valve port gradually increases. By precisely adjusting the opening of the pilot valve, the performance of the relief valve can be optimized and accurately controlled;

(2) The mapping relationship curves between the opening-pressure difference of the relief valve and the key design parameters of the elastic valve plate have been established. It shows that the pressure difference of the relief valve reaches maximum with min angle, max width, most arc plates, and the opening reaches maximum with max angle, min width, fewest plates. By adjusting the arc angle, width, and number of arc plates of the elastic valve plate, the relief valve port's opening and pressure difference characteristics can be optimized. These findings provide a foundation for the serialization design of the pilot relief valve and their damping compatibility with dampers;

(3) The study revealed the change rules of the inlet pressure, opening pressure, pressure difference, and opening of the relief valve with the structure of the guide flow disc. When the structure of the guide disc is irregular-shaped type, the inlet pressure and pressure difference of the relief valve reach the maximum; when the structure of the guide disc is round-hole type, the opening of the relief valve reaches maximum. The analysis of pressure difference-opening characteristics of the relief valve provides theoretical support for the precise design of the pilot-relief valve.

Author Contributions: Conceptualization, J.C. and P.H.; methodology, J.C.; software, P.H.; validation, P.H., X.X. and C.G.; formal analysis, P.H.; investigation, H.L.; resources, J.C.; data curation, P.H.; writing—original draft preparation, P.H.; writing—review and editing, X.X.; visualization, P.H.;

supervision, J.C.; project administration, J.C.; funding acquisition, J.C. All authors have read and agreed to the published version of the manuscript.

Funding: This research was funded by Jiangxi Provincial Natural Science Foundation, grant number 20224BAB204039.

Data Availability Statement: The original contributions presented in the study are included in the article; further inquiries can be directed to the corresponding author.

Acknowledgments: The authors express thanks sincerely to Changyu Guan, Zhejiang Leap Technology Co., Ltd., for deformation model of annular elastic valve plate. The authors also thank sincerely the editors and reviewers for their delicate work.

Conflicts of Interest: Author Changyu Guan was employed by the company Zhejiang Leap Technology Co., Ltd., and author Hao Liu was employed by the company Hunan Yunjian Group Co., Ltd. The remaining authors declare that the research was conducted in the absence of any commercial or financial relationships that could be construed as a potential conflict of interest.

References

1. Zhang, X. Transient Flow Characteristics of a Pressure Differential Valve with Different Valve Spool Damping Orifice Structures. *J. Mech. Eng.* **2024**, *73*, 141–158. [CrossRef]
2. Song, X.; Cui, L.; Cao, M. A CFD analysis of the dynamics of a direct-operated safety relief valve mounted on a pressure vessel. *Energy Convers. Manag.* **2014**, *81*, 407–419. [CrossRef]
3. Beune, A.; Kuerten, J.; van Heumen, M. CFD analysis with fluid-structure interaction of opening high-pressure safety valves. *Comput. Fluids* **2012**, *64*, 108–116. [CrossRef]
4. Finesso, R.; Rundo, M. Numerical and experimental investigation on a conical poppet relief valve with flow force compensation. *Int. J. Fluid Power* **2017**, *18*, 111–122. [CrossRef]
5. Yuan, X.; Guo, K. Non-linear dynamic characteristic of a relief valve controlled by a thin annular plate. *J. Ital. Assoc. Theor. Appl. Mech.* **2016**, *51*, 1141–1156. [CrossRef]
6. Yuan, X.; Tian, T.; Zhou, H.; Zhou, J. Comparisons of methods for solving static deflections of a thin annular plate. *Appl. Math. Comput.* **2018**, *127*, 266–279. [CrossRef]
7. Zhang, G.; Wang, J. A non-classical model for circular Kirchhoff disc incorporating microstructure and surface energy effects. *Acta Mech.* **2015**, *226*, 4073–4085. [CrossRef]
8. Hedayati, H.; Aragh, B.S. Influence of graded agglomerated CNTs on vibration of CNT-reinforced annular sectorial plates resting on Pasternak foundation. *Appl. Math. Comput.* **2012**, *218*, 8715–8735. [CrossRef]
9. Tornabene, F.; Viola, E.; Inman, D.J. 2-D differential quadrature solution for vibration analysis of functionally graded conical, cylindrical shell and annular plate structures. *J. Sound Vib.* **2009**, *328*, 259–290. [CrossRef]
10. Xu, J.; Chu, J. Hybrid modeling and verification of disk-stacked shock absorber valve. *Adv. Mech. Eng.* **2018**, *10*, 419–430. [CrossRef]
11. Bao, C.; Xu, S.; Long, X. Research on the dynamic characteristic of a Seawater-Hydraulic-Relief-Valve with a damping orifice. *Proc. Inst. Mech. Eng. Part C J. Mech. Eng. Sci.* **2023**, *237*, 1344–1356.
12. Yin, Y.; Wang, D.; Fu, J. Effect of Dynamic Pressure Feedback Orifice on Stability of Cartridge-Type Hydraulic Pilot-Operated Relief Valve. *J. Mech. Eng.* **2023**, *36*, 245–259. [CrossRef]
13. Chen, J.; Guan, C.; Liu, H.; Yuan, X. Deformation Model of Ring-shaped Elastic Valve Plate and Analysis of the Effect of Thickness on Relief Valve Opening. *Mod. Manuf. Eng.* **2024**, 110–118. [CrossRef]

actuators

MDPI

Article

Observer-Based Sliding Mode Control for Vehicle Way-Point Tracking with Unknown Disturbances and Obstacles

Jiacheng Song, Mingjie Shen and Yanan Zhang *

College of Mechanical and Electronic Engineering, Northwest A&F University, Yangling 712000, China; jiacheng.song@nwafu.edu.cn (J.S.); shenmingjie@nwafu.edu.cn (M.S.)
* Correspondence: ynzhang@nwafu.edu.cn

Abstract: In this paper, an advanced vehicle way-point tracking control method, including kinematic control, dynamic control and an obstacle avoidance strategy, is introduced. In the kinematic part, a vehicle kinematic model is established, along with the coordinate transformation between the vehicle and its target. A way-point tracking control law is developed to optimize the vehicle's movement along predefined way-points. In the dynamic part, a dynamic model considering the actual disturbances and losses is established. An observer compensation technique is utilized to monitor and mitigate disturbances, while sliding mode control, enhanced by a HyperSpiral algorithm, ensures accurate and stable tracking performance. Furthermore, to tackle real-world path planning challenges, an improved way-point tracking obstacle-avoidance algorithm is developed to generate effective way-points for navigating around obstacles. Finally, simulation results validate that the vehicle consistently tracks target way-points in complex scenarios, highlighting the robustness and effectiveness of the proposed method.

Keywords: nonlinear observer; sliding mode control; way-point tracking control; way-point obstacle avoidance

1. Introduction

In recent years, way-point tracking systems have been notably enhanced by significant technological advancements in intelligent control [1]. Substantial progress in multi-robot tracking control has been documented [2–4], while tracking control has also been successfully applied to unmanned aerial vehicles (UAVs) and unmanned surface vehicles (USVs), as shown in [5–7]. The challenge of implementing tracking control for robotic arms has been addressed in [8]. An H-Infinity control for path tracking with fuzzy hyperbolic tangent model was developed in [9]. Way-point tracking for vehicles is a crucial method for enabling patrol inspections in complex ground environments. Compared to operations in open-field environments, vehicles often operate in environments characterized by complexity and numerous obstacles. The practical deployment of vehicle way-point tracking in real-world environments remains complex due to challenges like navigating complex environments and counteracting unknown disturbances, with the presence of obstacles adding further difficulty.

The precision and effectiveness of way-point tracking control are inherently tied to the robustness of the control laws employed. Notable contributions, such as an adaptive PID-like controller for quadrotor unmanned aerial vehicle systems in [10] and the disturbance observer-based control, combined with sliding mode laws for tracking control of remotely operated vehicles (ROVs) discussed in [11], emphasize the critical role of these control

https://doi.org/10.3390/act14020089

strategies in achieving precise way-point tracking. Additionally, distributed tracking control mechanisms, as examined in [12], and the use of backstepping-based six-degrees-of-freedom adaptive control for spacecraft tracking of non-cooperative targets, as presented in [13], highlight the versatility and importance of advanced control laws. The tracking control of land vehicles necessitates high robustness, rapid response, and precise accuracy, which are essential for effective operation in real-world terrestrial environments. However, the aforementioned control laws are primarily applicable to specific environments, such as the seabed or aerial space, where the surroundings are typically open with few obstacles, and the requirements for tracking accuracy are relatively low. The unique challenges of ground vehicle way-point tracking often render conventional methods computationally burdensome, limiting their practical applicability. This underscores the necessity of developing control laws specifically designed to address the intricacies of way-point tracking in ground vehicles.

In the domain of tracking control, it is often assumed that scenarios are ideal, with load variations due to terrain and specific uncertainties and disturbances being ignored. However, the practical application of tracking control necessitates the incorporation of observers to manage the impact of these uncertain disturbances on motion. For instance, the study in [14] employed an adaptive fuzzy sliding mode control method that included an observer for omnidirectional robot control, focusing on the mitigation of uncertainties. Similarly, the control strategy for a towing vehicle in [15] accounted for disturbances through the use of an observer. Further research in [16–18], and Ref. [19] has addressed disturbances in motors by implementing observers to monitor and compensate for these uncertainties. The role of observers in ensuring stability and precision is exemplified by [20], which addressed lateral instability in electric vehicles under extreme conditions through observer-based control, and [21], in which the nonlinear extended-state observer and prescribed performance fault-tolerant control is used in quadrotor unmanned aerial vehicles against compound faults. These examples underscore the critical role of observers in enhancing the robustness and effectiveness of vehicle tracking control systems. However, the estimation of disturbances for most vehicles is typically performed under the assumption that the disturbances are predictable and the computational complexity is low. Disturbance estimation using observers, however, tends to rely heavily on the model and involves high computational complexity, making it challenging to implement in real-world environments. Therefore, it is necessary to develop an observer with low computational complexity that accounts for various disturbances.

Sliding mode control stands out in the field of vehicle control due to its robustness, adaptability, and rapid response capabilities. Its effectiveness is well documented across various applications, such as the acceleration tracking of double-link manipulators in [22] and the tracking of autonomous vehicles' expected velocity and angular velocity in [23], and the tracking of parameters in PMa-SynRM Drive Systems in [24,25]. For ground vehicles, the ability to quickly and stably track desired angular velocities and speeds is of the utmost importance. However, in complex terrain, the vehicle needs to respond swiftly and in real time to changes in the external environment. This necessitates the use of a sliding mode control law capable of fast convergence to ensure the vehicle can track the target speed and angular velocity promptly.

In the practical operation of vehicles, particularly in multi-obstacle and relatively narrow environments, obstacle avoidance is a crucial challenge. This task involves detecting both static obstacles and adjusting the vehicle's path to avoid collisions while maintaining progress toward predefined waypoints or goals. Various studies have tackled the issue of obstacle avoidance in specific domains. For example, reference [26] addresses obstacle-avoidance algorithms for orchard weeders, where the environment is relatively structured and obstacles are known in advance. Similarly, references [27,28] focus on obstacle avoidance for unmanned aerial vehicles (UAVs), which face the additional challenge of three-dimensional space navigation. References [29,30], on the other hand, propose solutions for robotic arms, which typically operate in controlled, semi-static environments. An adaptive network-based fuzzy inference system and particle swarm optimization algorithm was developed in [31]. The obstacle-avoidance algorithm for way-point-based vehicles is crucial for real-world operations in relatively fixed and complex environments. However, most of the existing studies focus on idealized environments, leaving significant challenges in the development of way-point obstacle-avoidance algorithms for complex terrestrial settings.

This study addresses way-point tracking control for vehicles in the presence of disturbances and obstacles. Firstly, a coordinate transformation is carried out to develop the vehicle kinematic model, and a way-point tracking law is employed to ensure high precision in point tracking. Then, a dynamic model considering various losses and uncertain disturbances is developed. To account for real-world uncertainties, an observer monitors disturbances, and sliding mode control is implemented to track the vehicle's desired speed. Finally, an algorithm for obstacle avoidance is developed to generate avoidance way-points and achieve the task of obstacle avoidance.

Considering these aspects, this study aims to explore a way-point tracking obstacle-avoidance control method for vehicles. The main contributions of this research can be summarized as follows:

- Based on the actual operational characteristics of the vehicular environment, a kinematic model of the vehicle is developed. Building upon this model, a dynamic controller for way-point tracking is designed. Structural analysis of the controller demonstrates that the vehicle is capable of accurately tracking designated target points. Furthermore, the stability of the system is analyzed and validated using the Lyapunov stability method.
- A motor model that incorporates various practical losses and uncertainties is developed. Based on this model, a rapid-response sliding mode control law is designed, utilizing the HyperSpiral algorithm as the reaching law. To mitigate chattering effects, a switching control strategy is introduced, and an observer is employed to provide compensation. The stability of the observer is verified using the pole placement method, while the stability of the sliding mode control is proven through the Lyapunov stability method.
- To enable the vehicle to perform obstacle avoidance in real-world environments, a way-point tracking-based obstacle-avoidance algorithm is developed. To address the issue of excessively dense way-points leading to inter-point distances smaller than the error tolerance, the algorithm is further optimized. Finally, the obstacle-avoidance challenge for a way-point tracking control vehicle is analyzed from two perspectives.
- A comprehensive obstacle-avoidance algorithm for vehicle way-point tracking is proposed, featuring fast response and disturbance tolerance. The effectiveness of the algorithm is validated through three simulations, demonstrating the vehicle's capabilities for high-precision target point tracking, disturbance rejection, and robust obstacle avoidance.

The organization of this paper is as follows. Section 2 describes the problem and introduces the control objects. Section 3 establishes the vehicle kinematic model and designs the kinematic controller. Section 4 describes the establishment of the dynamic model and the sliding mode control with observer. Section 5 presents the way-point obstacle-avoidance algorithm. Section 6 presents the system way-point tracking control algorithm. Section 7 presents the simulation experiment and analysis, and Section 8 presents the conclusion.

2. Problem Description and Control Objective

2.1. Way-Point Tracking Control Vehicle Problem Description

The way-point tracking control ensures that for a given way-point on an ideal two-dimensional plane, the vehicle can accurately track the target point. First, a kinematic analysis is performed to plan the vehicle's desired velocity and angular velocity. The kinematic controller uses these velocities to ensure stable tracking through a way-point tracking control law. In the dynamic part, the actual dynamics of the vehicle are considered, where the vehicle's forward and rotational motors must track the target velocities despite uncertainties. Sliding mode control is applied to achieve precise tracking of the target speed and angular velocity. A nonlinear observer is used to estimate and compensate for unmeasured disturbances, ensuring stable performance. For obstacle avoidance, the algorithm relies on pre-modeling the obstacle conditions of the work area, generating obstacle-avoidance way-points. A greedy algorithm is used to optimize the number of way-points, ensuring they are not too dense or fall below the set error tolerance. This allows the vehicle to effectively avoid obstacles while maintaining accurate tracking of the optimized path. To better represent the overall operational process of the system, Figure 1 illustrates the system architecture for tracking multiple target points.

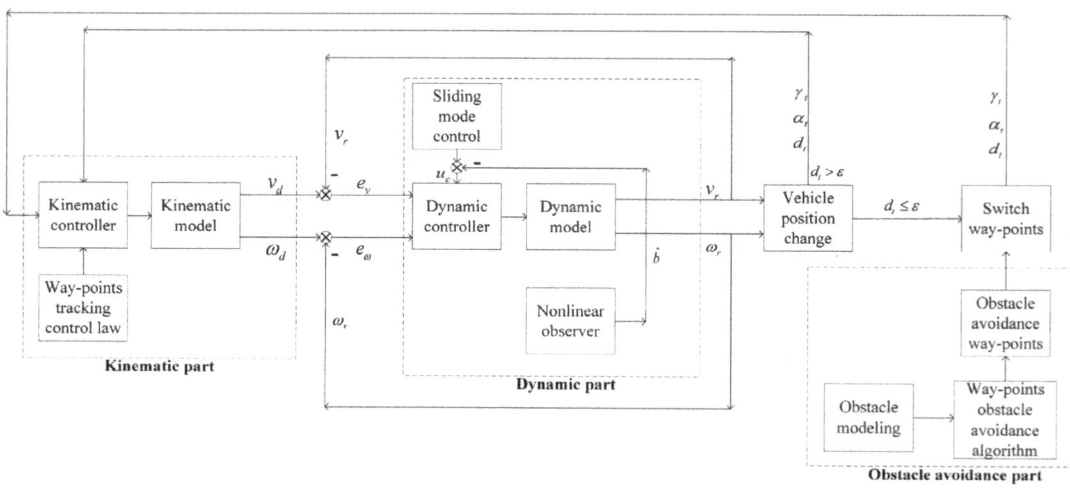

Figure 1. Control architecture for tracking multiple target points.

2.2. Control Objective of Way-Point Tracking Control Vehicle

The control objective for the kinematic model is to ensure that the vehicle moves towards the target point as described by the kinematic model. Additionally, the vehicle must continue moving towards the target point until the desired accuracy is achieved. The desired performance is expressed by the following equations:

$$\begin{cases} \lim\limits_{t\to\infty} \alpha_t \to 0 \\ \lim\limits_{t\to\infty} d_t \to 0 \end{cases} \tag{1}$$

where α_t is the difference angle between the current direction of the vehicle and the direction of the vehicle towards the target point, and d_t is the distance between the vehicle and the target point.

In the dynamic model, the primary control objective is to ensure that the actual velocity tracks the target velocity and the actual angular velocity tracks the target angular velocity. Furthermore, observer compensation is utilized to minimize the impact of uncertainty disturbances. The corresponding performance criteria are given by:

$$\begin{cases} e_v \to 0 \\ e_\omega \to 0 \\ b - \hat{b} \to 0 \end{cases} \tag{2}$$

where $e_v = v_d - v_r$ represents the velocity error, $e_\omega = \omega_d - \omega_r$ denotes the angular velocity error, and b and $\hat{b}$ are the actual and estimated disturbances, respectively.

3. Establishment of Kinematic Model and Kinematic Controller

In this section, the kinematic behavior of the vehicle is analyzed on an ideal two-dimensional plane. A way-point tracking control law is developed to ensure the vehicle accurately tracks the target point. The stability of the proposed control law is verified using the Lyapunov method, demonstrating its ability to guarantee convergence and robustness in tracking performance.

3.1. Vehicle Kinematics Analysis

In the way-point tracking control of the vehicle, the first thing is to establish a suitable kinematic model. In the establishment of the kinematic model, in order to highlight the problem, the slope and obstacles existing in the plane are ignored, and the establishment of the entire kinematic model is considered in the ideal two-dimensional plane, so that the distribution of coordinate points can be evenly distributed throughout the coordinate plane. According to the above requirements, the vehicle kinematics analysis diagram as shown in Figure 2 is established.

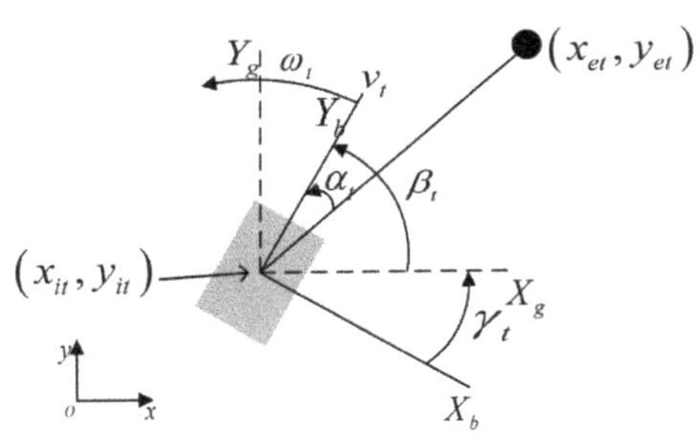

Figure 2. Kinematic model of vehicle in two-dimensional plane.

According to the kinematic analysis in Figure 2, the following formula can be obtained.

Considering the coordinate movement of the vehicle in a two-dimensional plane coordinate system, there is the following formula:

$$\begin{aligned} \dot{x}_{it} &= v_t \cos \beta_t \\ \dot{y}_{it} &= v_t \sin \beta_t \end{aligned} \tag{3}$$

where (x_{it}, y_{it}) represent the initial coordinates of the vehicle, and (x_{et}, y_{et}) denote the target point coordinates. The symbol β_t corresponds to the angle between the y-axis of the body coordinate system and the x-axis of the fixed coordinate system, and v_t corresponds to the forward speed of the vehicle.

As can be seen from Figure 2, the angle and distance between the vehicle and the target point are as follows:

$$d_t = \sqrt{(x_{it} - x_{et})^2 + (y_{it} - y_{et})^2}$$
$$\alpha_t = \begin{cases} -\beta_t + \arctan\frac{(y_{it} - y_{et})}{(x_{it} - x_{et})} & x_{it} \neq x_{et} \\ -\beta_t + \frac{\pi}{2} & x_{it} = x_{et}, y_{it} = y_{et} \end{cases} \quad (4)$$
$$\gamma_t = \frac{\pi}{2} - \beta_t$$

where the symbol d_t corresponds to the distance from the initial point to the target point, α_t corresponds to the angle between the y-axis of the body coordinate system and the line connecting the initial and target points, and γ_t corresponds to the angle between the x-axis of the fixed axis and the x-axis of the body coordinate axis.

Taking the updating of the angle and distance of the vehicle moving to the target point into account, the following can be obtained:

$$\begin{aligned} \frac{d\gamma_t}{t} &= -\omega_t \\ \dot{d}_t &= -v_t \cos \alpha_t \\ \dot{\alpha}_t &= k_{\text{sgn}t} \frac{v_t \sin \alpha_t}{d_t} - \omega_t \end{aligned} \quad (5)$$

For easy representation, the sign function is defined as follows:

$$k_{\text{sgn}t} = \text{sign}(\frac{\pi}{2} - |\alpha_t|) \quad (6)$$

where the symbol ω_{dt} corresponds to the angular velocity; all angles are measured counterclockwise. To facilitate expression, the symbolic function is represented as $k_{\text{sgn}t}$.

3.2. Establishment of Kinematic Controller

In order to enact precise way-point tracking control within the designated coordinate area, it is essential to devise a suitable kinematic controller that guarantees the vehicle's navigation to the specified coordinate points.

The equations governing the kinematic controller are delineated as follows:

$$\begin{cases} v_{dt} = k_\alpha d_t \\ \omega_{dt} = k_\beta \alpha_t + k_{\text{sgn}t} k_\gamma \gamma_t \end{cases} \quad (7)$$

where k_α, k_β and k_γ are the control gains.

Combined with Formulas (5) and (7), the following formula can be obtained:

$$\begin{cases} \dot{d}_t = -v_{dt} \cos \alpha_t = -k_\alpha d_t \cos \alpha_t \\ \dot{\alpha}_t = -k_\beta \alpha_t - k_{\text{sgn}t} k_\gamma \gamma_t + k_\alpha \sin \alpha_t \\ \dot{\gamma}_t = -k_\beta \alpha_t - k_{\text{sgn}t} k_\gamma \gamma_t \end{cases} \quad (8)$$

Theorem 1. *For a system defined by Equations (3)–(6), under the action of the control law (7), the system is asymptotically stable in the Lyapunov sense. If* $k_\gamma k_{\text{sgnt}} < 0, k_\alpha > 0, k_\beta > k_\alpha - k_\gamma k_{\text{sgnt}}$ *holds. Specifically, as* $t \to \infty$ *, the system states satisfy* $\lim_{t \to \infty} \alpha_t = 0$.

Proof of Theorem 1. Using the Lyapunov method to prove the stability of Formula (7), the following is the defined Lyapunov function:

$$V_k = \left[(k_\gamma k_{\text{sgn}t})^2(-\alpha_t + \gamma_t)^2 + 2k_\alpha k_\gamma k_{\text{sgn}t}(\cos \alpha_t - 1)\right] \quad (9)$$

Within the formula, the following restrictions are given:

$$\begin{cases} k_\gamma k_{\text{sgn}t} < 0 \\ k_\alpha > 0 \end{cases} \tag{10}$$

The following conclusions can be drawn:

$$V_k > 0, \forall \alpha_t, \gamma_t \neq 0 \tag{11}$$

From Formula (9), the following formula is derived:

$$\begin{aligned}
\dot{V}_k &= k_\gamma k_{\text{sgn}t}\left[2k_\gamma k_{\text{sgn}t}(-\alpha_t+\gamma_t)(-\dot{\alpha}_t+\dot{\gamma}_t) - 2k_\alpha \sin\alpha_t \dot{\alpha}_t\right] \\
&= k_\gamma k_{\text{sgn}t}\left[\begin{array}{l} 2k_\gamma k_{\text{sgn}t}(-\alpha_t+\gamma_t)\left[-(-k_\beta\alpha_t - k_{\text{sgn}t}k_\gamma\gamma_t + k_\alpha \sin\alpha_t) + (-k_\beta\alpha_t - k_{\text{sgn}t}k_\gamma\gamma_t)\right] \\ -2k_\alpha \sin\alpha_t(-k_\beta\alpha_t - k_{\text{sgn}t}k_\gamma\gamma_t + k_\alpha\sin\alpha_t) \end{array}\right] \\
&= k_\gamma k_{\text{sgn}t}\left[2k_\gamma k_{\text{sgn}t}(-\alpha_t+\gamma_t)(-k_\alpha\sin\alpha_t) - 2k_\alpha\sin\alpha_t(-k_\beta\alpha_t - k_{\text{sgn}t}k_\gamma\gamma_t + k_\alpha\sin\alpha_t)\right] \\
&= k_\gamma k_{\text{sgn}t}\left[\begin{array}{l} (2k_\alpha k_\gamma k_{\text{sgn}t}\alpha_t\sin\alpha_t - 2k_\alpha k_\gamma k_{\text{sgn}t}\gamma_t\sin\alpha_t) - 2k_\alpha\sin\alpha_t(-k_\beta\alpha_t - k_{\text{sgn}t}k_\gamma\gamma_t \\ +k_\alpha\sin\alpha_t) \end{array}\right] \\
&= k_\gamma k_{\text{sgn}t}\left[\begin{array}{l} (2k_\alpha k_\gamma k_{\text{sgn}t}\alpha_t\sin\alpha_t - 2k_\alpha k_\gamma k_{\text{sgn}t}\gamma_t\sin\alpha_t) + (2k_\alpha k_\beta\alpha_t\sin\alpha_t + 2k_\alpha k_{\text{sgn}t}k_\gamma\gamma_t\sin\alpha_t \\ -2k_\alpha^2\sin^2\alpha_t) \end{array}\right] \\
&= k_\gamma k_{\text{sgn}t}\left[2k_\gamma k_\alpha k_{\text{sgn}t}\sin\alpha_t\alpha_t + 2k_\alpha k_\beta\sin\alpha_t\alpha_t - 2k_\alpha^2\sin^2\alpha_t\right] \\
&= 2k_\gamma k_\alpha k_{\text{sgn}t}\sin\alpha_t\alpha_t\left(k_\beta + k_{\text{sgn}t}k_\gamma - k_\alpha\frac{\sin\alpha_t}{\alpha_t}\right)
\end{aligned} \tag{12}$$

For Formula (12), $k_\beta + k_\gamma k_{\text{sgn}t} - k_\alpha\frac{\sin\alpha_t}{\alpha_t} > 0$ is required; the restrictions in Formula (10) are given as $k_\gamma k_{\text{sgn}t} < 0$ and $k_\alpha > 0$. For $\alpha_t \in \left(-\pi \quad \pi\right)$, $k_\alpha\frac{\sin\alpha_t}{\alpha_t} > 0$, $\frac{\sin\alpha_t}{\alpha_t} < 1 - \frac{\alpha_t^2}{3} < 1$ and $\sin\alpha_t\alpha_t \geq 0$. Thus, if $k_\beta > k_\alpha - k_\gamma k_{\text{sgn}t}$, the following conclusions can be drawn:

$$\dot{V}_k \leq 0 \tag{13}$$

In Formula (13), $\dot{V}_k$ is negative semi-definite. Taking the derivative of (13):

$$\begin{aligned}
\ddot{V}_k &= (\dot{\alpha}_t\cos\alpha_t\alpha_t + \sin\alpha_t\dot{\alpha}_t)(2k_\gamma k_\alpha k_\beta k_{\text{sgn}t} + 2k_\gamma^2 k_\alpha k_\beta k_{\text{sgn}t}^2) - 4k_\gamma k_\beta k_\alpha^2 k_{\text{sgn}t}\sin\alpha_t\dot{\alpha}_t \\
&= (-k_\beta\alpha_t - k_{\text{sgn}t}k_\gamma\gamma_t + k_\alpha\sin\alpha_t)\left[(\cos\alpha_t\alpha_t + \sin\alpha_t)(2k_\gamma k_\alpha k_\beta k_{\text{sgn}t} + 2k_\gamma^2 k_\alpha k_\beta k_{\text{sgn}t}^2)\right. \\
&\quad \left.-4k_\gamma k_\beta k_\alpha^2 k_{\text{sgn}t}\sin\alpha_t\right]
\end{aligned} \tag{14}$$

as α_t, γ_t are boundedness, $\ddot{V}_k$ is boundedness, and $\dot{V}_k$ is uniformly continuous. According to the Barbalat lemma [32], $\lim_{t\to\infty}\dot{V}_k = 0$. Since $k_\beta + k_\gamma k_{\text{sgn}t} - k_\alpha\frac{\sin\alpha_t}{\alpha_t} > 0$ and $2k_\gamma k_\alpha k_{\text{sgn}t} < 0$, $\lim_{t\to\infty}\alpha_t = 0$. Therefore, the effectiveness of $\lim_{t\to\infty}\alpha_t \to 0$ is proven.

Then, the local stability of Formula (8) around $\alpha_t = 0$ is analyzed as follows:

$$\begin{bmatrix}\dot{d}_t \\ \dot{\alpha}_t \\ \dot{\gamma}_t\end{bmatrix} = \begin{bmatrix} -k_\alpha & 0 & 0 \\ 0 & k_\alpha - k_\beta & -k_{\text{sgn}t}k_\gamma \\ 0 & -k_\beta & -k_{\text{sgn}t}k_\gamma \end{bmatrix}\begin{bmatrix} d_t \\ \alpha_t \\ \gamma_t\end{bmatrix} \tag{15}$$

The eigenvalues of (15) can be presented as follows:

$$p = (\lambda + k_\alpha)(\lambda^2 + (k_{\text{sgn}t}k_\gamma + k_\alpha - k_\beta)\lambda - k_\alpha k_\gamma k_{\text{sgn}t}) \tag{16}$$

where λ is the eigenvalue of matrix (15). Then, the following formula can be obtained:

$$\begin{cases} \lambda_1 = -k_\alpha < 0 \\ \lambda_2 + \lambda_3 = -(k_{\text{sgn}t} k_\gamma + k_\alpha - k_\beta)/2 < 0 \\ \lambda_2 \lambda_3 = -k_\alpha k_\gamma k_{\text{sgn}t} > 0 \end{cases} \tag{17}$$

The above formula leads to $\lambda_1 < 0, \lambda_2 < 0, \lambda_3 < 0$, which indicates the local stability of the system. □

Theorem 2. *For the system model defined by Equations (3)–(6), under the application of the control law (7), if* $\lim_{t\to\infty} \alpha_t = 0$ *holds, then* $\lim_{t\to\infty} d_t = 0$ *is asymptotically stable in the Lyapunov sense.*

Proof of Theorem 2. At this point, the stability of the system's angle control is verified. Furthermore, considering the distance control objective, the following Lyapunov function is constructed:

$$V_d = \frac{1}{2} d_t^2 \tag{18}$$

From Equation (18), it is clear that $V_d > 0, \forall d_t \neq 0, V_d(0) = 0$. Taking its derivative of time, the following formula can be obtained:

$$\dot{V}_d = d_t \dot{d}_t = -k_\alpha d_t^2 \cos \alpha_t \tag{19}$$

If the initial angle α_t lies within $\left[\frac{\pi}{2} \quad \pi\right)$ or $\left(-\pi \quad -\frac{\pi}{2}\right]$, it will first be adjusted to fall within the range $\left[-\frac{\pi}{2}, \frac{\pi}{2}\right]$. Subsequently, with a positive velocity directed towards the target point, the vehicle will move towards the target. Considering the vehicle's turning process as a transitional phase, when α_t is within $\left[-\frac{\pi}{2}, \frac{\pi}{2}\right]$, the vehicle will align its orientation to effectively approach the target. The following conclusions can be drawn:

$$\dot{V}_d \leq 0 \tag{20}$$

Since $\dot{V}_d$ is negative semi-definite, taking the derivative of (19) we obtain:

$$\begin{aligned} \ddot{V}_d &= -2k_\alpha \dot{d}_t \cos \alpha_t + k_\alpha d_t^2 \dot{\alpha}_t \sin \alpha_t \\ &= 2k_\alpha^2 d_t \cos^2 \alpha_t + k_\alpha d_t^2 \sin \alpha_t (-k_\beta \alpha_t - k_{\text{sgn}t} k_\gamma \gamma_t + k_\alpha \sin \alpha_t) \end{aligned} \tag{21}$$

Since d_t, α_t, γ_t are bounded, $\ddot{V}_d$ is also bounded. Therefore, $\dot{V}_d$ is uniformly continuous. For $\lim_{t\to\infty} \alpha_t = 0$, according to the Barbalat lemma [32], $\lim_{t\to\infty} \dot{V}_d = 0$, $\lim_{t\to\infty} d_t = 0$. Therefore, the effectiveness of $\lim_{t\to\infty} d_t \to 0$ is proven. □

4. Establishment of Dynamics Model and Dynamics Controller

In this part, the dynamic model is first established. A permanent magnet synchronous motor is used as the dynamic model, and the final dynamic model is given, considering the disturbance, friction, torque loss, etc., existing in practice. Secondly, the dynamic model is rewritten to establish an observer to estimate the uncertain disturbance. Then, the final compensation control law is given by using the sliding mode control method.

4.1. Establishment of Dynamic Model

In the pursuit of effective way-point tracking control for the vehicle within the complex environment, it is crucial to establish a suitable dynamic model as a foundational step.

The dynamic model is simplified by using the equation of a three-phase permanent magnet synchronous motor in per-unit value. First, the time base values and torque base values are defined as follows:

$$\begin{aligned} T_\partial &= \frac{P_\partial}{\omega_\partial / p} \\ t_\partial &= \frac{1}{\omega_\partial} \end{aligned} \tag{22}$$

where P_∂ is the base value of the power, and ω_∂ is the base value of the angular frequency.

For the motor rotor equation, there is the following formula:

$$\begin{aligned} J\frac{d^2\beta_t}{dt^2} &= T_e - T_m \\ \beta_t &= \frac{\gamma_t}{p} \end{aligned} \tag{23}$$

The following formula can be obtained by Formula (15).

$$\frac{J}{P}\frac{d^2\gamma_t'}{dt} = T_e - T_m \tag{24}$$

By dividing the torque base value and combining the basic equation $\omega_\partial = 2\pi f$, $f = \frac{pn}{60}$, $P_\partial = S_N \times 10^3$, $J = mr^2 = \frac{1}{4}GD^2 \times 10^3$, the following formula can be obtained:

$$\frac{\frac{1}{4}GD^2}{p^2 S_N}\frac{\omega_\partial}{t_\partial^2}\frac{d^2\gamma_t'}{d\left(\frac{t}{t_\partial}\right)^2} = \frac{T_e}{T_\partial} - \frac{T_m}{T_\partial} \tag{25}$$

The following formula is obtained:

$$\begin{aligned} H\frac{d\omega_\sigma}{dt_\sigma} &= T_{e\sigma} - T_{m\sigma} \\ H = \frac{J\omega_\partial^2}{P} &= 2\pi^3 f\frac{GD^2}{S_N}\left(\frac{n}{60}\right)^2 \end{aligned} \tag{26}$$

The provided formula represents an ideal motor model, where T_m denotes the mechanical torque, T_e represents the electromagnetic torque of the motor, J corresponds to the moment of inertia, β_t signifies the mechanical angle of the rotor x' axis preceding the stator coordinate axis, and γ' represents the electrical angle of the rotor x' axis preceding the stator coordinate axis. Furthermore, p indicates the number of poles, and H is the inertia constant of the rotor at per-unit value. P is the standard value of moment of inertia. In expression (24), the constant GD^2 is the flywheel torque, f is grid frequency, n is synchronous speed, and S_N is the apparent power.

The preceding formula represents the ideal model for a permanent magnet synchronous motor. In complex scenarios, the model can be refined as follows:

$$\begin{aligned} 2\pi^3 p_v P_v \frac{G_v D_v^2}{\omega_{\partial v}^2 S_{Nv} r_c}\left(\frac{n_v}{60}\right)^3 \frac{dv_m}{dt} &= T_{ev} - T_{mv} - B_m v_m - \tau_{1v} \\ 2\pi^3 p_\omega P_\omega \frac{G_\omega D_\omega^2}{\omega_{\partial\omega}^2 S_{N\omega}}\left(\frac{n_\omega}{60}\right)^3 \frac{d\omega_m}{dt} &= T_{e\omega} - T_{m\omega} - B_m \omega_m - \tau_{1\omega} \end{aligned} \tag{27}$$

where ω_m denotes the angular velocity of the motor, B_m represents the damping coefficient, and τ_1 accounts for uncertain factors. In addition, in order to simplify the problem, only the angular velocity after the motor transfer of the trolley is considered, and the mechanical equation of the velocity and angular velocity of the vehicle can be obtained by the formula of the change of angular velocity and velocity.

In practical scenarios, there exists a torque loss in the angular speed coupling between the motor and the vehicle, denoted by $\omega_r = \xi_\omega \eta_\omega \omega_m$. Similarly, the torque loss of the forward motor can be obtained by combining the conversion formula of speed and angular speed. Subsequently, the following formula can be derived:

$$
\begin{aligned}
2\pi^3 p_v P_v \frac{G_v D_v^2}{\omega_{\partial v}^2 S_{Nv} r_c} \left(\frac{n_v}{60}\right)^3 \frac{dv_r}{dt} &= \xi_v \eta_v (T_{ev} - T_{mv}) - \tau_{2v} \\
2\pi^3 p_\omega P_\omega \frac{G_\omega D_\omega^2}{\omega_{\partial \omega}^2 S_{N\omega}} \left(\frac{n_\omega}{60}\right)^3 \frac{d\omega_r}{dt} &= \xi_\omega \eta_\omega (T_{e\omega} - T_{m\omega}) - \tau_{2\omega}
\end{aligned}
\tag{28}
$$

where ξ represents the gear ratio, η accounts for torque loss, r_c is the radius of the vehicle wheel, and $\tau_{2\omega} = B_\omega \omega_r + f(\omega_r, i_q)$ introduces the uncertainty factor.

A three-phase permanent magnet synchronous motor has the following formula:

$$
\begin{aligned}
u_d &= R_s i_d + L_d \frac{di_d}{dt} - \omega_e L_q i_q \\
u_q &= R_s i_q + L_q \frac{di_q}{dt} + \omega_e L_d i_d + \omega_e \psi_f \\
T_e &= \frac{3}{2} p_n \left[\psi_f i_q + (L_d - L_q) i_d i_q\right]
\end{aligned}
\tag{29}
$$

where R_s represents the equivalent resistance, and L_d and L_q denote the inductance of the d-axis and q-axis, respectively. ψ_f stands for the magnetic flux of the permanent magnet, ω_e is the electric angular speed of the motor, u_d and u_q are the voltages of the d-axis and q-axis, and i_d and i_q represent the currents of the d-axis and q-axis, respectively.

For the sake of discussion, if the q-axis inductance of the motor is equal to the d-axis inductance, the following formula applies:

$$
T_e = \frac{3}{2} p \psi_f i_q \tag{30}
$$

In this case, Formula (28) can be represented as:

$$
\begin{aligned}
2\pi^3 p_v P_v \frac{G_v D_v^2}{\omega_{\partial v}^2 S_{Nv} r_c} \left(\frac{n_v}{60}\right)^3 \frac{dv_r}{dt} &= \frac{3}{2} \xi_v \eta_v p_v \psi_{fv} i_{qv} - \xi_v \eta_v T_{mv} - \tau_{2v} \\
2\pi^3 p_\omega P_\omega \frac{G_\omega D_\omega^2}{\omega_{\partial \omega}^2 S_{N\omega}} \left(\frac{n_\omega}{60}\right)^3 \frac{d\omega_r}{dt} &= \frac{3}{2} \xi_\omega \eta_\omega p_\omega \psi_{f\omega} i_{q\omega} - \xi_\omega \eta_\omega T_{m\omega} - \tau_{2\omega}
\end{aligned}
\tag{31}
$$

The resulting formula is as follows:

$$
\begin{aligned}
\dot{v}_r &= \frac{3\xi_v \eta_v \psi_{fv} \omega_{\partial v}^2 S_{Nv} r_c}{4\pi^3 P_v G_v D_v^2 \left(\frac{n_v}{60}\right)^3} i_{qv} - \frac{\omega_{\partial v}^2 S_{Nv} r_c}{2\pi^3 P_v p_v G_v D_v^2 \left(\frac{n_v}{60}\right)^3} \left[B_v v_r + \xi_v \eta_v T_{mv} + f(v_r, i_{qv})\right] \\
\dot{\omega}_r &= \frac{3\xi_\omega \eta_\omega \psi_{f\omega} \omega_{\partial \omega}^2 S_{N\omega}}{4\pi^3 P_\omega G_\omega D_\omega^2 \left(\frac{n_\omega}{60}\right)^3} i_{qv} - \frac{\omega_{\partial \omega}^2 S_{N\omega}}{2\pi^3 P_\omega p_\omega G_\omega D_\omega^2 \left(\frac{n_\omega}{60}\right)^3} \left[B_\omega \omega_r + \xi_\omega \eta_\omega T_{m\omega} + f(\omega_r, i_{q\omega})\right]
\end{aligned}
\tag{32}
$$

At this point, the basic motor model is established.

4.2. Establishment of Nonlinear Observation for Way-Point Tracking Control Vehicle

In this part, nonlinear observation is to estimate the uncertain disturbance existing in the running process of the vehicle. Through the difference between the estimated value and the actual value, the uncertain disturbance of the vehicle can be compensated for and the stability of the system can be strengthened.

To establish the observer, rewrite Formula (32) as:

$$
\begin{aligned}
\dot{\omega}_r &= a_\omega i_{q\omega} + b_\omega \\
\dot{v}_r &= a_v i_{qv} + b_v
\end{aligned}
\tag{33}
$$

In Formula (33) $a_\omega = \frac{3\xi_\omega \eta_\omega \psi_{f\omega}}{4\pi^3 P_\omega \frac{G_\omega D_\omega^2}{\omega_{\partial\omega}^2 S_{N\omega}} \left(\frac{n_\omega}{60}\right)^3}$, $b_\omega = -\frac{\xi_\omega \eta_\omega T_{m\omega} + B_\omega \omega_r + f(\omega_r, i_{q\omega})}{2\pi^3 P_\omega \frac{G_\omega D_\omega^2}{\omega_{\partial\omega}^2 S_{N\omega}} \left(\frac{n_\omega}{60}\right)^3}$, $a_v = \frac{3\xi_v \eta_v \psi_{fv}}{4\pi^3 P_v \frac{G_v D_v^2}{\omega_{\partial v}^2 S_{Nv} r_c} \left(\frac{n_v}{60}\right)^3}$, $b_v = -\frac{\xi_v \eta_v T_{mv} + B_v v_r + f(v_r, i_{qv})}{2\pi^3 P_v \frac{G_v D_v^2}{\omega_{\partial v}^2 S_{Nv} r_c} \left(\frac{n_v}{60}\right)^3}$.

Since the rotating motor of the vehicle is decoupled from the forward motor, but the same set of observers is used in the end, the following formula is constructed using the rotating motor as an example. Formula (33) can be rewritten as:

$$\begin{cases} \dot{x} = Ax + Bu \\ y = Cx \end{cases} \tag{34}$$

where $x = \begin{pmatrix} \omega_r \\ b_\omega \end{pmatrix}$, $A = \begin{pmatrix} 0 & 1 \\ 0 & 0 \end{pmatrix}$, $B = \begin{pmatrix} a_\omega \\ 0 \end{pmatrix}$, $C = \begin{pmatrix} 1 & 0 \end{pmatrix}$, $u = i_{q\omega}$, $y = \omega_r$.

Define the following equations:

$$\begin{aligned} \dot{\hat{x}} &= A\hat{x} + Bu + L(y - \hat{y}) \\ \hat{y} &= C\hat{x} \end{aligned} \tag{35}$$

where $\hat{x} = \begin{pmatrix} \hat{\omega}_r \\ \hat{b}_\omega \end{pmatrix}$, $L = \begin{pmatrix} l_1 \\ l_2 \end{pmatrix}$.

$$\hat{e} = \dot{x} - \dot{\hat{x}} = (A - LC)e \tag{36}$$

Theorem 3. *For the system corresponding to Formula (33), $b - \hat{b} \to 0$ holds on the premise of an established observer (34)–(36).*

Proof of Theorem 3. To ensure the dynamic stability of the observer, it is essential that the eigenvalues of the matrix $(A - LC)$ reside on the left-hand side of the imaginary plane. The gain matrix $L = (24\ 128)^T$ is derived by placing the poles at $[-8, -16]$ using the pole placement method.

The error component, as defined by Equation (27), is $f(\omega_r, i_{q\omega}) = \mu_1 \Delta i_q - \mu_2 \Delta\omega_r + \tau$. τ is a disturbance caused by multiple factors. Given the bounded nature of the input, it can be asserted that $f(\omega_r, i_q) < \rho_1$ and $f'(\omega_r, i_q) < \rho_2$.

In the estimation of uncertainty factors, b_ω includes the computable part $\frac{B_\omega}{2\pi^3 P_\omega \frac{G_\omega D_\omega^2}{\omega_{\partial\omega}^2 S_{N\omega}} \left(\frac{n_\omega}{60}\right)^3}$ and the non-computable part $\frac{\xi_\omega \eta_\omega T_{m\omega} + f(\omega_r, i_{q\omega})}{2\pi^3 P_\omega \frac{G_\omega D_\omega^2}{\omega_{\partial\omega}^2 S_{N\omega}} \left(\frac{n_\omega}{60}\right)^3}$. This uncertainty estimation enables effective compensation within the sliding mode control framework.

At this juncture, the establishment of the nonlinear observer has been essentially completed. □

4.3. Sliding Mode Control for Way-Point Tracking Control Vehicle

In this part, the approach law of the integral sliding mode and HyperSpiral algorithm are used in combination with the dynamic model to obtain the system control input. In order to accelerate the approach speed and reduce the buffering during the approach, a switching control is added to the control input. At the same time, the observer is used to compensate for and strengthen the stability of the system.

Formula (32) can be rewritten as follows:

$$\begin{aligned} \dot{\omega}_r &= k_{m\omega} i_{q\omega} - k_{f\omega}\omega_r - \tau_\omega^p \\ \dot{v}_r &= k_{mv} i_{qv} - k_{fv} v_r - \tau_v^p \end{aligned} \tag{37}$$

where $k_{mv} = \frac{3\xi_v \eta_v \psi_{fv}}{4\pi^3 P_v \frac{G_v D_v^2}{\omega_{\partial v}^2 S_{Nv} r_c} \left(\frac{n_v}{60}\right)^3}$, $k_{fv} = \frac{B_v}{2\pi^3 p_v P_v \frac{G_v D_v^2}{\omega_{\partial v}^2 S_{Nv} r_c} \left(\frac{n_v}{60}\right)^3}$, $\tau_v^p = \frac{f(v_r, i_q) + \xi_v \eta_v T_{mv}}{2\pi^3 p_v P_v \frac{G_v D_v^2}{\omega_{\partial v}^2 S_{Nv} r_c} \left(\frac{n_v}{60}\right)^3}$,
$k_{m\omega} = \frac{3\xi_\omega \eta_\omega \psi_{f\omega}}{4\pi^3 P_\omega \frac{G_\omega D_\omega^2}{\omega_{\partial \omega}^2 S_{N\omega}} \left(\frac{n_\omega}{60}\right)^3}$, $k_{f\omega} = \frac{B_\omega}{2\pi^3 p_\omega P_\omega \frac{G_\omega D_\omega^2}{\omega_{\partial \omega}^2 S_{N\omega}} \left(\frac{n_\omega}{60}\right)^3}$, $\tau_\omega^p = \frac{f(\omega_r, i_{q\omega}) + \xi_\omega \eta_\omega T_{m\omega}}{2\pi^3 p_\omega P_\omega \frac{G_\omega D_\omega^2}{\omega_{\partial \omega}^2 S_{N\omega}} \left(\frac{n_\omega}{60}\right)^3}$.

Define the error as follows:

$$\begin{cases} e_v = v_d - v_r \\ e_\omega = \omega_d - \omega_r \end{cases} \tag{38}$$

where ω_d is the expected angular speed and ω_r is the actual angular speed of the motor.

Define the sliding surface as follows:

$$s = k_p e + k_i \int e dt \tag{39}$$

where k_p is the proportional gain used to adjust for the direct effect of the error. k_i is the integral gain, which adjusts for the integral effect of the error.

The reaching law is shown in the following formula:

$$\begin{cases} \dot{s} = -\lambda |s|^{\frac{1}{2}} sign(s) + v \\ \dot{v} = -\alpha_k sign(s) \end{cases} \tag{40}$$

In sliding mode control, the sliding mode control method applied to both the forward motor and the rotating motor follows the same principles; only the parameters used may differ in magnitude. Therefore, in the derivation process, the sliding mode control of the rotating motor will be used as a representative example for the derivation.

The following equation can be obtained:

$$\begin{aligned} \dot{s} &= k_{p\omega} \dot{e}_\omega + k_{i\omega} e_\omega \\ &= k_{p\omega}(\dot{\omega}_d - k_{m\omega} i_{q\omega} + k_{f\omega}\omega_r + \tau_\omega^p) + k_{i\omega}(\omega_d - \omega_r) \\ &= -\lambda_\omega |s|^{\frac{1}{2}} sign(s) - \alpha_{k\omega} \int sign(s) dt \end{aligned} \tag{41}$$

The following equation can be solved by Equation (41).

$$i_q^{eq} = \frac{1}{k_{m\omega}} \left[\frac{1}{k_{p\omega}} \left[\begin{array}{l} \lambda_\omega |s|^{\frac{1}{2}} sign(s) + \alpha_{k\omega} \int sign(s) dt \\ + k_{i\omega}(\omega_d - \omega_r) \end{array} \right] + \tau_\omega^p + k_{f\omega}\omega_r + \dot{\omega}_d \right] \tag{42}$$

The control quantity consists of equivalent control and switching control:

$$u_c = i_q^{eq} + i_q^{sw} \tag{43}$$

Let i_q^{sw} be $k_{s\omega}|s|^{\frac{1}{2}} \tanh\left(\frac{s}{\varepsilon_\omega}\right)$. The control quantity can be obtained as follows:

$$u_c = \frac{1}{k_{m\omega}} \left[\frac{1}{k_{p\omega}} \left[\begin{array}{l} \lambda_\omega |s|^{\frac{1}{2}} sign(s) + \alpha_{k\omega} \int sign(s) dt \\ + k_{i\omega}(\omega_d - \omega_r) \end{array} \right] + \tau_\omega^p + k_{f\omega}\omega_r + \dot{\omega}_d \right] + k_{s\omega}|s|^{\frac{1}{2}} \tanh\left(\frac{s}{\varepsilon_\omega}\right) \tag{44}$$

Add the estimation $\hat{b}_\omega$ of the uncertain disturbance in Equation (44).

$$u_c = \frac{1}{k_{m\omega}} \left[\frac{1}{k_{p\omega}} \left[\begin{array}{l} \lambda_\omega |s|^{\frac{1}{2}} sign(s) + \alpha_{k\omega} \int sign(s) dt \\ + k_{i\omega}(\omega_d - \omega_r) \end{array} \right] + \tau_\omega^p + k_{f\omega}\omega_r + \dot{\omega}_d + \hat{b}_\omega \right] + k_{s\omega}|s|^{\frac{1}{2}} \tanh\left(\frac{s}{\varepsilon_\omega}\right) \tag{45}$$

By combining Formula (33) with Formula (37), $k_{f\omega}\omega_r + \tau_\omega^p = -b_\omega$, $b_\omega - \hat{b}_\omega \to 0$ can be obtained, and the following formula can be obtained:

$$\tau_\omega^p + k_{f\omega}\omega_r + \hat{b}_\omega \to 0 \tag{46}$$

Theorem 4. *The system model described by Equation (32), under the action of the control law defined in Equations (37)–(40), is asymptotically stable in the Lyapunov sense, if* $k_p k_m k_s > 0, \lambda > 0, \alpha_k > 0, \varepsilon > 0$ *holds. Specifically, as* $t \to \infty$ *with* $e_v \to 0,\ e_\omega \to 0$.

Proof of Theorem 4. The Lyapunov method will then be used to prove its stability.

The Lyapunov function is defined as follows:

$$V_s = \frac{1}{2}s^2 \tag{47}$$

Obviously, the following can be obtained:

$$V_s > 0, \forall s \neq 0, V_s(0) = 0 \tag{48}$$

Taking its derivative of time, the following formula can be obtained:

$$\dot{V}_s = s\dot{s} = s(k_{p\omega}\dot{e} + k_{i\omega}e)$$

Then, the following formula is obtained.

$$\begin{aligned}
\dot{V}_s &= s\left[k_{p\omega}(\dot{\omega}_d - k_{m\omega}i_q + k_{f\omega}\omega_r + \tau_\omega^p) + k_{i\omega}(\omega_d - \omega_r)\right] \\
&= s\left[\begin{array}{l} k_{p\omega}\left(\dot{\omega}_d - \left[\begin{array}{l}\left[\begin{array}{l}\frac{1}{k_{p\omega}}\lambda_\omega |s|^{\frac{1}{2}} sign(s) + \tau_\omega^p + \frac{1}{k_{p\omega}}\alpha_{k\omega}\int sign(s)dt + k_{f\omega}\omega_r + \frac{1}{k_{p\omega}}k_{i\omega}(\omega_d - \omega_r) \\ +\dot{\omega}_d\end{array}\right] \\ +k_{s\omega}k_{m\omega}|s|^{\frac{1}{2}}\tanh\left(\frac{s}{\varepsilon_\omega}\right)\end{array}\right]\right. \\ \left. +k_{F\omega}\omega_r + \tau_\omega^p\right) + k_{i\omega}(\omega_d - \omega_r)\end{array}\right] \\
&= s\left[\begin{array}{l} k_{p\omega}\dot{\omega}_d - \left[\begin{array}{l}\left[\begin{array}{l}\lambda_\omega |s|^{\frac{1}{2}} sign(s) + \alpha_{k\omega}\int sign(s)dt + k_{i\omega}(\omega_d - \omega_r) + k_{p\omega}\tau_\omega^p + k_{p\omega}k_{f\omega}\omega_r + k_{p\omega}\dot{\omega}_d\end{array}\right] \\ +k_{p\omega}k_{s\omega}k_{m\omega}|s|^{\frac{1}{2}}\tanh\left(\frac{s}{\varepsilon_\omega}\right)\end{array}\right] \\ +k_{p\omega}k_{f\omega}\omega_r + k_{p\omega}\tau_\omega^p) + k_{i\omega}(\omega_d - \omega_r)\end{array}\right]
\end{aligned}$$

A further formula derivation can be obtained.

$$\begin{aligned}
\dot{V}_s &= s\left[\begin{array}{l} k_{p\omega}\dot{\omega}_d - \lambda_\omega |s|^{\frac{1}{2}} sign(s) - \alpha_{k\omega}\int sign(s)dt - k_{i\omega}(\omega_d - \omega_r) - k_{p\omega}\tau_\omega^p - k_{p\omega}k_{f\omega}\omega_r - k_{p\omega}\dot{\omega}_d - \\ k_{p\omega}k_{m\omega}k_{s\omega}|s|^{\frac{1}{2}}\tanh\left(\frac{s}{\varepsilon_\omega}\right) + k_{p\omega}k_{f\omega}\omega_r + k_{p\omega}\tau_\omega^p + k_{i\omega}(\omega_d - \omega_r)\end{array}\right] \\
&= s\left[-\lambda_\omega |s|^{\frac{1}{2}} sign(s) - \alpha_{k\omega}\int sign(s)dt - k_{p\omega}k_{m\omega}k_{s\omega}|s|^{\frac{1}{2}}\tanh\left(\frac{s}{\varepsilon_\omega}\right)\right]
\end{aligned}$$

Then, the following is obtained:

$$\dot{V}_s = -\lambda_\omega s|s|^{\frac{1}{2}} sign(s) - \alpha_{k\omega} s \int sign(s)dt - k_{p\omega}k_{m\omega}k_{s\omega}|s|^{\frac{1}{2}} \text{stanh}\left(\frac{s}{\varepsilon_\omega}\right) \tag{49}$$

For the above formula, the following requirements must be met:

$$\begin{cases} k_p k_s > 0 \\ k_m > 0 \\ \lambda > 0 \\ \alpha_k > 0 \\ \varepsilon > 0 \end{cases} \tag{50}$$

The following conclusions can be drawn.

$$\dot{V}_s < 0, \forall s \neq 0 \tag{51}$$

At this point, the proof of stability has been basically completed. □

4.4. Analysis of the System Singularity

Considering Formula (5), when $t \to \infty$, it holds that $d_t \to 0,\ v_r \to 0$. So, there is a singularity in $\dot{\alpha}_t = k_{\text{sgnt}} \frac{v_t \sin \alpha_t}{d_t} - \omega_t$.

Define the following equation:

$$e_t = \frac{v_r}{d_t} - c \geq 0 \tag{52}$$

where c is a sufficiently small positive number.

The function is defined as follows:

$$V_e(t) = \frac{1}{2}e_t^2 = \frac{1}{2}\left(\frac{v_r}{d_t} - c\right)^2 \geq 0 \tag{53}$$

Then, take the derivative of Equation (53):

$$\dot{V}_e(t) = e_t\dot{e}_t = \left(\frac{v_r}{d_t} - c\right)\left(\frac{\dot{v}_r d_t - v_r \dot{d}_t}{d_t^2}\right) \tag{54}$$

Considering the dynamic control law in Formulas (7) and (8), the simplified dynamic model in Formula (37) and the control input in Formula (44), we proceed as follows.

Firstly, we substitute Formula (44) into the model in (37):

$$\begin{aligned} \dot{v}_r &= k_{mv}\frac{1}{k_{mv}}\left[\frac{1}{k_{pv}}\left[\begin{array}{l} \lambda_v|s|^{\frac{1}{2}} sign(s) + \alpha_{kv}\int sign(s)dt \\ +k_{iv}(v_d - v_r) \end{array}\right] + \tau_v^p + k_{fv}v_r + \dot{v}_d\right] \\ &+k_{sv}|s|^{\frac{1}{2}} \tanh\left(\frac{s}{\varepsilon_v}\right) - k_{fv}v_r - \tau_v^p \\ &= \left[\frac{1}{k_{pv}}\left[\begin{array}{l} \lambda_v|s|^{\frac{1}{2}} sign(s) + \alpha_{kv}\int sign(s)dt \\ +k_{iv}(v_d - v_r) \end{array}\right] + \dot{v}_d\right] + k_{sv}|s|^{\frac{1}{2}} \tanh\left(\frac{s}{\varepsilon_v}\right) \end{aligned} \tag{55}$$

Thus, Formula (54) can be simplified as follows:

$$\begin{aligned}
\dot{V}_e(t) &= (\frac{v_r}{d_t} - c)[(\frac{1}{k_{pv}}\left[\frac{\lambda_v|s|^{\frac{1}{2}}sign(s)}{d_t} + \frac{\alpha_{kv}\int sign(s)dt}{d_t} + k_{iv}(k_\alpha - \frac{v_r}{d_t})\right] - k_\alpha^2\cos\alpha_t) \\
&+ \frac{k_{sv}|s|^{\frac{1}{2}}\tanh\left(\frac{s}{\varepsilon_v}\right) + v_r k_\alpha \cos\alpha}{d_t}] \\
&= (\frac{v_r}{d_t} - c)[\frac{\lambda_v|s|^{\frac{1}{2}}sign(s)}{k_{pv}d_t} + \frac{\alpha_{kv}\int sign(s)dt}{k_{pv}d_t} + \frac{k_{iv}}{k_{pv}}(k_\alpha - \frac{v_r}{d_t}) - k_\alpha^2\cos\alpha_t \\
&+ \frac{k_{sv}|s|^{\frac{1}{2}}\tanh\left(\frac{s}{\varepsilon_v}\right)}{d_t} + \frac{v_r}{d_t}k_\alpha\cos\alpha] \\
&= (\frac{v_r}{d_t} - c)[\frac{[\frac{\lambda_v}{k_{pv}}sign(s) + k_{sv}\tanh\left(\frac{s}{\varepsilon_v}\right)]|s|^{\frac{1}{2}} + \frac{\alpha_{kv}}{k_{pv}}\int sign(s)dt}{d_t} \\
&+ k_\alpha(\frac{k_{iv}}{k_{pv}} - k_\alpha\cos\alpha_t) - \frac{v_r}{d_t}(\frac{k_{iv}}{k_{pv}} - k_\alpha\cos\alpha)] \\
&= (\frac{v_r}{d_t} - c)[\frac{[\frac{\lambda_v}{k_{pv}}sign(s) + k_{sv}\tanh\left(\frac{s}{\varepsilon_v}\right)]|s|^{\frac{1}{2}} + \frac{\alpha_{kv}}{k_{pv}}\int sign(s)dt}{d_t} + (k_\alpha - \frac{v_r}{d_t})(\frac{k_{iv}}{k_{pv}} - k_\alpha\cos\alpha_t)]
\end{aligned} \tag{56}$$

Since v_d, as defined in Equation (5), is a constant positive quantity along the vehicle body coordinate axis Y_b, therefore, in the whole tracking process of the vehicle for a certain point, the change in speed is a process of increase and then decrease, and for $t \to \infty$, the actual speed at the previous moment is greater than the expected speed at the later moment, making $v_d - v_r < 0$ valid, and the motor model in Equation (37) is a first-order system. Adjusting to monotonically without overshoot, it follows that when $t \to \infty$ for $e_v \le 0$ the model approaches zero from the negative direction, and the accumulation of errors is greater than the error at this point. Thus, $s = k_p e + k_i \int edt < 0 k_p > 0, k_i > 0$ holds true. Therefore, the above equation can be simplified as follows:

$$\dot{V}_e(t) = (\frac{v_r}{d_t} - c)[\frac{[-\frac{\lambda_v}{k_{pv}} + k_{sv}\tanh\left(\frac{s}{\varepsilon_v}\right)]|s|^{\frac{1}{2}} - \frac{\alpha_{kv}}{k_{pv}}\int dt}{d_t} + (k_\alpha - \frac{v_r}{d_t})(\frac{k_{iv}}{k_{pv}} - k_\alpha\cos\alpha_t)] \tag{57}$$

In Formula (50), $k_p k_m k_s > 0$ is required. Thus, set the $k_p > 0$, $k_s > 0$, $k_m > 0$. Then, in Formula (57), $-\frac{\lambda_v}{k_{pv}} < 0, k_{sv}\tanh(\frac{s}{\varepsilon_v}) < 0, -\frac{\alpha_{kv}}{k_{pv}}\int dt < 0, k_\alpha > 0, d_t > 0$.

When $t \to \infty$, $\frac{v_r}{d_t} \ge c, v_r \to 0, d_t \to 0$, there are three cases:

Case 1. *$c < k_\alpha$, at first states, $c < k_\alpha < \frac{v_r}{d_t}$; as time goes on, d_t is small enough that $\frac{[-\frac{\lambda_v}{k_{pv}} + k_{sv}\tanh\left(\frac{s}{\varepsilon_v}\right)]|s|^{\frac{1}{2}} - \frac{\alpha_{kv}}{k_{pv}}\int dt}{d_t} < -(k_\alpha - \frac{v_r}{d_t})(\frac{k_{iv}}{k_{pv}} - k_\alpha\cos\alpha_t)$. Then, the following formula can be obtained:*

$$\begin{cases} V_e(t) > 0 \\ \dot{V}_e(t) = (\frac{v_r}{d_t} - c)[\frac{[-\frac{\lambda_v}{k_{pv}} + k_{sv}\tanh\left(\frac{s}{\varepsilon_v}\right)]|s|^{\frac{1}{2}} - \frac{\alpha_{kv}}{k_{pv}}\int dt}{d_t} + (k_\alpha - \frac{v_r}{d_t})(\frac{k_{iv}}{k_{pv}} - k_\alpha \cos\alpha_t)] < 0 \end{cases} \tag{58}$$

Then, at the second state, $c \leq \frac{v_r}{d_t} < k_\alpha$*. In this state, it holds that* $V_e(t) \geq 0$*,* $\dot{V}_e(t) < 0$*. Until when* $c = \frac{v_r}{d_t}$*, it can be concluded that* $V_e(t) = 0, \dot{V}_e(t) = 0$*. Thus, finally,* $\lim_{t\to\infty} \frac{v_r}{d_t} = c \geq 0$*.*

Case 2. $c > k_\alpha$ *, at first state,* $k_\alpha < c \leq \frac{v_r}{d_t}$*. As time goes on,* d_t *is small enough that* $\frac{[-\frac{\lambda_v}{k_{pv}} + k_{sv}\tanh\left(\frac{s}{\varepsilon_v}\right)]|s|^{\frac{1}{2}} - \frac{\alpha_{kv}}{k_{pv}}\int dt}{d_t} < -(k_\alpha - \frac{v_r}{d_t})(\frac{k_{iv}}{k_{pv}} - k_\alpha \cos\alpha_t)$ *is true. Thus, it holds that* $V_e(t) \geq 0, \dot{V}_e(t) < 0$*. Until* $c = \frac{v_r}{d_t}$*, it can be concluded that* $V_e(t) = 0, \dot{V}_e(t) = 0$*. Thus, finally,* $\lim_{t\to\infty} \frac{v_r}{d_t} = c > k_\alpha$

Case 3. $c = k_\alpha$*, then it can get* $k_\alpha = c \leq \frac{v_r}{d_t}$*. Thus, when* $\lim_{t\to\infty} \frac{v_r}{d_t} = k_\alpha, V_e(t) = 0 \dot{V}_e(t) = 0$*.*

Therefore, while the vehicle tracks the target points, v_r and d_t are all decreasing, which always keeps $\lim_{t\to\infty} \frac{v_r}{d_t} = c$. As the three cases discuss above, the system proposed will remain non-singular.

5. Way-Point Obstacle-Avoidance Algorithm

In this section, the path planning and obstacle avoidance for vehicle navigation is considered. To ensure proper vehicle operation within complex environments, an improved Rapidly Exploring Random Tree algorithm is used to generate way-points and a greedy algorithm to optimize the way-points. The overall process is outlined as follows:

First, a random point, *sample,* is generated in each iteration to guide the expansion direction of the tree. This is calculated as:

$$sample = \begin{cases} rand(x,y) \times searchsize + start(x,y) \\ goal(x,y) \end{cases} \tag{59}$$

where *sample* represents a random point used to guide the tree's growth, *searchsize* is the defined search space, and $rand(x,y)$ generates two values between 0 and 1. Multiplying by *searchsize* ensures the randomly generated way-points fall within the defined search area; *start* denotes the starting point, and *goal* represents the target point.

Next, within the path tree, identify the node closest to the generated sample by computing its Euclidean distance:

$$dist = \sqrt{(x_{sample} - x_{tree})^2 + (y_{sample} - y_{tree})^2} \tag{60}$$

where *dist* is the Euclidean distance between two points, (x_{smple}, y_{sample}) is the sample point, and (x_{tree}, y_{tree}) is the point in the path tree.

A new node, *newpoint,* is then generated in the direction of *sample* from the nearest node, *closestnode.* The step size is denoted as *stepsize,* and the direction vector, *growvec,* is unitized:

$$growvec = \frac{sample - closestnode}{||sample - closestnode||} \tag{61}$$

The new node is computed as:

$$newpoint = closestnode + growvec \times stepsize \tag{62}$$

Subsequently, collision detection is performed. If the new node collides with an obstacle, it is discarded. To check if the path from node $P_{closestnode}(x_{closestnode}, y_{closestnode})$ to node $P_{new}(x_{new}, y_{new})$ intersects with a circular obstacle, describe the path equation as:

$$P(t) = (1-t)P_{closestnode} + tP_{new},\ t \in [0,1] \tag{63}$$

where $c_i(x_{ci}, y_{ci})$ is the center and r_i is the radius of the $i-th$ obstacle. The obstacle equation is:

$$\|P(t) - c_i\|^2 = r_i^2 \tag{64}$$

Substituting the path equation:

$$\begin{bmatrix} (1-t)P_{closestnode}+ \\ tP_{new} - c_i \end{bmatrix}^{\mathrm{T}} \begin{bmatrix} (1-t)P_{closestnode}+ \\ tP_{new} - c_i \end{bmatrix} = r_i^2 \tag{65}$$

Expanding and simplifying yields the quadratic form:

$$At^2 + Bt + C = 0 \tag{66}$$

where $A = \|P_{new} - P_{closestnode}\|^2$, $B = 2(P_{closestnode} - c_i)^{\mathrm{T}}$, $(P_{new} - P_{closestnode})$ and $C = \|P_{closestnode} - c_i\|^2 - r_i^2$. The discriminant of this quadratic equation determines whether an intersection occurs, indicating a collision.

The next step is reconnection, aimed at reducing the path cost from the root node to other nodes by reconnecting nodes in the tree for optimization. For each adjacent node x_j with index j, the cost $K_{new}(x_j)$ of the new path from x_{last} to x_j is computed as:

$$K_{new}(x_j) = K(x_{last}) + dist(x_{last}, x_j) \tag{67}$$

If the new path cost $K_{new}(x_j)$ is less than the current path cost $K(x_{last})$ and no collision occurs from x_{last} to x_j, the following updates are made:

$$\begin{cases} x_{parent(j)} = x_{last} \\ K(x_j) = K_{new}(x_j) \end{cases} \tag{68}$$

where x_{last} is the last generated node, x_j represents the path cost of node x_i, and $x_{parent(i)}$ denotes the parent node.

Finally, path backtracking is performed to generate the full path from the target node to the start node by tracing the parent nodes. To minimize the number of way-points, reduce unnecessary velocity and angular velocity oscillations caused by frequent way-point replacements, and enhance system stability, a greedy algorithm is applied for path optimization.

The core idea is to attempt direct connections from the current node to further nodes in the path. If a successful connection is made without collisions, intermediate nodes are skipped. The original path is $P = [p_1, p_2, p_3 \ldots, p_n]$, where $p_i = (x_i, y_i)$ represents the $i-th$ waypoint. The optimized path, P_{opt}, starts with the initial waypoint $P_{opt} = [p_1]$, and the current point is $p_{cur} = p_1$.

For each point in the path, starting from P_{cur}, we attempt to connect directly to further points P_j in the path, checking for collisions. If no collision is detected, P_{cur} is updated to P_j, and the optimized path is updated as follows:

$$\begin{cases} p_{\text{cur}} = p_j \\ P_{\text{opt}} = P_{\text{opt}} \cup \{p_j\} \end{cases} \tag{69}$$

At the current juncture, the development of the obstacle-avoidance algorithm for way-point tracking control is substantially finalized. However, it is important to take into account certain practical considerations.

Remark 1. *In real-world applications, the distribution, size, and shape of obstacles must be preemptively modeled and computed. Although the algorithm operates under the assumption of circular obstacles for the sake of simplicity, in practical scenarios, it is only necessary to approximate the contours of the obstacles. This approach allows for a more accurate representation of the environment without overcomplicating the computational process.*

Remark 2. *The control law that has been implemented decouples the angular velocity from the linear velocity, which leads to discrepancies between the vehicle's actual path and the straight-line connections between way-points. This potential path deviation can be identified and addressed through rigorous simulation testing. Consequently, the arrangement of target waypoints should be fine-tuned based on empirical observations and practical constraints. The root cause of this path deviation stems from the vehicle's high speed in conjunction with the relatively slow changes in angular velocity. This dynamic can result in the vehicle heading a considerable distance before the vehicle is aligned with the intended target point. To mitigate this issue, careful adjustment of the algorithm's parameters is required. By doing so, it can ensure that the vehicle's path more closely mirrors the planned path, thereby enhancing the overall accuracy and efficiency of the waypoint tracking system.*

6. Design of Sliding Mode Control with Observer Compensation and Intelligent Obstacle Avoidance Way-Point Tracking Control Algorithm

The vehicle way-point tracking logic includes an obstacle-avoidance algorithm that generates safe way-points around obstacles, an observer for real-time state estimation, a sliding mode control law that tracks speed and angular velocity while ensuring robustness against disturbances, and updates for the vehicle's kinematics and dynamics. The pseudocode outlines the algorithm flow, and Figure 3 illustrates the structural logic.

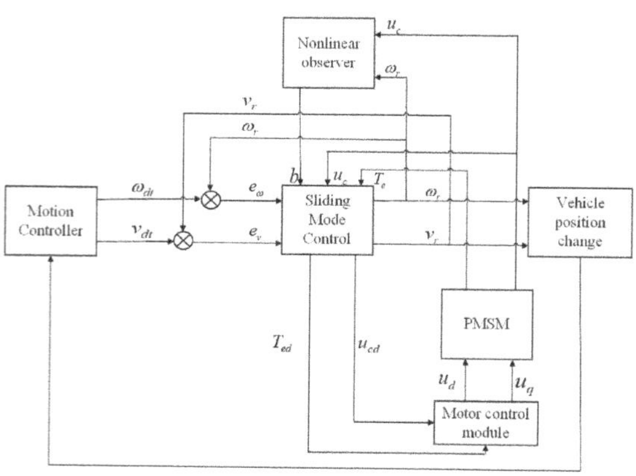

Figure 3. Pseudo-code structure logic diagram.

Algorithm 1 is given based on Figure 3.

Algorithm 1: Vehicle way-point tracking control algorithm.

1. Vehicle Way-Points Obstacle Avoidance

Initialize obstacles and goal:

Basic modeling based on the actual environment, define obstacle positions and radii, and set start and goal points.

From Formulas (59)–(62) and (63)–(66), RRT* Path Generation:

Randomly select sample points, extend the tree, check for collisions, and stop when a path to the goal is found.

Greedy optimization:

From Formulas (67)–(69). Simplify the path by removing unnecessary intermediate nodes if no collisions occur.

2. Construction of Observer for Way-Point Tracking Control Vehicle

Define parameters and system matrix through Formulas (34)–(36), obtain matrix gain L using pole placement method, and initialize variables.

For $i = 1 :$ length(*time*) **do**

Define system dynamics $\dot{x}$, x by Formula (34).

Update $x(1, k+1)$, $x(2, k+1)$ through $\dot{x}$.

Define observer dynamics $\hat{y}$, $\hat{x}$, $\dot{\hat{x}}$ by Formula (35).

Update $\hat{x}$ through $\dot{\hat{x}}$.

Extract $\hat{x}(2, k)$ and use it as the estimator of uncertainty in sliding control to compensate.

End for

3. Establishment of Sliding Mode Control for Way-Point Tracking Control

Define parameters and initialize variables.

For $i = 1 :$ length(*time*) **do**

Define error $e(i)$ by Formula (38).

Define control input $u(i)$ by Formula (45).

The estimation error of the observer from **part2** is added to compensate.

The motor dynamics $T_e(i)$, $\dot{\omega}_r(i)$ $\dot{v}_r(i)$ are expressed by Formulas (25) and (27).

Update $\dot{\omega}_r(i+1)$, $\dot{v}_r(i+1)$ and update $\omega_r(i+1)$ $v_r(i+1)$.

End for

4. Establishment of Vehicle Motion Controller and Way-Point Tracking Control

The values of k_α, k_γ, and k_β must satisfy the inequalities in (7) and (9).

Define the initial position (x_{it}, y_{it}), and the target point is obtained according to the obstacle-avoidance algorithm (x_{et}, y_{et}).

For $i = 1$: length(*time*) **do**

According to Formulas (3)–(6), represent ω_{dt}, d_t, v_{dt}, α_t, β_t, γ_t, $\dot{\gamma}_t$, $\dot{\alpha}_t$, T_e, $\dot{v}_r$, $\dot{\omega}_r$, v_r, ω_r are obtained by the sliding mode control program and the motor program from **part2** and **part3**.

Update α_t, γ_t by $\dot{\gamma}_t$, $\dot{\alpha}_t$.

Update (x_{it}, y_{it}) by Formula (3).

If $\|d_t(i)\| \leq \varepsilon$ **then**

Update the next way-points from **part1** (x_{et}, y_{et}).

End if

#ε is the error precision set.

End for

7. Simulation Experiment and Analysis

In this part, the previously designed code is mainly used to conduct simulation experiments. The whole simulation process is divided into three parts: first, the tracking of a single point to verify the tracking accuracy of a single point; then, the tracking of multiple points to verify the effectiveness of the operation under the disturbance in the actual environment; and finally, the obstacle avoidance simulation to verify the operation capability under the complex environment.

7.1. Tracking Simulation for Single Point

Firstly, the simulation of single-point tracking is carried out, which aims to verify the accuracy of waypoint tracking.

The parameters of the rotating motor are defined as follow: $\eta_\omega = 0.8$, $\xi_\omega = 0.75$, $p_{n\omega} = 2$, $\psi_{f\omega} = 0.5$, $B_{f\omega} = 0.03$, $f_\omega = 50$, $G_\omega = 0.15$, $D_\omega = 0.1$, $P_\omega = 40$, $n_\omega = 1000$, $S_{N\omega} = 20$, $k_{p\omega} = 50$, $k_{i\omega} = 5$, $\lambda_\omega = 2$, $\alpha_{k\omega} = 10$, $k_{s\omega} = 0.8$, $\delta_\omega = 0.2$. The parameters of the forward motor are defined as follows: $\eta_\omega = 0.8$, $\xi_\omega = 0.85$, $p_{n\omega} = 2$, $\psi_{f\omega} = 0.5$, $B_{f\omega} = 0.05$, $f_\omega = 50$, $G_\omega = 0.15$, $D_\omega = 0.05$, $P_\omega = 50$, $n_\omega = 1500$, $S_{N\omega} = 175$, $k_{p\omega} = 10$, $k_{i\omega} = 2$, $\lambda_\omega = 0.2$, $\alpha_{k\omega} = 5$, $k_{s\omega} = 1$, $\delta_\omega = 0.15$, $r_c = 0.15$.

In addition, set the target point to $(10, 20)$, the time step to $dt = 0.1$, and the tracking accuracy to $\varepsilon = 0.1$, $v_d(1) = 0$, $\omega_d(1) = 0$, $k_\alpha = 0.75$, $k_\gamma = 13$, $k_\beta = -3.7$. At the same time, the vehicle starts at zero point.

Figure 4 presents the path diagram for the single-point simulation, illustrating the vehicle's path. Figure 5 depicts the tracking performance of the sliding mode control for the single-point tracking task, showcasing how effectively the control method follows the target point. Lastly, Figure 6 provides the control input diagram, highlighting the system's inputs during the tracking process and offering insights into the control strategy's dynamics.

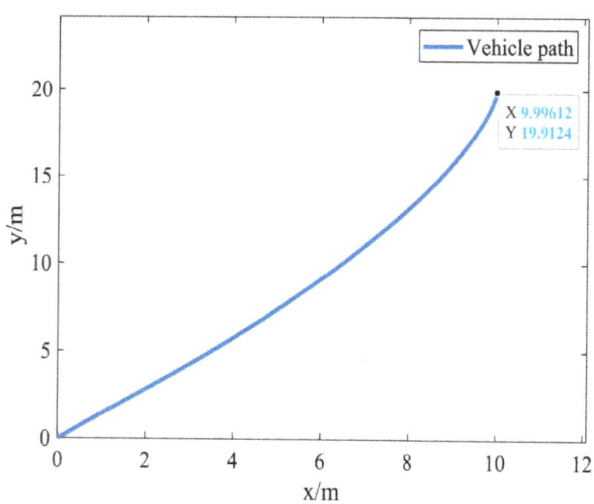

Figure 4. Single-point tracking simulation image.

From Figure 4, it is evident that during the single-point tracking, the vehicle successfully tracks the target point at coordinates with an accuracy error of less than or equal to 0.1. This demonstrates the effectiveness of the tracking system, as the vehicle achieves precise alignment with the target point within a very small margin of error.

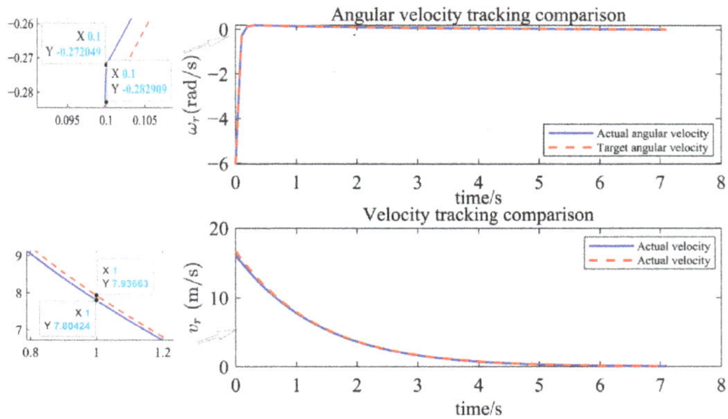

Figure 5. Comparison image of velocity and angular velocity tracking under single-point tracking.

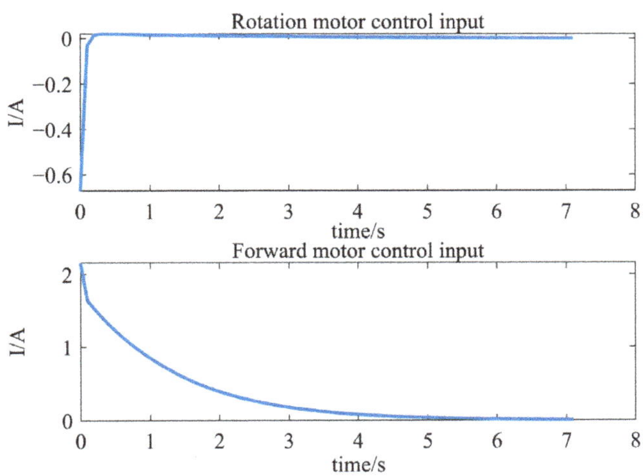

Figure 6. Single-point tracking system control input.

Through the analysis of these graphs and simulation results, it is evident that the way-point tracking control can accurately track the target point within the specified error margins. This approach not only achieves precise tracking of a single target point but also demonstrates the capability to track any effective point within the designated region in an ideal environment. The simulation results effectively validate the proposed method, showcasing its reliability. This successful demonstration underscores the robustness and precision of the tracking control strategy, providing a solid foundation for subsequent simulation implementations.

The demonstration of the tracking control strategy indicates its potential to achieve precise and reliable tracking within the specified error margins. It is important to note that the method was evaluated under ideal conditions. In real-world applications, factors such as sensor noise, system disturbances, and environmental unpredictability may present additional challenges that could affect performance.

7.2. Tracking Simulation for Multiple Points

First, set the way-points as $(0,0)\rightarrow(20,50)\rightarrow(20,50)\rightarrow(20,43)\rightarrow(20,50)\rightarrow(30,53)$, and define the error tolerance ε as 0.1. The parameters used are the same as those in the single-point simulation. As shown in Figure 7, the vehicle tracks the target points with stable accuracy, completing the way-point tracking as set, and successfully follows the specified coordinate points.

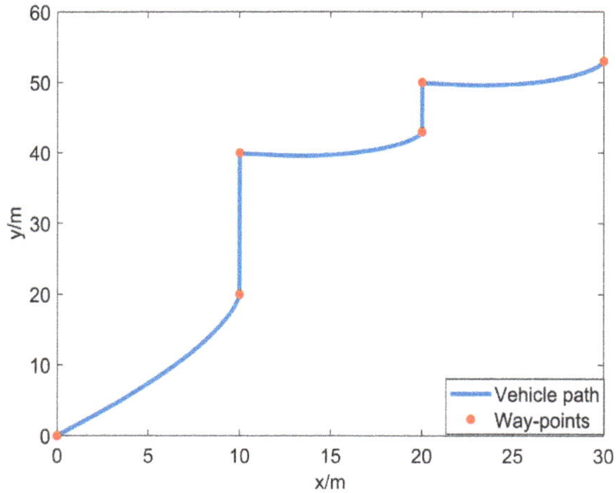

Figure 7. Multi-point tracking simulation path.

In Figure 7, the vehicle undergoes five steering processes, which correspond to the changes in angular velocity, as seen in Figures 8 and 9, as well as the control inputs to the rotating motor. Additionally, the vehicle experiences five point transitions throughout the tracking process, indicating five distinct speed adjustment phases, consistent with the data shown in Figures 7 and 8. This demonstrates that the control system effectively manages both direction and speed during the way-point tracking.

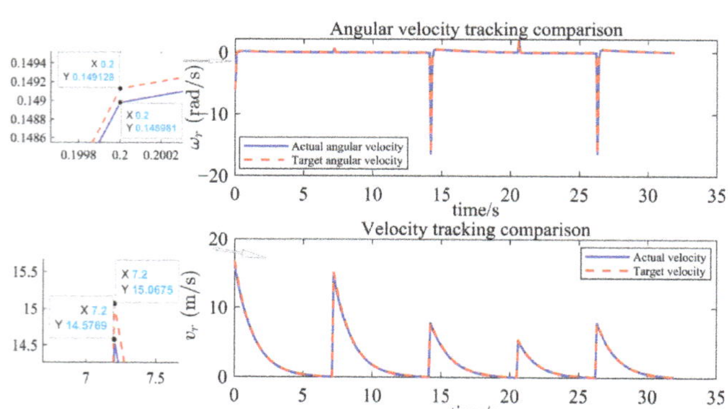

Figure 8. Comparison of velocity and angular velocity tracking under multi-point point tracking.

Figures 10 and 11 are the simulation images of the observer of the forward motor and the rotating motor, which contain the estimated image of the uncertain disturbance and the control input after the compensation image, respectively.

In Figure 10, the disturbance during the vehicle's operation was artificially introduced. The disturbance was defined by the function $b = 0.07\sin(t) + 0.18\cos(3t)$. The simulation results show that the observer effectively estimates the uncertain disturbance under these conditions.

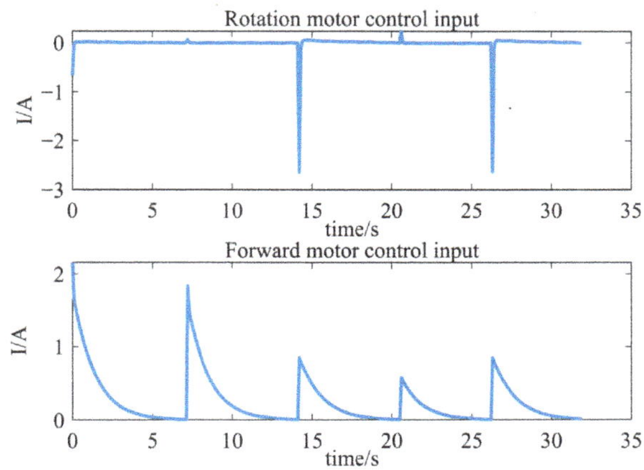

Figure 9. System control input imitation under multi-point.

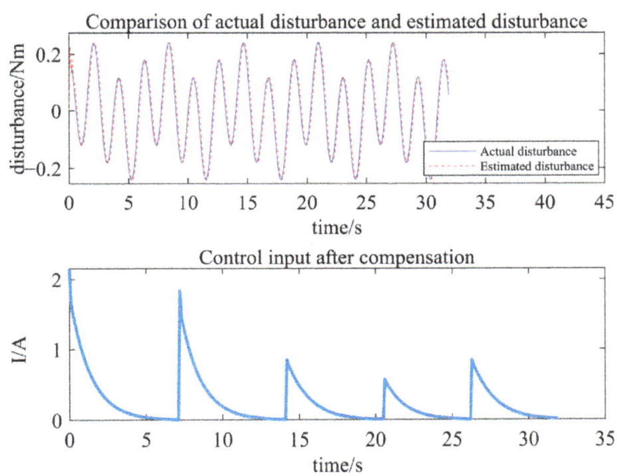

Figure 10. Forward motor observer simulation.

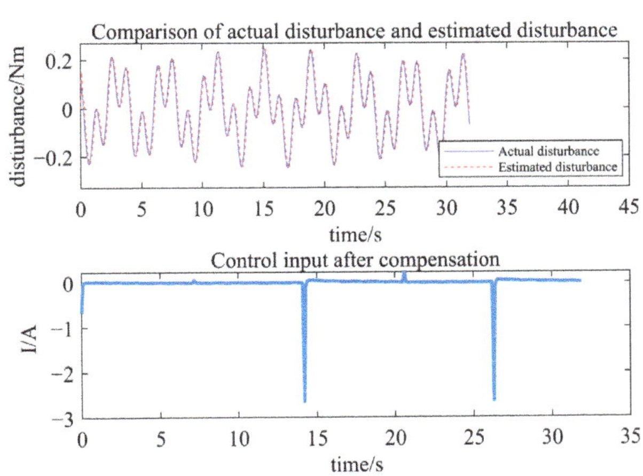

Figure 11. Rotating motor observer simulation.

In Figure 11, the disturbance during the vehicle's operation was artificially introduced. The disturbance was modeled using the function $b = 0.13\sin[\frac{\pi(t-2)}{2}] + 0.12\cos(5t)$. The simulation results demonstrate that the observer is capable of accurately estimating the uncertain disturbance, even under these dynamic conditions.

7.3. Tracking Simulation for Obstacle-Avoidance Algorithm

In this subsection, the obstacle-avoidance algorithm simulation is carried out. In this algorithm, the starting point of the car is set to $(0,0)$, and the destination is set to $(57,97)$. The algorithm generates a path between these two points, guiding the car to navigate around obstacles. In the simulation, the obstacles are modeled as circles with the following parameters: a center at $(10,20)$ and radius 18.5, a center at $(20,50)$ and radius 25, a center at $(80,70)$ and radius 30, and a center at $(70,20)$ and radius 30.

Using the obstacle-avoidance algorithm, the overall way-points from the start to the target are as follows: $(0,0) \rightarrow (19.9515, 1.3914) \rightarrow (39.5405, 28.9161) \rightarrow (46.3067, 47.7368) \rightarrow (46.5937, 87.3197) \rightarrow (57,95)$.

A comparison between Figures 12 and 13 demonstrates that, despite some deviations between the actual and target paths, the vehicle successfully completes the obstacle avoidance task. The figures also highlight that, in complex operational environments, allowing for a margin of error is often essential. This margin ensures the vehicle has adequate space to navigate, perform effective obstacle avoidance, and carry out coverage monitoring tasks. Additionally, Figure 14 illustrates the control input applied to the vehicle during this phase, providing further insights into the system's response and performance under these conditions.

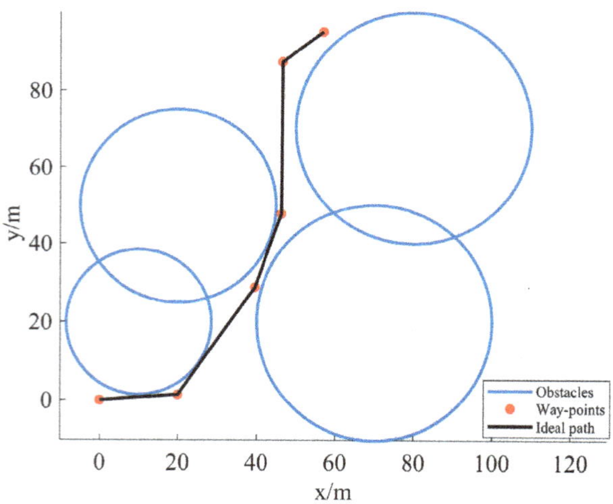

Figure 12. Obstacle-avoidance algorithm generates target path.

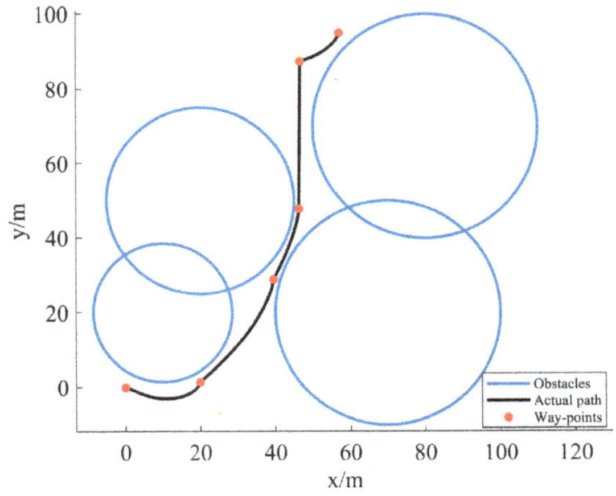

Figure 13. The actual path of the vehicle tracking.

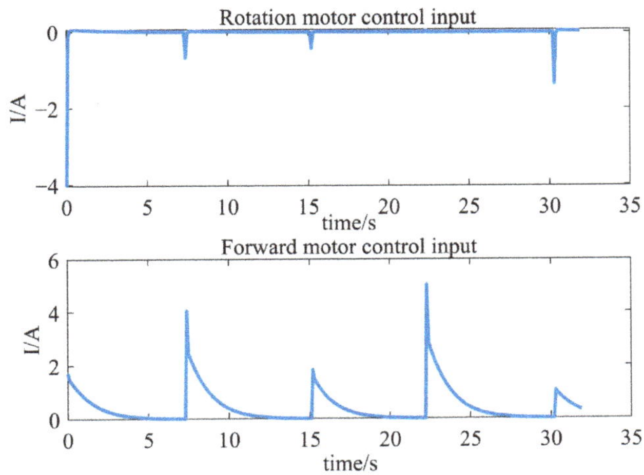

Figure 14. Control input of vehicle tracking under obstacle-avoidance algorithm.

8. Conclusions

This paper investigates the problem of way-point tracking control for vehicles in complex environments. Way-point tracking control is employed to ensure accurate navigation through a set of predefined coordinates, while observer-compensated sliding mode control is used to enable stable tracking of the vehicle's desired speed, ensuring effective performance. During the tracking process, an obstacle-avoidance algorithm was utilized to generate way-points, allowing the vehicle to efficiently bypass obstacles and reach target points. In the simulation, obstacles were modeled as circular regions with different radii, and the obstacle-avoidance algorithm generated safe way-points to guide the vehicle. The final simulation results demonstrate that the vehicle can accurately track target points and, by following the desired speed with the aid of sliding mode control compensated by an observer, successfully complete way-point tracking in complex environments. This approach provides valuable theoretical guidance for practical way-point tracking applications.

Future work will focus on: (1) Extending the obstacle-avoidance process from a static approach, which involves pre-modeling obstacles, to a dynamic one that incorporates full closed-loop observation, perception, and control. This enhancement will make the system more suitable for operations in complex environments. (2) Building upon the expanded obstacle avoidance functionality, future research will also address practical challenges in positioning errors and path planning, including considerations of various types of errors, motor control, and dynamic path planning. These improvements aim to achieve more robust tracking control by taking into account dynamic environmental changes, ensuring better adaptability in complex and dynamic environments.

Author Contributions: Conceptualization, J.S.; Methodology, M.S.; Software, M.S.; Validation, J.S. and M.S.; Writing original draft, M.S.; Funding acquisition, J.S. and Y.Z. All authors have read and agreed to the published version of the manuscript.

Funding: This work was supported by the Fundamental Research Funds for the Central University of China under Grants Z1090122053 and Z1090124102, and the Shaanxi Postdoctoral Science Foundation under Grant 2023BSHYDZZ63.

Data Availability Statement: Data are contained within the article.

Conflicts of Interest: The authors declare no conflicts of interest.

References

1. Zuo, L.; Ma, W.; Yan, M.; Yang, P. High precision tracking control for a sequence of way-points using dynamic unicycles. *J. Frankl. Inst.* **2019**, *356*, 5104–5121. [CrossRef]
2. Liu, L.; Zhang, L.; Hou, Y.; Tang, D.; Liu, H. Implementation of adaptive fault-tolerant tracking control for robot manipulators with integral sliding mode. *Int. J. Robust Nonlinear Control.* **2023**, *33*, 5337–5364. [CrossRef]
3. Meng, Q.; Yan, L. Adaptive admittance tracking control for interactive robot with prescribed performance. *J. Syst. Eng. Electron.* **2024**, *35*, 444–450. [CrossRef]
4. Peng, Z.; Cheng, H.; Shi, K.; Zou, C.; Huang, R.; Li, X.; Ghosh, B.K. Optimal tracking control for motion constrained robot systems via event-sampled critic learning. *Expert Syst. Appl.* **2023**, *234*, 1–11. [CrossRef]
5. Qi, W.; Tong, M.; Li, X.; Wang, Q.; Song, W. Linear disturbance observer-enhanced continuous-time predictive control for straight-line path-following control of small unmanned aerial vehicles. *Aerospace* **2024**, *11*, 902. [CrossRef]
6. Li, W.; Wang, Y.; Ning, J.; Li, Z. Unmanned ship heading tracking control strategy with state quantization and input quantization. *Chin. J. Ship Res.* **2024**, *19*, 111–118.
7. Liu, Y.; Jia, H.; Liu, L.; Wang, D.; Peng, Z. Anti-jamming optimal coverage control for unmanned ship cluster. *China Ship Res.* **2019**, *18*, 67–77.
8. Guo, G.; An, X.; Sun, J.; Ji, Z.; Zhao, Z. Observer-based event-triggered sliding mode tracking control for uncertain robotic manipulator systems. *J. Braz. Soc. Mech. Sci. Eng.* **2023**, *45*, 453. [CrossRef]
9. Shi, G.; Yang, J.; Zhao, X.; Li, Y.; Zhao, Y.; Li, J. A H-Infinity control for path tracking with fuzzy hyperbolic tangent model. *J. Control Sci. Eng.* **2016**, *2016 Pt 2*, 8. [CrossRef]
10. Boubakir, A.; Souanef, T.; Labiod, S.; Whidborne, J.F. A robust adaptive PID-like controller for quadrotor unmanned aerial vehicle systems. *Aerospace* **2024**, *11*, 980. [CrossRef]
11. Luo, G.; Gao, S.; Jiang, Z.; Luo, C.; Zhang, W.; Wang, H. ROV trajectory tracking control based on disturbance observer and combinatorial reaching law of sliding mode. *Ocean Eng.* **2024**, *15*, 304. [CrossRef]
12. Xing, G.; Chen, H.; Aghayan, Z.; Jiang, J.; Guirao, J.L.G. Tracking control for a class of fractional order uncertain systems with time-delay based on composite nonlinear feedback control. *AIMS Math.* **2024**, *9*, 13058–13076. [CrossRef]
13. Zhu, Z.; Gao, Z.; Guo, Y. Backstepping-based six degrees of freedom adaptive control for spacecraft tracking a noncooperative target. *J. Circuits Syst. Comput.* **2024**, *33*, 2450004. [CrossRef]
14. Ye, C.; Shao, J.; Liu, Y.; Yu, S. Fuzzy active disturbance rejection control method for an omnidirectional mobile robot with MY3 wheel. *Ind. Robot.* **2023**, *50*, 706–716. [CrossRef]
15. Oei, M.; Sawodny, O. Quaternion attitude and disturbance observer mass estimation with low-cost sensors for industrial vehicles with trailers. *Mechatronics* **2023**, *93*, 102991. [CrossRef]
16. Yang, N.; Zhang, S.; Li, X.; Li, X. A new model-free deadbeat predictive current control for PMSM using parameter-free Luenberger disturbance observer. *IEEE J. Emerg. Sel. Top. Power Electron.* **2022**, *11*, 407–417. [CrossRef]
17. Yuan, L.; Jiang, Y.; Xiong, L.; Wang, P. Sliding mode control approach with integrated disturbance observer for PMSM speed system. *CES Trans. Electr. Mach. Syst.* **2023**, *7*, 118–127. [CrossRef]
18. Gil, J.; You, S.; Lee, Y.; Kim, W. Nonlinear sliding mode controller using disturbance observer for permanent magnet synchronous motors under disturbance. *Expert Syst. Appl.* **2023**, *214*, 119085. [CrossRef]

19. Yin, Y.; Liu, L.; Vazquez, S.; Xu, R.; Dong, Z.; Liu, J.; Leon, J.I.; Wu, L.; Franquelo, L.G. Disturbance and uncertainty attenuation for speed regulation of PMSM servo system using adaptive optimal control strategy. *IEEE Trans. Transp. Electr.* **2022**, *9*, 3410–3420. [CrossRef]
20. Guo, G.; Yang, Z.; Chen, D.; Xing, X.; Houzhong, Z. Observer-based lateral stability adaptive control method for distributed drive electric vehicles. *Proc. Inst. Mech. Eng. Part D J. Automob. Eng.* **2023**, *237*, 1277–1289.
21. Mai, G.; Wang, H.; Wang, Y.; Wu, X.; Jiang, P.; Feng, G. Nonlinear extended state observer and prescribed performance fault-tolerant control of quadrotor unmanned aerial vehicles against compound faults. *Aerospace* **2024**, *11*, 903. [CrossRef]
22. Hernandez-Sanchez, A.; Chairez, I.; Poznyak, A.; Andrianova, O.; Chertopolokhov, V. Cueing end-effector acceleration of a two-link robotic arm by dynamic averaged sub-gradient integral sliding mode control. *Asian J. Control* **2023**, *25*, 2577–2587. [CrossRef]
23. Alika, R.; Mellouli, E.M.; Tissir, E.H. Adaptive modified super-twisting sliding mode control based on PSO with neural network for lateral dynamics of autonomous vehicle. *Int. J. Model. Identif. Control* **2023**, *42*, 296–311. [CrossRef]
24. Ahangarkolaei, J.M.; Izadi, M.; Nouri, T. Applying sliding mode control to suppress double frequency voltage ripples in single-phase quasi-z-source inverters. *J. Chin. Soc. Electr. Eng. (Engl. Ed.)* **2023**, *9*, 671–681.
25. Aziz, J.G.A.A.; Khan, R.A. Encoderless five-phase PMA-SynRM drive system based on robust torque-speed estimator with super-twisting sliding mode control. *J. Chin. Soc. Electr. Eng. (Engl. Ed.)* **2023**, *7*, 54–62. [CrossRef]
26. Jia, W.; Tai, K.; Wang, X.; Dong, X.; Ou, M. Design and simulation of intra-row obstacle avoidance shovel-type weeding machine in orchard. *Agriculture* **2024**, *14*, 1124. [CrossRef]
27. Cheng, K.; Hu, T.; Wu, D.; Li, T.; Wang, S.; Liu, K.; Tian, Z.; Yi, D. Heterogeneous UAV swarm collaborative search mission path optimization scheme for dynamic targets. *Int. J. Aerosp. Eng.* **2024**, *2024*, 6643424. [CrossRef]
28. He, Y.; Shi, X.; Lu, J.; Zhao, C.; Zhao, G. Grouping formation and obstacle avoidance control of UAV swarm based on synchronous DMPC. *Int. J. Aerosp. Eng.* **2024**, *2024*, 4934194. [CrossRef]
29. Hasanbasri, F.H.; Said, I.A.; Gururajan, S. Obstacle avoidance algorithm for circular objects during multiple target waypoints navigation mission of a fixed-wing UAV using polar coordinates. In Proceedings of the AIAA AVIATION 2023 Forum, San Diego, CA, USA, 12–16 June 2023; p. 3335.
30. Ye, B.; Liu, C.; Han, F.; Hu, X.; Chen, L. Obstacle avoidance motion planning for space redundant manipulator based on improved RRT algorithm. *J. Phys. Conf. Ser.* **2024**, *2764*, 012052. [CrossRef]
31. Li, J.; Yin, S.W.; Shi, G.S.; Wang, L. Optimization of indoor thermal comfort parameters with the adaptive network-based fuzzy inference system and particle swarm optimization algorithm. *Math. Problems Eng.* **2017**, *2017*, 1–13. [CrossRef]
32. Athans, M.; Stein, G.; Valavani, L. *Nonlinear and Adaptive Control*; Springer: Berlin/Heidelberg, Germany, 1986.

actuators

MDPI

Article

A New Dual Steering System in a Compact Tractor

Giorgio Paolo Massarotti [1,*], German Filippini [2], Gustavo Raush Alviach [1], Pedro Javier Gamez-Montero [3,*] and Esteban Codina Macia [1]

1 Department of Fluid Mechanics, CATMech-LABSON, Universitat Politecnica de Catalunya, Colom 11, 08222 Terrassa, Spain; gustavo.raush@upc.edu (G.R.A.); esteban.codina@upc.edu (E.C.M.)
2 Escuela de Ingeniería Mecánica, Facultad de Ciencias exactas, Ingeniería y Agrimensura, Universidad Nacional de Rosario, Rosario S2000CGK, Argentina; germanf@fceia.unr.edu.ar
3 Department of Fluid Mechanics, Universitat Politecnica de Catalunya, Campus Terrassa, Colom 7, 08222 Terrassa, Spain
* Correspondence: giorgio.paolo.massarotti@upc.edu (G.P.M.); pedro.javier.gamez@upc.edu (P.J.G.-M.)

Abstract: To achieve optimal controllability in a dual steering tractor (a four-wheel, iso-diametric tractor equipped with a dual-hydraulic steering system), this study proposes a coordinated approach that combines experimental testing (using a special agricultural tractor) with numerical analysis of the entire vehicle, developed in Bond Graph-3D. For certain crops, a dual steering vehicle is used to meet the needs of professionals who require easy maneuverability in narrow spaces and/or reduced steering time. This study aims to explore the reasons behind the need for dual steering tractors, highlighting the advantages and disadvantages of these two different configurations and ultimately focusing on the combined benefits of both. Based on an extensive review of the literature and drawing from previous studies, this paper analyzes aspects such as the variation in noise levels (or comfort level) experienced at the steering wheel when switching from Ackermann steering to a dual steering system. After outlining the theoretical methodology used to describe the model, both experimental and numerical analyses of a vineyard tractor in operation are presented. The goal of this work is to provide guidance on design methods and, through the Bond Graph-3D model, suggest the best control algorithms to minimize steering noise and enhance driving comfort. This research aims to pave the way for future control strategies in electrohydraulic steering systems.

Keywords: dual steering; compact tractor; hydraulic system; Ackermann steering; articulated steering; Bond Graph-3D

1. Introduction

The main goal of this research is to identify an optimal control algorithm that enhances maneuverability at low speeds, with a reduced tractor turning radius, while maintaining stability and comfort at high speeds, similar to a classic Ackermann vehicle. Dual steering systems integrate both front and articulated steering methods, aiming to improve stability and optimize operational efficiency by reducing cycle times. Among early studies, researchers such as Horton and Crolla [1] have investigated lateral stability using planar linearized three-degree-of-freedom (3-DOF) models, yielding results consistent with the issues reported by manufacturers. For articulated vehicles, they used a torsional spring and damper at the articulation joint to model the hydraulic cylinders between the front and rear sections. Other researchers have taken a different approach by integrating 3-DOF vehicle models with the inclusion of normalized equations of the steering valve and hydraulic cylinder, as proposed by Merritt [2] and proved by Azad [3]. Figure 1 shows various configurations, along with simplified expressions, that allow for the estimation of the inner and outer radii for each configuration. The vehicle being analyzed is a tractor designed for agricultural use, specifically in vineyards. The tractor's ultra-compact design requires an independent steering system for both the front axle and central articulation to ensure consistent chassis behavior at high speeds and to enhance steering capabilities during maneuvers. With specific reference to Figure 1, Table 1 presents the main geometric data of the tractor used in the experimental tests.

$$R_O = \frac{L}{\sin\alpha} \tag{1}$$

$$R_i = \frac{L}{\tan\alpha} - T \tag{2}$$

$$\alpha = \arctan\left(\frac{T}{2F}\right) \tag{3}$$

$$R_O = \frac{F+x}{\tan\alpha} + \frac{T}{2} = \frac{F + \frac{R}{\cos\alpha}}{\tan\alpha} + \frac{T}{2} \tag{4}$$

$$R_i = \frac{F+x}{\sin\alpha} - y - \frac{T}{2} = \frac{F + \frac{R}{\cos\alpha}}{\sin\alpha} - R\cdot\tan\alpha\frac{T}{2} \tag{5}$$

$$R_i = \frac{\frac{T\sin\alpha}{2} + F\cdot\cos\alpha + R}{\sin(\theta+\alpha)} \tag{6}$$

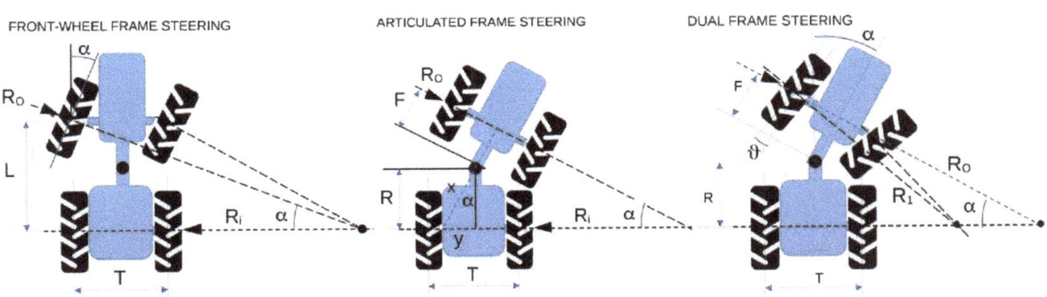

(a) (b) (c)

Figure 1. Schematics of front steering (**a**), articulated steering (**b**), and dual steering systems (**c**).

Table 1. Comparative analysis for one specific tractor (based on geometric layouts).

Variables	Description	Akermann Steering System	Articulated Steering System	Dual Steering System
Ro [m]	Outer diameter	3.7	3.8	3.8
Ri [m]	Inner diameter	2.2	3.48	2.2 (R1)
α	Angle on outer diameter	24°	26°	26°
L [mm]	Wheelbase	1495	1495	1496 (R1)
T [mm]	Median roadway	1150	1150	1150
F [mm]	Central pin—front axle	-	650	650
R [mm]	Central pin—rear axle	-	845	845
ϑ	Steering angle	-	-	24
% Ro	Percentage on Ro (ref. Ackermann)	-	+4.2	−40.3

Equations (1) and (2) refer to Figure 1a. Equations (3) and (5) refer to Figure 1b, and Equation (6) refers to Figure 1c. Equation (4) is the same for Figure 1b,c. It is noteworthy that the turning radius of the system with central articulation is greater than the corresponding turning radius of a front-wheel steering system, such as the Ackermann. However, it is important to mention that tractors come in various widths (median roadway), and in this case, the smallest value has been used. This study primarily focuses on enhancing the driver's steering experience to minimize unnecessary corrections during operation (as clearly highlighted in Section 2) and, secondarily, on a numerical model that encompasses all of the vehicle's components and systems. To achieve the main goal, it was necessary to highlight the advantages and disadvantages of these different steering architectures, first, by reviewing the literature on similar problems and conditions studied by other researchers, and second, by employing a technical approach. To describe the vehicle behavior in each part (or sub-system), a numerical approach is used. Using a bond graph methodology, the system demonstrates a thorough integration and understanding of the physical principles governing it. This process allows researchers to "read" and conceptualize the system holistically, fostering a detailed understanding of its internal interactions. Multidomain simulation tools offer preprogrammed libraries for mechanical and hydraulic systems, simplifying the modeling process. However, they abstract key physical phenomena within prebuilt components, which can limit users' insights into the underlying physics. This approach prioritizes convenience over a deeper, conceptual understanding, with users typically focusing on assembling predefined modules rather than fully engaging with system interactions. A bond graph is a graphical language that allows researchers to systematically derive the state equations of the system from the diagram itself. Using bond graphs ensures that no aspects of the system's dynamics are overlooked, providing a solid foundation for implementing control strategies and regulatory systems. For these reasons, in this work, 20-Sim software was used. An overview of important software tools specifically designed for multibody and multiphysics simulations is presented in Table 2.

Table 2. Main software for multibody and multiphysics simulations.

Category	Software Name	Characteristics
A	ADAMS, SIMPACK, RECURDyn, ALTAIR MS	General purpose software, often part of larger platforms, supporting complex multiphysics simulations
B	MODELICA, SimCenter AMESim, MapleSim, 20-Sim, MAthWorks (MATLAB, Simulink)	Programs focused on the dynamics of physical or multidomain systems with extended functionality (e.g., bond graph methods)

The literature concerns off-road machinery simulation. This study identified two trends in this field: the first includes systems such as the MODELICA Multibody library with HyLib (Otther P. and Beather M., 2003 [4]; Patil A. and Radle M., 2021 [5]), AMESim with the Planar Mechanical library (Altare G., 2009 [6] and Altare et al. 2012 [7]; Casoli P. and Alvin A., 2011 [8]), and MATLAB/Simulink with SimMechanics (Simscape) (Prabhu S.M., 2007 [9]; Jhala H.S., 2023 [10]; Hong T.D. et al., 2024 [11]); the second encompasses systems that require specialized interfaces to exchange variables across domains during simulation. Examples include co-simulations such as ADAMS and AMESim (Roccatello et al., 2007 [12]), ADAMS and MATLAB (Sapietova A. et al., 2012 [13]), and AMESim and Virtual Lab Motion (Prescott, 2009 [14]).

On the introduction of the tractor virtual model in 3D-BG, Xu and others [15] advocate for the incorporation of cylinder speeds as inputs, yet they subsequently employ algorithms based on the principle of minimum resistance within the Lagrangian framework. Figure 2 presents a block diagram representing the Lagrange approach. It is important to note a discrepancy: while the depicted inputs are forces, this contradicts the physical principle that the input variables from the hydraulic actuators, applied to the multibody model, should be velocities rather than forces. Similarly, the inputs from the tire/wheel interaction are forces exerted by the ground, and they impose a certain kinematic constraint dictated by the cylinders. Specifically, the mechanical domain is modeled using a 3D-BG vectorial domain, while the hydraulic domain is represented through a 1D-BG scalar domain. De las Heras and Codina [16] have a proven track record and extensive experience employing this methodology. This article focuses on the structural baseline part, consisting of a finite number of rigid bodies and joints. Several modules of this toolbox have demonstrated successful application in vehicle modeling to predict dynamic behavior, as shown by Filippini [17,18], and have been effectively employed in another project related to vehicle fault diagnosis by Silva L. et al. [19]. The spatial motion of the rigid bodies is determined using the well-established Euler equations. Moreover, it is highly capable of interacting with multiple energy domains, as indicated by Borutzky [20] and Thoma J. [21]. Regarding multibody systems, a variety of bond graph techniques have been introduced: Karnopp and Rosenberg [22] have outlined a strategy for generating bond graphs using Lagrange's equations, and Bos [23] has presented a modular strategy rooted in Newton–Euler equations. Various experimental and numerical tests have been conducted by the authors to establish a clear path for investigating the optimal approach to achieving controllability in travel and drivability in dual steering mode. The first technical experiment, carried out using a tractor prototype, aimed to transition from an Ackermann steering system to a dual steering system. This was followed by a model of the entire tractor, created using 20SIM v5.1. [24]. This article demonstrated a possible control strategy for a steering system.

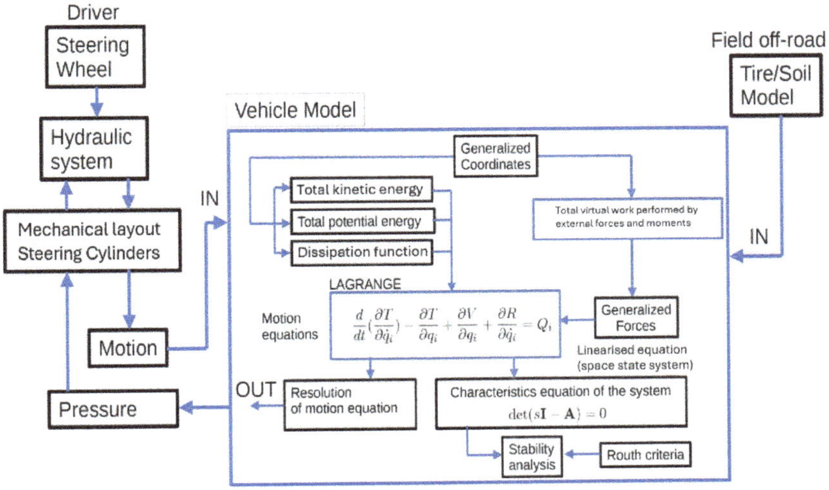

Figure 2. Block diagram schematizing the Lagrange approach [6].

2. Methodologies

Initially, an experimental design was produced to highlight the differences in instability between the Ackermann steering mode and the dual steering mode. The differences are discussed in Section 4. Following this, a numerical analysis with a 3D simulation was performed to compare the experimental results with all available (and unavailable) data introduced into the simulation. The tractor used in the field experiments is an isodiametric tractor (a tractor with all wheels of the same diameter), equipped with a 73 kW engine and a new electrohydraulic dual steering system. Figure 3 shows the prototype. The chassis used for the experiments is based on a common dual steering system tractor (with mechanical links), which was modified to incorporate the aforementioned hydraulic system. Due to confidentiality, detailed descriptions of the hydraulic components used cannot be disclosed in this article.

(a) (b)

Figure 3. (**a**) Prototype tractor with a new ECU and new hydraulic system under analysis; (**b**) roto transducer installed under the steering wheel to log the steering movement in the two working conditions under investigation: traveling in dual steering mode and traveling in Ackermann mode.

Figure 4 depicts the specific track used for conducting field tests and evaluating both stability and vehicle trajectories. The letters a–e depicts the speed trend. The track was asphalted (not tarmac), dry (on a sunny day at 28 °C), and in a test area closed track in northern Italy. The test track consisted of multiple segments, starting at point "a", which marks the beginning of the trial. Prior to the start of the experiment, a briefing session was held, involving full steering maneuvers to the left and right. This was carried out to demonstrate the functionality of each sensor and the data-logging equipment. No ground experiment was performed during this period. The main sensors used for the experiment are described in Table 3.

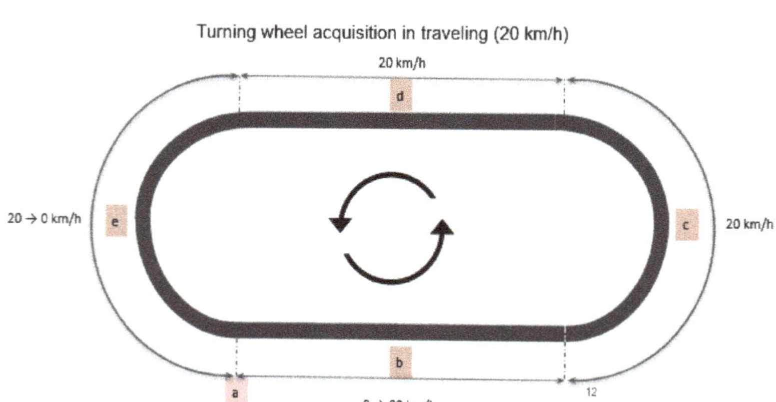

Figure 4. Circuit for the stability experiment in dual steering and Ackermann mode.

The experiments were conducted using a portable HYDAC HMG 4000 data recorder, a powerful mobile handheld measuring device with a 5.7″ color touchscreen, designed for recording and storing measurement data. The J1939 CAN protocol was used for the experiment, connecting to the ECU for central steering control and the WDGA encoder. This model aims to encompass the specifications of both steering systems, whether synchronized or not, in conjunction with a dynamic model of the articulated vehicle. The proposed methodology, which considers cylinder velocities as inputs, provides a more straightforward and clear approach for dynamic modeling.

Table 3. Main sensors used in tractor experiments.

Sensor Type	Sensor Nomenclature	Accuracy/Main Characteristic
Rotational transducer, i_s	Wachendorff WDGA 58E-12-1213-COA-B00-CC5-AHW Magnetic Absolute Encoder Binary/CANopen	12-bit ST/13-bit MT 10 . . . 32 V 6000 min^{-1}
Angular sensor	ELEN Angular sensor ±60° 10–30 V ball bearings	Linearity < ±1%
Logging instrument	HYDAC HMG 4000	J1939 CAN Protocol

Before analyzing the dynamic model, it is important to provide an overview of the hydraulic circuit used to control the dual steering system in the tractor under analysis. The vehicle under study is a tractor with isodiametric wheels, central articulation, and directional front wheels, as depicted in Figure 5. Additionally, Figure 6 presents a simplified diagram of the hydraulic circuit with a description for each component shown in Table 4. The approach encompasses two primary modules: the hydraulic domain model and the mechanical domain model. Firstly, this study addresses the actuators for the implements, which include both the steering system and its associated hydraulic drive system. These systems are treated as one-dimensional (1D) systems in our modeling approach, utilizing fundamental 1D bond graph elements. Finally, the validity of this approach will be confirmed by comparing its numerical results with a series of experimental data obtained from field tests.

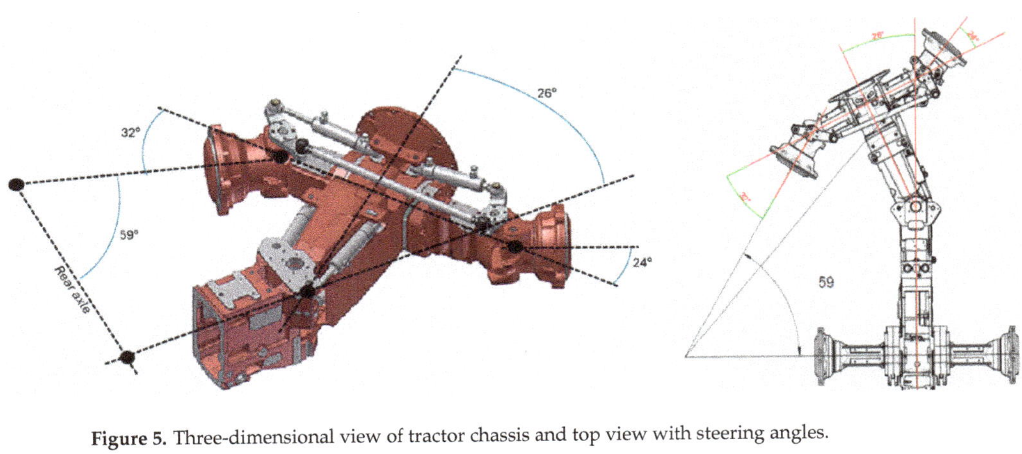

Figure 5. Three-dimensional view of tractor chassis and top view with steering angles.

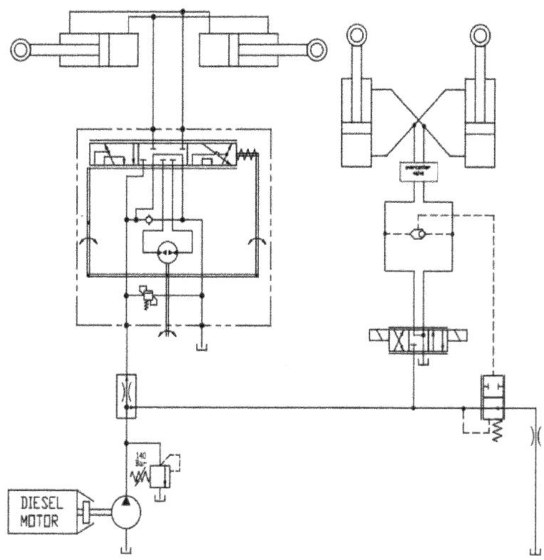

Figure 6. Simplified diagram of hydraulic circuit.

Table 4. Hydraulic components.

Symbol	Description/Characteristics
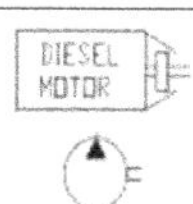	Diesel engine 75 kW, 2100 rpm Pump Cv = 15 cm^3/rev
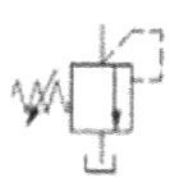	Relief valve 140 bar
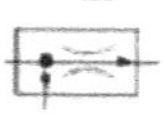	Control flow valve (3 ways) 13 L/min regulated
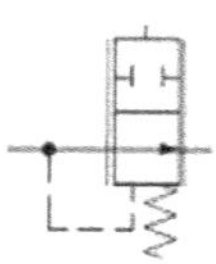	Load sensing pressure compensator Dp = 5 bar pressure setting
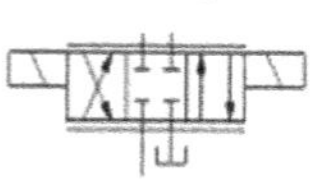	4/3 proportional directional control valve (30 L/min nominal) 12 V electric actuator
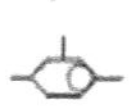	Shuttle valve cartridge
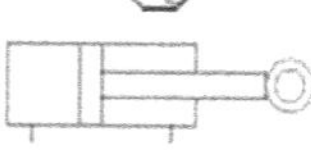	Double-effect cylinder 48 × 22 × 130 front—42 × 22 × 87 central
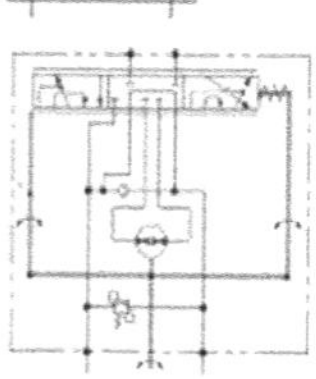	Hydrostatic steering unit 100 cm^3/rev (Gerotor + rotary valve)
	Tank
	Hydraulic load 18–20 bar

The dual steering tractor model, depicted in Figure 7, was designed to emulate a tractor with its two rigidly joined bodies (rear and front sections), with the pivot point located near the geometric center of the vehicle. Alongside the pivot point system, the vehicle's two bodies are connected by hydraulic actuators integrated into the central steering system. It is worth noting that this type of vehicle is not equipped with suspensions, and the cantilever engine is mounted in a forward position, affecting the center of gravity (CoG) and the distribution of mass. More than 50% of the tractor's weight is supported by the front axle, mainly due to the weight of the diesel engine. The vehicle's sections are modeled as rigid bodies, with a local coordinate system (x, y, z) attached to their center of mass and aligned with their principal axes. Each section has a mass mmm, and moments of inertia JrJ_rJr (roll) about the x-axis, JpJ_pJp (pitch) about the y-axis, and JyJ_yJy (yaw) about the z-axis. To determine the motion of a rigid body in space, a set of relations known as Euler equations is used. Additionally, it is necessary to model the interconnections between rigid bodies (joints) and the actions of the external environment on them. In the model, a joint represents the force relationship between the rigid bodies it connects, depending on their position and relative orientation. In summary, the dual steering tractor is composed of multiple interconnected rigid bodies, including the rear- and front-section bodies (orange box), tie rod (red box), knuckles (green box), and wheels and wheel/tire assemblies (yellow box), as can be seen in the 3D-BG of Figure 7.

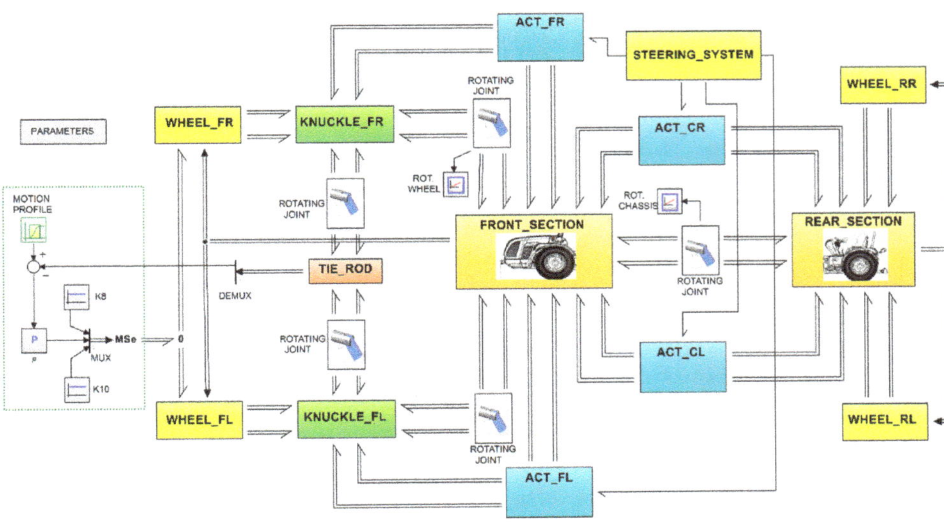

Figure 7. Bond graph virtual model of articulated tractor (3D-GB upper layer).

Given the hierarchical structure—model, submodels, BG basic elements, and constitutive equations—Figure 7 presents the bond graph submodel schematic implicit in the front-section icon. To establish the connection between the rigid body of the front section and the central rotating joint (white box), a translation is required. Additionally, a series of translations (T elements) are necessary to facilitate the interconnections between this rigid body and its interactions with external environmental forces, linking them to the steering hydraulic actuators (blue box) and the wheels (orange box).

Next, this study will focus specifically on the wheel submodule, which, as shown in Figure 8, incorporates the behavior of the wheel as a rigid solid (green box) and the tire/soil interaction (yellow box).

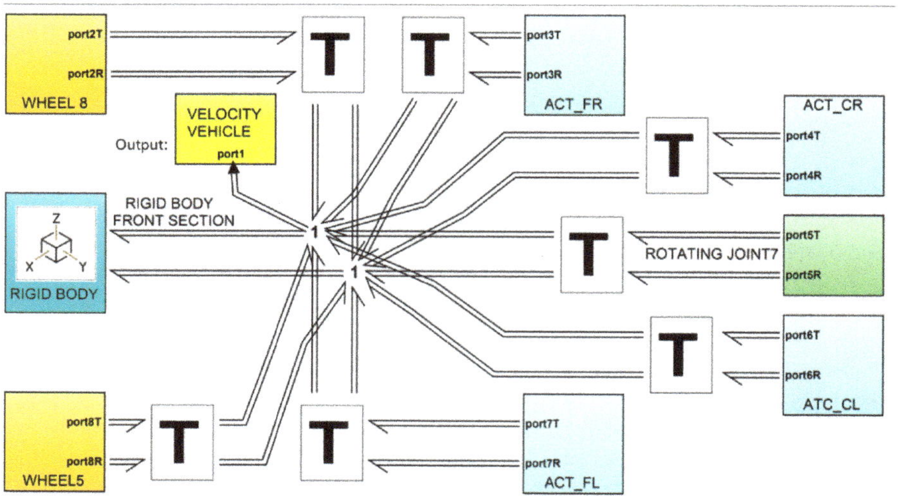

Figure 8. Front-section platform submodel (3D-BG scheme—sublayer n°1).

In Figure 9, which shows a part of the scheme from Figure 6 in the bottom left corner, the icons of the hydraulic actuators (blue box) responsible for the kinematics of the operator-imposed steering are visible. These submodels are connected to the hydraulic system submodel (yellow box in Figure 6). Additionally, Figure 9 illustrates the hierarchical structure of the hydraulic cylinder, highlighting the connection between the mechanical domain and the hydraulic domain.

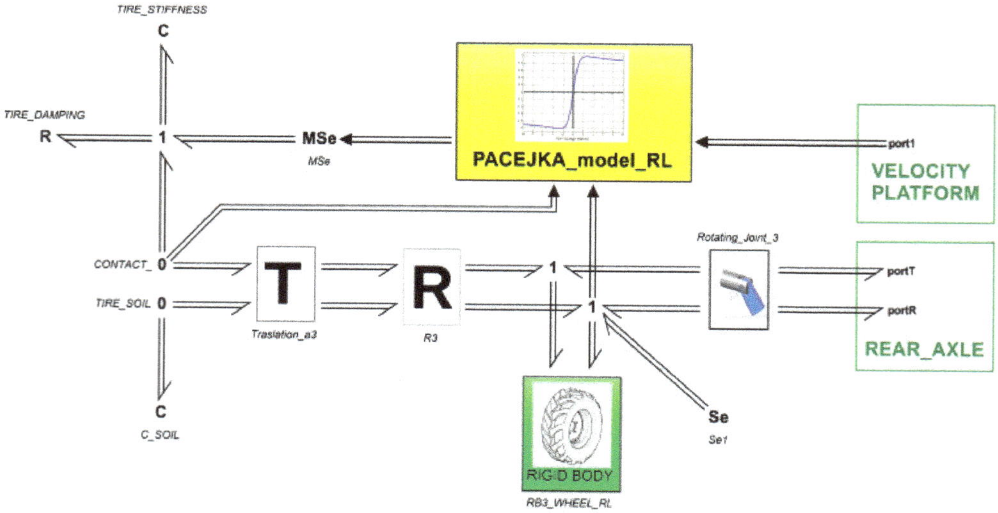

Figure 9. Wheel submodel tire/soil interaction (3D-BG scheme—sublayer n°1).

The four hydraulic actuator submodels (RL, RR, FL, FR), which are responsible for the dual steering system, are configured by combining the mechanical domain (a hydraulic cylinder consisting of two mechanical elements, the housing or cylinder, and the piston/rod assembly that slides inside the former) and the hydraulic domain (which, powered by energy from the ICE engine to the hydraulic group, generates relative motion between the two mechanical parts). In other words, the flow provided by the hydraulic group is transformed into rod velocity, which in turn serves as input for the multibody system. This motion is then returned to the hydraulic system as resistive force and working pressure, ultimately requiring a certain torque from the ICE engine.

Hydraulic Domain

As described in the previous section and outlined in Figure 6, the four hydraulic actuators are responsible for the dual steering system implemented in the tractor. Building on this, using the hydraulic domain as a reference point, Figure 10 presents the scalar bond graph structure developed for the proposed hydraulic circuit to simulate the dual steering system (front and central steering). Additionally, the 3D prismatic joint is connected to a 1D bond graph diagram corresponding to the fluid mechanics domain, incorporating equations of mass conservation (including oil compressibility) and the conservation of momentum.

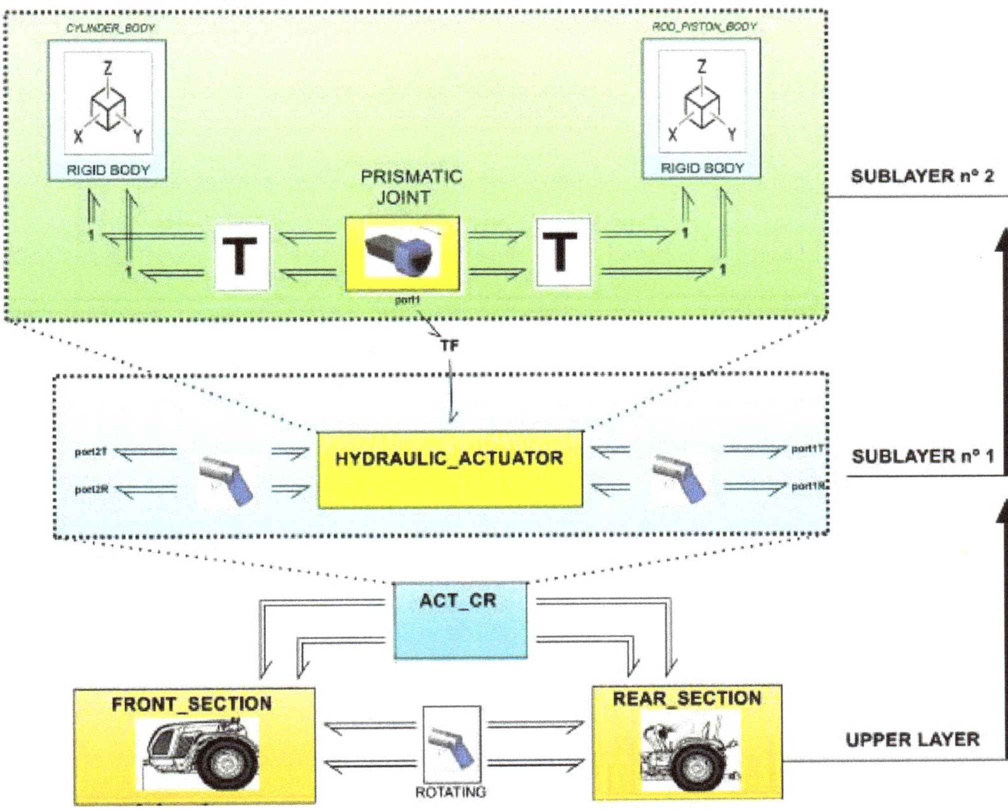

Figure 10. Hierarchical structure (by layer) of the hydraulic cylinder.

Other submodels worth mentioning in Figure 10 are those corresponding to the hydraulic block (green box), called orbitrol, and the 3-way valve (yellow boxes). The scalar bond graph diagrams for these two components (orbitrol) are shown in Figures 11 and 12. The 3-way valve is shown in Figure 13.

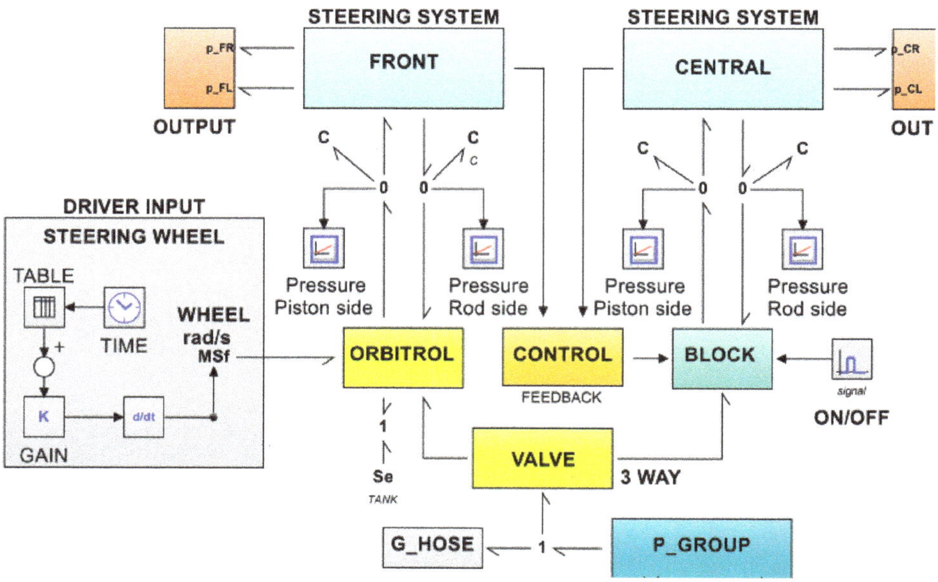

Figure 11. Steering hydraulic system (1D-BG scheme—sublayer n°1).

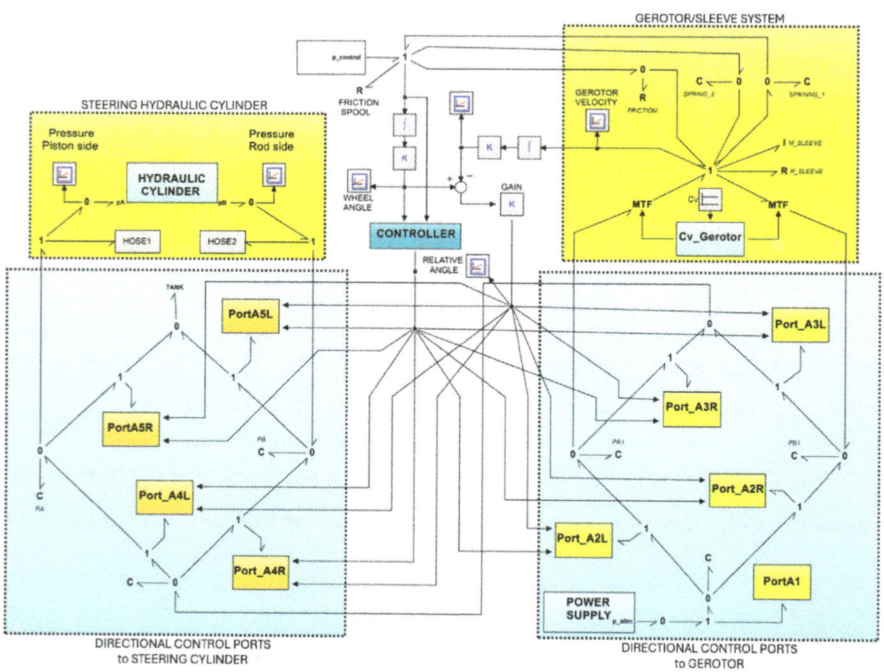

Figure 12. HSU orbitrol hydraulic component (1D-BG scheme—sublayer n°2).

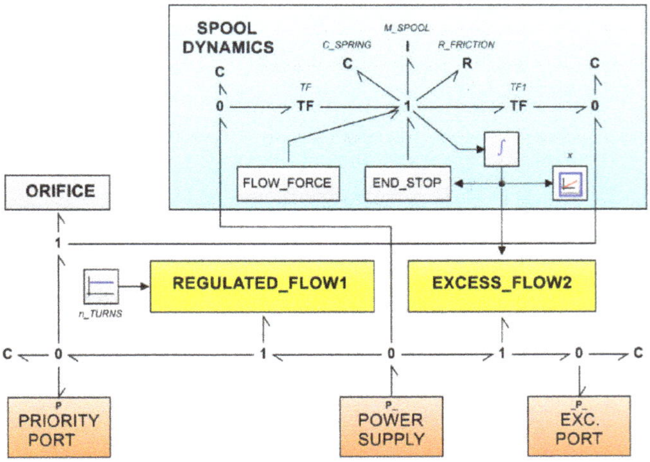

Figure 13. Three-way valve (priority concept) (1D-BG scheme—sublayer n°2).

The key submodels in these two bond graph diagrams correspond to the icons labeled "Port_iii", which physically represent the nine variable orifices of the rotary valve (spool/sleeve assembly) in the case of the orbitrol element, and the corresponding orifices of the directional control valve in the case of the three-way valve. To conclude this section, it is necessary to describe the submodel referred to as the block (green box in Figure 10). In this submodel, it is important to highlight the presence of the proportional valve with pressure compensation, for which the corresponding bond graph diagrams are illustrated in Figures 14 and 15.

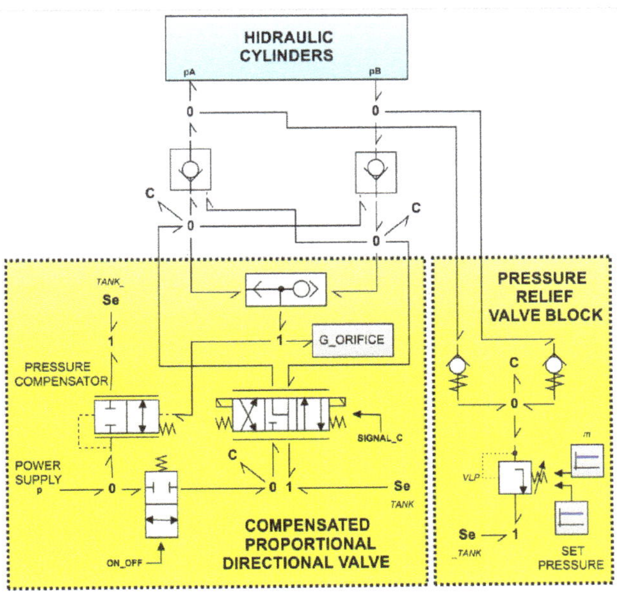

Figure 14. Central steering system hydraulic circuit (submodel n°2).

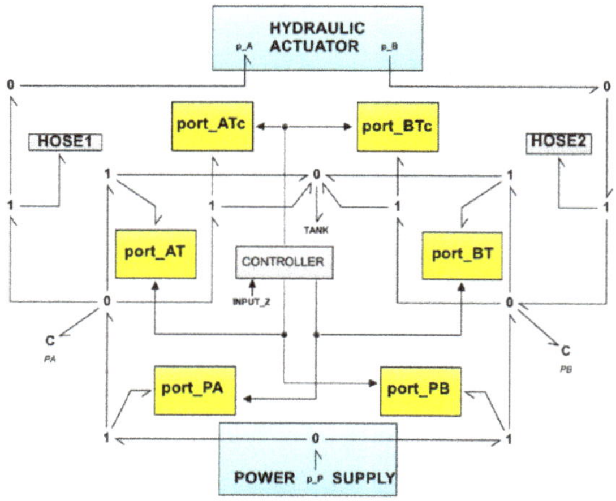

Figure 15. Proportional directional control valve (1D-BG scheme—sublayer n°3).

In these diagrams, the presence of variable-section throttles, also known as metering, is once again emphasized; they control the flow of oil feeding the hydraulic actuators of the steering systems. For this reason, it is pertinent to briefly explain how they have been simulated and the basic equations required for their modeling. On the other hand, it is unnecessary to investigate the description of the other hydraulic elements, as they are more conventional and well known to those familiar with fluid power technology and scalar bond graph diagrams. Table 5 provides a summary of the most significant parameters of the modeled hydraulic components, offering a comprehensive overview of their key characteristics and attributes.

Table 5. Main simulation factors used in the numerical analysis.

Hydraulic Parameters	Value	Units
Oil density	875	kg/m^3
Oil bulk modulus	17.500	bar
Oil kinematic viscosity	46	mm^2/s
Pump maximum displacement	6.50	cm^3/rev
Pump volumetric efficiency	0.93	
Pump hydraulic–mechanical efficiency	0.96	
Engine speed	2.300	rpm
Relief valve cracking pressure	18	MPa
Proportional directional control valve 4/3	SP10-47D	serie
Flow control valve pressure compensated	BFCV12	serie
Pressure compensating element	EC16-32	serie
Viscous friction coefficient	50	N/m/s
Cylinder leakage coefficient	0.001	L/min/MPa
Piston diameter of front steering hydraulic cylinder	48	mm
Rod diameter of front steering hydraulic cylinder	22	mm
Travel of front hydraulic cylinder	131	mm
Piston diameter of central articulation hydraulic cylinder	42	mm
Rod diameter of central articulation hydraulic cylinder	22	mm
Travel of central articulation hydraulic cylinder	87	mm
Steering unit (HSU) displacement	80	cm^3/rev
HSU spring rate (estimated)	6.40	N∗cm/rad
Steering wheel angles	±700	degree

This structure model and its characteristics started with the modelization of all information available (experimental data and bibliography) when this study started. The first improvement aims to make it easier to structure each submodel in order to reduce the simulation time. The identification of minor aspects that do not aid the quality of the results is the right way to proceed.

3. Numerical Results and Validation Procedure

The trends displayed in Figure 16 represent the steering displacement logged by the rotary sensor, as reported in Table 2. Sector "a" corresponds to the warm-up phase, where the tractor is stopped and a full steering displacement is applied. This action is used to identify any issues before the tests. Sectors "c" and "d" highlight a significant shift in the tractor's behavior. A noticeable trend can be observed at the corners, where the external forces are higher. The tendency is an increase in steering noise, especially in dual steering mode. Finally, sector "e" represents the last corner before the finish line. As shown in Figure 16 (8860 samples every 0.02 s as the logging time), the variability between the maximum and minimum values changes as the system switches from dual steering mode to Ackermann mode.

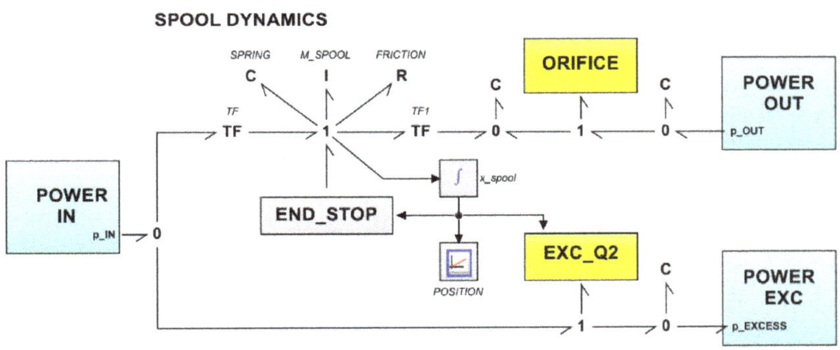

Figure 16. Hydraulic pressure compensator (1D-BG scheme—sublayer n°3).

Parameter uncertainties are the main source of error in the simulation (real viscous friction coefficients, etc.). Inertial and geometrical parameters of the mechanical system, including the hydraulic actuator components, can be easily obtained from the CAD model. However, other parameters are more elusive, such as the elastic-damping parameters of the tires and, in particular, those related to verifying specific parameters of the hydraulic system and its components. In some cases, confidence has been placed in the technical expertise and professional experience of the laboratory technicians involved in this work, particularly in the art of simulating components and estimating more subjective parameters. Nonetheless, this result represents a starting point.

4. Discussion

The tractor accelerates from 0 to 20 km/h while operating in dual steering mode. During the turn, the tractor maintains a constant speed of 20 km/h, a pattern that continues through sector d. Upon entering sector e, the tractor decelerates until it reaches a speed of zero. Once it reaches point a, there is a switch to Ackermann mode, and another turn is executed under the same acceleration and speed conditions. As described in the following data acquisition process, the first turn is performed in dual steering mode, followed by the second turn in Ackermann mode. When traversing the circuit at the same speed and vehicle condition in Ackermann mode, the steering noise lower, and the feeling of controllability is improved. According to the experimental data, it is evident that steering ability improves by 37.5% during cornering and by 40% in a straight-line direction, passing through the Ackermann configuration to dual steering configuration. This can be explained by the low impact of the tractor's dynamics, and consequently, a smaller number of driving factors. In dual steering mode, the tendency difference is more pronounced. This can be attributed to both the vehicle's dynamics (where some damping parameters have been hypothesized) and the metering devices (flow control valve and orbitrol model), where the metering function was not fully known, as indicated by the hysteresis in the experimental graph. This supports the conclusion that the concept of having a vehicle capable of switching between an Ackermann configuration and dual steering mode makes sense, as it can be adapted to the different tasks required by the tractor (maneuvers in tight spaces vs. high-speed transport operations). The primary focus was to maintain tractor stability and ensure as linear a trajectory as possible. Due to the significant gain associated with even a small movement of the steering wheel in dual mode, corrective counter-steering was required whenever snaking behavior occurred to realign the tractor. The results from simulations conducted using 20-SIM reveal differences in offset values. The input in the tractor model was the steering wheel acquisition (Figures 17 and 18) performed during the experiment with the tractor. Practically, in running the model, the goal was to achieve the same trend shown by the sensor acquisition on the tractor. These variations are due to the zero value assumed by the position transducers located at the rotation points near the wheels/articulation. Due to space constraints, the sensors could not be oriented identically along the tractor's axis due to the bulk of electrical connections and the space needed for cables (right/left steering). The values and trends in the graph shown in Figure 19 suggest that the numerical results closely follow the experimental trend in the Ackermann configuration and in a straight-line direction. However, the results differ for the dual steering system. Despite using the same control algorithm for both the front steering system and the central system, there is evidence of a lack of synchronization in the control system. A specific point worth mentioning is the difference observed in the central articulation (Figure 19, right), where a significant disparity is visible between the experimental trend and the numerical analysis. The input for the numerical analysis is the movement of the steering wheel, and the trends in the experimental and numerical data align for this aspect. However, the experimental data reveal a slight lag between the steering wheel movement and the movement of the front wheels, indicating elastic effects along the control chain (e.g., flexible hydraulic pipes, clearances between mechanical parts). The lag in the movement of the central articulation is more noticeable due to the combination of elastic elements (similar to the front steering) and the time required for the front steering variation detection, which is measured by the rotary sensor on the front wheels. The calculation took several hours (four) for just 10 s of simulation. The issues with the match between numerical and experimental data can be attributed to the absence of the central actuator in the numerical model. This could be identified as a delay in the actuation due to a restriction or an element that stiffens either the hydraulic model or the mechanical model. The data then undergo processing by the ECU before the hydraulic actuation reaches the cylinders. This results in a "congenital" actuation delay, which can only be reduced with a more rigid system (hydromechanical) and a faster electronic system. Analyzing the data further, the numerical model's complexity arises from the large number of operations and the time required to obtain numerical results. The four-wheel submodels (RL, RR, FL, FR in Figure 7) play a crucial role in determining the vehicle's behavior, as they are the only components directly interacting with the ground. In agricultural research, understanding the interaction between implements and unpaved soft soil is critical. Due to the soil's low tensile strength, it deforms permanently, making any modeling effort involving soil interaction nonlinear and subject to significant material variability. Empirical approaches and field tests have been the primary methods for analyzing tire/soil interaction. The Pacejka tire models [25] are commonly used in professional vehicle dynamics simulations due to their reasonable accuracy, ease of implementation, and quick solution times. In this study, while the wheel is typically modeled as a rigid body, dynamic effects such as friction between the tire and the ground and the tire's response to compression are also considered. The tire's response to compression is modeled using a spring and damper mechanism operating vertically with respect to the global coordinate system, with parameters derived from relevant references [26,27]. To analyze the forces along the axes (Fx for longitudinal and Fy for lateral) and moments about these axes (Mx for overturning, My for rolling resistance, and Mz for aligning torque), the general form of the Magic Formula, introduced by Pacejka, is adopted. This formula uses empirical approaches to model force or moment as functions of a carefully selected "slip" variable representing the interaction between the tire and the ground. This issue, common in articulated vehicles [28–33], is largely mitigated when the Ackermann configuration is activated [34]. It offers a valuable perspective for discussing the need to analytically describe the physical model in each part and provide crucial information regarding driving strategy.

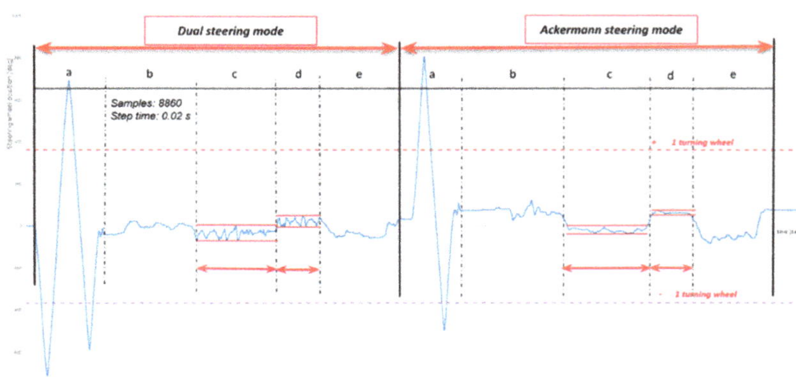

Figure 17. Circuit path for the stability experiment in dual steering and Ackermann mode where a–e are the speed segments of the Figure 4 (time vs. steering grade displacement).

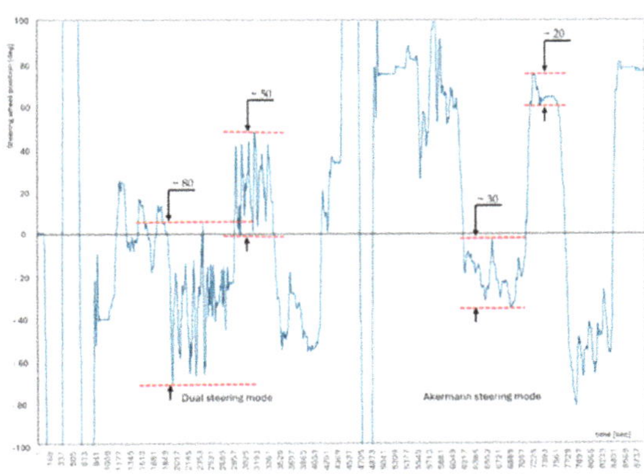

Figure 18. Zoomed-in view of the variations in the experiment stability in dual steering mode and Ackermann mode (time vs. steering grade displacement).

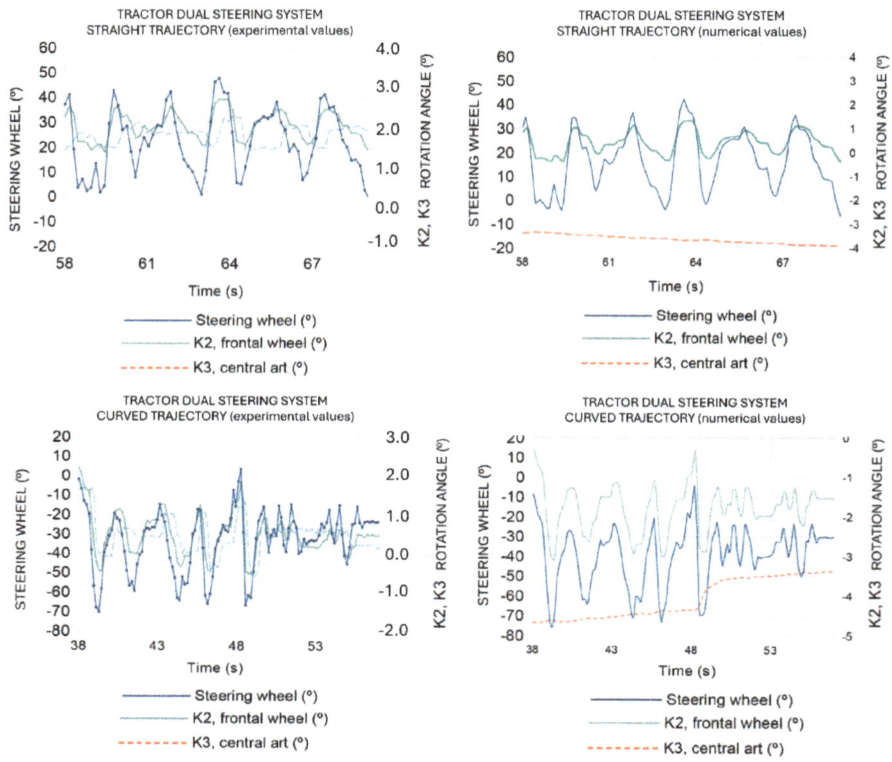

Figure 19. Experimental and numerical results regarding the dual steering systems (involving corrective counter of steering wheel movement in dual-mode steering) during curved and straight trajectory at 20 km/h.

5. Conclusions

This work has demonstrated a difference in terms of lateral stability between Ackermann and dual steering architectures under the same operating conditions, quantifying the variation between them. The results validate the large body of data available in the literature, which is only partially referenced in this work. The difference in terms of operator sensitivity lies in the increased attention required when using the tractor at a speed of 20 km/h, as the recall and correction operations are greater in terms of amplitude. The numerical analysis has highlighted the inherent difficulty of fully describing all aspects present in the dynamic field, noting that control in dual steering mode is complex and challenging to implement. The acquisition data presented above show a difference in terms of the vehicle's steering ability, with the Ackermann mode being considered the most effective at high speeds. For this reason, the Ackermann mode can be switched automatically once a certain speed threshold is exceeded. This speed threshold can be chosen based on the maximum speed usable in dual steering mode, which typically corresponds to the end-field maneuverer speed. In the near future, the time response between the front and central steering systems will be investigated to identify delays in electronic and hydraulic signals within the "master and slave" dual steering system. A threshold can be set, relative to the vehicle speed, to disable dual steering mode when a certain speed is exceeded. Based on experimental data from experienced operators, this threshold is set to around 8.5 km/h. This value is specific to the case in question (type of soil, type of tractor wheelbase and length, type of tires and diameters, hydraulic circuit setup, etc.), should be based on expert opinions, and should not be generalized to other applications. The switch speed must be taken into account in different environmental conditions, and this aspect is under study in relation to the tractor model. For future work, different terrains will also be investigated to evaluate the repeatability regarding comfortability under each condition.

6. Patents

The system analyzed in this work has been patented as EP3871950A1-HIGH-MANEUVERABILITY VARIABLE-STEERING TRACTOR.

Author Contributions: The investigation was led and supervised by E.C.M. Experimental works and data processing were completed by G.P.M.; bond graph simulation models and illustrations were finalized by G.F., G.R.A., P.J.G.-M., G.P.M. and E.C.M. The manuscript was finalized by G.P.M. and P.J.G.-M. All authors have read and agreed to the published version of the manuscript.

Funding: This research received no external funding.

Data Availability Statement: The data in this study are available on request from the corresponding author. The data are not publicly available due to privacy reasons.

Acknowledgments: The authors would like to express their heartfelt thanks to Technical Director of BCS S.p.A. (BCS Group), for the exceptional development work carried out by him and his team on the tractor steering system. We also extend our gratitude to Albert Porta Forradelles for his significant contributions to the development of the numerical model at the beginning of this project.

Conflicts of Interest: The funding sponsor had no role in the design of this study; in the collection, analysis, or interpretation of data; or in the writing of the conclusions and in the decision to publish the results. The authors declare no conflicts of interest.

References

1. Horton, D.N.L.; Crolla, D.A. Theoretical Analysis of the Steering Behaviour of Articulated Frame Steer Vehicles. *Veh. Syst. Dyn.* **1986**, *15*, 211–234. [CrossRef]
2. Merrit, H.E. *Hydraulic Control Systems*; Wiley: New York, NY, USA, 1967; 358p.
3. Lashgarian Azad, N. Dynamic Modelling and Stability Controller Development for Articulated Steer Vehicles [Internet]. University of Waterloo. 2007. Available online: http://hdl.handle.net/10012/2633 (accessed on 26 November 2024).
4. Beater, P.; Otter, M. Multi domain simulation: Mechanics and hydraulics of an excavator. In Proceedings of the Conference: 3rd International Modelica Conference, Linkoping, Sweden, 3–4 November 2003; The Modelica Association: Linkoping, Sweden, 2003.
5. Patil, A.; Radle, M. Hydraulic and multibody combined simulations for electric forklift design using modelica. *Int. J. Eng. Sci. Technol.* **2022**, *10*, 27–35.
6. Altare, G. Analisi e Modellazione del Circuito Idraulico di un Miniescavatore. Master's Thesis, Politecnico di Torino, Torino, Italy, 2009.
7. Altare, G.; Lovuolo, F.; Nervegna, N.; Rundo, M. Coupled Simulation of a Telehandler Forks Handling Hydraulics. *Int. J. Fluid Power* **2012**, *13*, 15–28. [CrossRef]
8. Casoli, P.; Alvin, A. Modelling of an Excavator Pump Nonlinear Model and Structural Linkage/Mechanical Model. In Proceedings of the 12th Scandinavian International Conference on Fluid Power, Tampere, Finland, 18–20 May 2011; pp. 25–40.
9. Prabhu, S.M. *Model-Based Design for Off-Highway Machine Systems Development*; SAE: Warrendale, PA, USA, 2007.
10. Jhala, H.S. A Multibody Simulation Approach to Identify Critical Instability Scenarios of a Forklift Truck. Master Automotive Technology. Master's Thesis, Eindhoven University Technology, Eindhoven, The Netherlands, 2023.
11. Hong, T.D.; Pham, M.Q.; Tram, S.C.; Tram, L.Q.; Nguyen, T.T. A comparative study on kinetics and dynamics of two dump truck lifting mechanisms using Matlab Simscape. *Theor. Appl. Mech. Lett.* **2024**, *14*, 100502. [CrossRef]
12. Roccatello, A.; Manc, S.; Nervegna, N. Modelling a Variable Displacement Axial Piston Pump in a Multibody Simulation Environment. *J. Dyn. Sys. Meas. Control* **2007**, *29*, 456–469. [CrossRef]
13. Sapietova, A.; Saga, M.; Novak, P. Multi software platform for solving of multibody systems synthesis. *Commun. Sci. Lett. Univ. Zilina* **2012**, *14*, 43–48. [CrossRef]
14. Prescot, W. Using multibody dynamics solvers in a Multiphysics environment. Multibody Dynamics. In Proceedings of the ECCOMAS Thematic Conference, Warsaw, Poland, 29 June–2 July 2009.

15. Xu, T.; Shen, Y.; Huang, Y.; Khajepour, A. Study of Hydraulic Steering Process for Articulated Heavy Vehicles Based on the Principle of the Least Resistance. *IEEE/ASME Trans. Mechatronics* **2019**, *24*, 1662–1673. [CrossRef]
16. de las Heras, S.; Codina, E. *Modelización de Sistemas Fluidos Mediante Bondgraph*; S. de las Heras: Barcelona, Spain, 1997.
17. Filippini, G.; Nigro, N.; Junco, S. Estudio del comportamiento dinamico de vehiculos terrestres utilizando la tecnica de bond graphs. *Mec. Comput.* **2005**, *XXIV*, 1559–1582.
18. Filippini, G.; Nigro, N.; Junco, S. Vehicle Dynamics Simulation Using Bond Graphs. December 2004.
19. Silva, L.; Delarmelina, D.; Junco, S.J.; M'Sirdi, N.K.; Noura, H. Bond Graph Based Fault Diagnosis of 4 W-Vehicles Suspension Systems I: Passive Suspensions. Engineering. 2007. Available online: https://www.google.it/url?sa=t&source=web&rct=j&opi=89978449&url=https://bicyt.conicet.gov.ar/fichas/produccion/1038957&ved=2ahUKEwif3Iu53viKAxVF8bsIHY9rCZMQFnoECBkQAQ&usg=AOvVaw0xTVpxpM8ouuZrtodG6tjX (accessed on 26 November 2024).
20. Borutzky, W. (Ed.) *Bond Graphs for Modelling, Control and Fault Diagnosis of Engineering Systems*; Springer International Publishing: Cham, Switzerland, 2017.
21. Thoma, J.U. *Simulation by Bondgraphs: Introduction to a Graphical Method*; Springer: Berlin/Heidelberg, Germany, 1990; p. 1.
22. Karnopp, D.; Margolis, D.L.; Rosenberg, R.C. *System Dynamics: Modeling and Simulation of Mechatronic Systems*, 5th ed.; Wiley: Hoboken, NJ, USA, 2012; p. 411.
23. Bos, A.M. *Modelling Multibody Systems in Terms of Multibond Graphs: With Applications to a Motorcycle*; Universiteit Twente: Enschede, The Netherlands, 1986.
24. 20-SIM Software Package for Mechatronic Systems [Internet]. Controllab Products. Available online: https://www.20sim.com/download/ (accessed on 26 November 2024).
25. Pacejka, H.B.; Besselink, I. *Tire and Vehicle Dynamics*, 3rd ed.; Butterworth-Heinemann Elsevier: Waltham, MA, USA, 2012.
26. Hao, L. Analysis of Off-Road Tire-Soil Interaction through Analytical and Finite Element Methods. Engineering, Environmental Science, Agricultural and Food Sciences. August 2013, p. 33. Available online: https://www.google.it/url?sa=t&source=web&rct=j&opi=89978449&url=https://kluedo.ub.rptu.de/files/3644/_Analysis+of+Off-Road+Tire-Soil+Interaction+through+Analytical+and+Finite+Element+Methods.pdf&ved=2ahUKEwj-1b2p4fiKAxWw3AIHHbmOIYgQFnoECB8QAQ&usg=AOvVaw2qganoqOMonCq3Uuojk2p_ (accessed on 26 November 2024).
27. Lines, J.A.; Murphy, K. The stiffness of agricultural tractor tyres. *J. Terramech.* **1991**, *28*, 49–64. [CrossRef]
28. Polotski, V. New reference point for guiding an articulated vehicle. In Proceedings of the 2000 IEEE International Conference on Control Applications Conference Proceedings (Cat No00CH37162), Anchorage, AK, USA, 27 September 2000; pp. 455–460.
29. Dudziński, P.; Skurjat, A. System for improving directional stability for articulated vehicles. *AIP Conf. Proc.* **2019**, *2078*, 020084.
30. Łopatka, M.J.; Rubiec, A. Concept and Preliminary Simulations of a Driver-Aid System for Transport Tasks of Articulated Vehicles with a Hydrostatic Steering System. *Appl. Sci.* **2020**, *10*, 5747. [CrossRef]
31. Alhimdani, F.F. Dynamic Analysis of Steering Articulated Tracked Vehicles. *J. Eng. Sustain. Dev.* **2006**, *10*, 32–44.
32. Delrobaei, M.; McIsaac, K.A. Design and Steering Control of a Center-Articulated Mobile Robot Module. *J. Robot.* **2011**, *2011*, 621879. [CrossRef]
33. Dudziński, P.; Skurjat, A. Directional dynamics problems of an articulated frame steer wheeled vehicles. *J. KONES Powertrain Transp.* **2015**, *19*, 89–98. [CrossRef]
34. Qiao, Y.; Chen, X.; Yin, D. Coordinated Control for the Trajectory Tracking of Four-Wheel Independent Drive–Four-Wheel Independent Steering Electric Vehicles Based on the Extension Dynamic Stability Domain. *Actuators* **2024**, *13*, 77. [CrossRef]

actuators

MDPI

Article

A Robust Path Tracking Controller for Autonomous Mobility with Control Delay Compensation Using Backstepping Control

Munjung Jang, Sehwan Kim, Byeonghun Yoo and Kwangseok Oh *

School of ICT, Robotics and Mechanical Engineering, Hankyong National University, Anseong-si 17579, Republic of Korea; munjung0831@hknu.ac.kr (M.J.); ksh49@hknu.ac.kr (S.K.); yoobh00@hknu.ac.kr (B.Y.)
* Correspondence: oks@hknu.ac.kr; Tel.: +82-31-670-5117

Abstract: Control delay phenomena, such as time delays and actuator lags, can compromise the control performance of autonomous mobility systems, leading to increased control errors. Therefore, it is essential to develop a control delay compensation algorithm. This paper proposes a Lyapunov-based backstepping steering control algorithm to compensate for control delays in autonomous mobility systems. To estimate the control delay in the steering system, the Recursive Least Squares (RLS) algorithm was employed to calculate the time constant in real time. The control delay was estimated using an RLS designed based on a first-order differential equation. A backstepping steering controller was developed to calculate the desired steering angle using simplified error dynamics for reference path tracking. The control errors, specifically the lateral preview and yaw angle errors, were derived by calculating the path error between the current position and the waypoint. The performance of the proposed control algorithm was evaluated using the DC motor and CarMaker software 8.1.1(IPG Automotive, Karlsruhe, Germany) under scenarios involving sinusoidal input and four-curved loop and S-curved paths respectively.

Keywords: control delay compensation; lyapunov stability; backstepping control; recursive least squares; autonomous mobility

1. Introduction

In recent decades, advancements in electrical and electronic components, battery technologies, and the development and structure of big data systems have become feasible. Additionally, an increase in computational power has enabled the application of advanced algorithms that consider the driving stability and energy efficiency of hybrid and autonomous vehicles [1,2]. Accordingly, the commercialization of autonomous mobility systems is accelerating in various areas, and mobility solutions capable of performing multiple functions are being developed across various industries using these systems [3,4]. In these systems, a significant number of sensors, actuators, and electronic control units (ECUs) are used, which can lead to control delays caused by hardware (such as physical loads) [5,6] and software (including computational loads and communication delays) [7,8]. When transmitting control signals over a network, data loss or signal delays may occur due to network traffic or limitation of network bandwidth. In these systems, the time delay within the control loop presents a significant challenge that can lead to stability issues. Additionally, these delays reduce the response speed of the controller, making it challenging to maintain the desired performance. This can result in system instability or quantization errors, which complicate real-time control [9,10]. Control delays in an over-actuated system, which include both delays and actuator lags, can negatively affect the control performance, stability, and operator safety. Therefore, to ensure optimal performance, it is essential to develop a control methodology that compensates for these delays. Additionally, when designing a control algorithm based on a mathematical model to ensure control robustness, a precise understanding of the system parameters can significantly influence control performance.

Studies are being conducted to estimate and compensate for delays in networks [11,12], software [13,14], and actuators [15,16] across various systems. These studies aim to enhance the stability of the systems and target performance by applying mathematical model-based control, optimal control, and adaptive theory. Therefore, we propose a methodology that estimates control delay as a system parameter in real time and considers it when deriving the steering control input for autonomous mobility. The main contributions of this study are as follows:

1. Develop a backstepping-based steering control algorithm for robust path tracking that satisfies Lyapunov stability conditions while considering control delays, through a straightforward computational process based on system dynamics.
2. Design of real-time control delay estimation algorithm using first-order differential equation of the steering system, using RLS with forgetting factor.
3. Validation of real-time control delay estimation algorithm on DC motor platform.

Ackermann kinematics-based simplified error dynamics were utilized to design a backstepping-based steering control algorithm to track the reference path. Control errors were derived based on the waypoints of the target path, considering the lateral preview and yaw angle errors. The steering angle, compensating for the control delay, was calculated using the system dynamics expressed by a first-order differential equation that included the control delay. The control delay was applied to the value estimated in real time using the RLS. The remainder of this paper is organized as follows:

Section 2 discusses related work, and Section 3 describes the design process of the proposed steering control algorithm. Section 4 presents a performance evaluation based on simulations, and Section 5 discusses the conclusions and future work.

2. Related Works

Intelligent programs and robots can enhance production because they only require electrical energy while humans need rest and food. For use in industrial fields such as factories, medical institutions, and farms, various types of mobility are being developed to suit specific purposes, allowing workers to operate more efficiently [17]. In factories, mobile robots can coexist with humans and navigate throughout the factory to assist in storage, transportation, and production processes. They utilize LiDAR and cameras to identify current and target locations based on precise maps, and execute the corresponding location or task through actuator control [18,19]. In hospitals, Internet of Things (IoT) modules are utilized to monitor and control autonomous mobility for transporting medicine and food to patients as well as to perform disinfection tasks within the facility [20,21]. In addition, farms use autonomous mobility to identify target crops, assess growth data, and facilitate harvesting. Based on various sensors and actuators, it detects target crops and obstacles, minimizing crop damage, and performs path planning to reach the target quickly and accurately to achieve a specific goal [22–24].

Robust control algorithms are essential for the effective and consistent execution of various types of mobility robots. Park et al. proposed a Linear Quadratic Regulator (LQR) method to enhance the safety of autonomous vehicles by executing drift maneuvers that induce rear-wheel slip under extreme conditions, such as on slippery roads. They calculated the drift equilibrium state using nonlinear vehicle dynamics and tire models, and maintained this equilibrium by simultaneously controlling the steering torque and rear-wheel torque based on the LQR [25]. Amertet et al. developed a kinematic model for a four-wheel skid-steering mobile robot and proposed an effective control methodology to address changes in the dynamics of the model caused by defects, uncertainties, and parameter variations. This was achieved by calculating the Q and R matrices of the LQR based on observability [26].

The over-actuated system must consider constraints on both the actuators and state variables. Accordingly, Model Predictive Control (MPC) is utilized, offering the advantage of deriving realistic optimal control inputs by considering the constraints [27,28]. Hadian and Zhang proposed an algorithm that predicts the robot's state using a recurrent neural network and incorporates this prediction into robust path tracking control based on MPC. They expressed the complex nonlinear system of a mobile robot as a quasi-linear parameter-varying model and approximated it with a linear model [29]. Additionally, it is utilized as a longitudinal and lateral control algorithm based on Sliding Mode Control (SMC), which is robust control methodology against external disturbances and system uncertainties [30,31]. Thus, it can be confirmed that control methodologies such as MPC, SMC, and LQR exist to secure reasonable control performance of the system. Furthermore, with the advancement of Artificial Intelligence (AI), Deep Learning techniques are increasingly being employed the self-tuned variables of control algorithms [32,33] or to derive control inputs [34,35].

To ensure and enhance the stable performance of the mathematical models and various AI-based control algorithms, it is essential that they rely on accurate system parameters and state variables. Accordingly, the uncertainty in the observer-based system and external disturbances [36–38] were addressed using the vehicle's internal sensors and estimation algorithms. These algorithms incorporate parameters such as the vehicle's center point, tire stiffness [39], and road friction coefficient [40], utilizing unscented, extended, and adaptive Kalman filters. Additionally, factors such as the body slip angle [41,42] and longitudinal velocity [43] were considered.

Autonomous mobility systems require additional sensors, actuators, and control devices. Therefore, random network delays may arise due to limitations in vehicle network bandwidth, computational load, and external disturbances. This may result in excessive performance degradation and instability within the system; therefore, an algorithm that estimates and compensates for control delay is necessary. Research has been conducted to estimate network delays using machine learning techniques [44] and mathematical modeling [45]. Additionally, efforts are underway to estimate Vehicle-to-Vehicle (V2V) communication delay [46] and incorporate this information into the controller for compensation. Furthermore, neglecting the steering delay, which is the time required to generate the tire force, can increase the model's uncertainty and degrade performance [47]. Consequently, research has focused on estimating and compensating for actuator delays in conjunction with network delays [48–51].

Through related research, various strategies to estimate and compensate for control delays have been proposed, but real time estimation and compensation algorithms have been insufficiently performed. Section 3 describes the design process for the proposed steering control algorithm.

3. Control Delay Compensated Path Tracking Control

Control delays, such as actuator lags and communication delays, can increase control errors and decrease control performance. In this study, to address the control performance degradation caused by the control delay of the steering system, a Lyapunov-based backstepping steering control algorithm is used to compensate for the control delay in the steering system by estimating the control delay in real time. A backstepping-based steering controller was designed to compensate for the control delay based on the Lyapunov stability condition. To estimate the control delay in real time, an RLS with a forgetting factor was applied. Figure 1 illustrates a block diagram of the proposed algorithm, which comprises backstepping-based steering control and RLS-based control-delay estimation for control delay compensation.

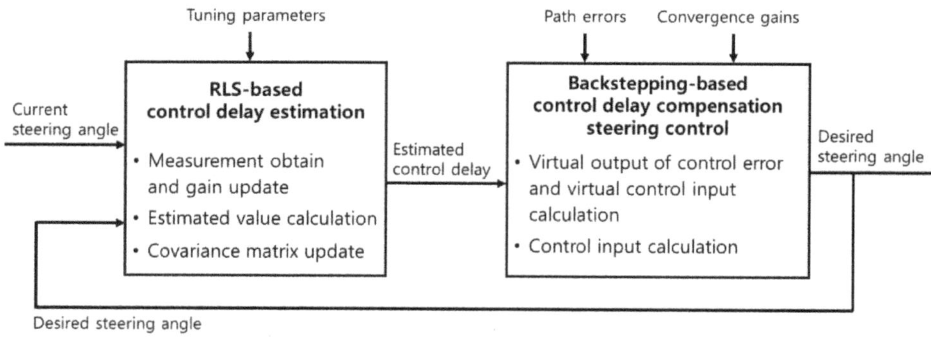

Figure 1. Block diagram of the proposed algorithm.

3.1. Lyapunov Stability-Based Backstepping Steering Controller

In this study, a steering control algorithm for tracking the reference path was deduced using simplified error dynamics based on Ackerman kinematics, and the control input was calculated using the Lyapunov approach-based backstepping control. To derive simplified error dynamics, a bicycle model based on Ackerman kinematics was applied, considering 3 DOF planar motion, as shown in Figure 2.

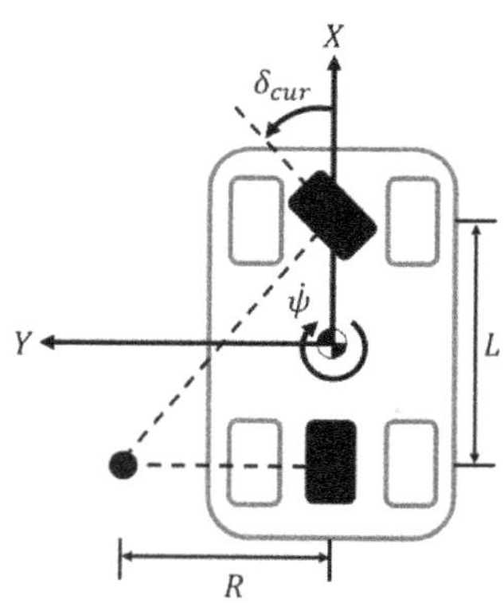

Figure 2. Ackerman kinematics-based bicycle model.

According to Ackerman kinematics, the longitudinal velocity is defined as Equation (1), assuming that the vehicle satisfies uniform circular motion.

$$v_x = R\dot{\psi} \tag{1}$$

where v_x represents the longitudinal velocity, R represents the radius, and $\dot{\psi}$ denotes the yaw rate respectively. The radius can be derived as follows:

$$R = \frac{L}{\delta} \tag{2}$$

where L is the wheel base and δ is the steering angle, respectively. The yaw rate can be derived using the longitudinal velocity, wheel base and steering angle, as shown in Equation (3), under the assumption of a small steering angle.

$$\dot{\psi} = \frac{v_x}{L}\delta \tag{3}$$

$$where\ tan(\delta) \approx \delta$$

The error dynamics for tracking the reference path were designed based on Ackerman kinematics-based simplified error dynamics, as follows:

$$\dot{e}_\psi = \dot{\psi}_{cur} - \dot{\psi}_{des} \tag{4}$$

$$\dot{e}_y = v_x e_\psi \tag{5}$$

where e_ψ and e_y denote the yaw angle error and lateral error respectively, and $\dot{\psi}_{cur}$ is the current yaw rate, $\dot{\psi}_{des}$ is the desired yaw rate, v_x denotes the longitudinal velocity.

To converge the lateral error to zero, the relationship between the lateral error and virtual control input for lateral error u_1 is defined by Equation (6). If the virtual control input for lateral error is designed as Equation (7), the lateral error satisfies the Lyapunov stability condition.

$$\dot{e}_y = v_x u_1 \tag{6}$$

$$u_1 = -\frac{\eta_1}{v_x} e_y\ (where,\ \eta_1 > 0) \tag{7}$$

where η_1 denotes the convergence rate of the lateral error. It is designed as a tuning parameter and determines reasonable value through evaluation.

Then, the virtual control output of the yaw angle error y_1 can be defined using the virtual control input for the lateral error u_1, as shown in Equation (8). By Equations (1), (2) and (8), the time-differential equation of the virtual control output for the yaw angle error can be rewritten as Equation (9).

$$y_1 = e_\psi - u_1 \tag{8}$$

$$\dot{y}_1 = \dot{e}_\psi - \dot{u}_1 = \frac{v_x}{L}\delta_{cur} - \dot{\psi}_{des} - \dot{u}_1 \tag{9}$$

where δ_{cur} denotes the current steering angle.

Then, the current steering angle can be derived using the convergence rate of the lateral error and the virtual control output of the lateral error, as in Equation (10).

$$\delta_{cur} = -\frac{L}{v_x}\alpha_1\eta_1 - \frac{L}{v_x}(\eta_1 + \alpha_1)e_\psi + \frac{L}{v_x}\dot{\psi}_{des} \tag{10}$$

$$where, \dot{y}_1 = -\alpha_1 y_1\ (\alpha_1 > 0)$$

where α_1 denotes the convergence rate of output of lateral error.

To derive the virtual control input for steering angle u_2, the time differential of the yaw angle error can be formulated as follows:

$$\dot{e}_\psi = \frac{v_x}{L}u_2 - \dot{\psi}_d \tag{11}$$

To converge the yaw angle error to zero, the convergence rate of the yaw angle error was defined to satisfy the Lyapunov stability condition.

$$\dot{e}_\psi = -\eta_2 e_\psi\ (where, \eta_2 > 0) \tag{12}$$

where η_2 represents the convergence rate of the yaw angle error and it is set at a reasonable value through the experiment. The virtual control input for the steering angle can be derived using Equation (13), by subtracting Equations (11) and (12), respectively.

$$u_2 = \frac{L}{v_x}\left(-\eta_2 e_\psi + \dot{\psi}_d\right) \tag{13}$$

To converge the virtual control input for the steering angle to the current steering angle, the output of the lateral error is defined as

$$y_2 = \delta_{cur} - u_2 \tag{14}$$

The first-order differential equation of the steering system can be expressed using the time constant in Equation (15).

$$\dot{\delta}_{cur} = \frac{1}{\tau_d}(\delta_{des} - \delta_{cur}) \tag{15}$$

where τ_d denotes the time constant of steering system. By subtracting Equations (14) and (15), Equation (16) can be derived.

$$\dot{y}_2 = \frac{1}{\tau_d}\delta_{des} - \frac{1}{\tau_d}\delta_{cur} - \dot{u}_2 \tag{16}$$

Using the Lyapunov stability condition, the time differential of the virtual output of the lateral error is defined in Equation (17) with the convergence rate.

$$\dot{y}_2 = -\alpha_2 y_2 \tag{17}$$

The desired steering angle can then be expressed using Equation (18) by Equation (15). The time differential of the virtual output of the lateral error can be expressed as shown in Equation (19).

$$\delta_{des} = \delta_{cur} + \tau_d \dot{y}_2 + \tau_d \dot{u}_2 \tag{18}$$

$$\dot{y}_2 = \left(\eta_2 - \frac{1}{\tau_d}\right)\delta_{cur} + \frac{1}{\tau_d}\delta_{des} - \frac{\eta_2}{v_x}L\dot{\psi}_{des} - \frac{L}{v_x}\ddot{\psi}_{des} \tag{19}$$

Finally, the desired steering angle is derived using Equation (20), by subtracting Equations (9), (18) and (19), respectively.

$$\delta_{des} = (1 - \tau_d(\alpha_2 + \eta_2))M - N \tag{20}$$

$$where,\ M = -\frac{L}{v_x}(\alpha_1\eta_1 e_y + (\alpha_1 + \eta_1)e_\psi),\ N = \frac{\tau_d L}{v_x}\alpha_2\eta_2 e_\psi$$

3.2. RLS-Based Control Delay Estimation

The control delay of an autonomous mobility steering system can vary not only physically but also computationally in real time. Therefore, it is necessary to develop a real time control delay estimation algorithm.

In this study, to estimate the control delay of the steering system of autonomous mobility in real time, an RLS with the forgetting factor methodology has been designed using the first-order differential equation of the steering system, as shown in Equation (15).

$$\delta_d - \delta_c = \dot{\delta}_c \hat{\tau}_d \tag{21}$$

where $\hat{\tau}_d$ is the estimated time constant. To calculate the control delay, the difference in the desired and current steering angles is considered as the output of the linear system (y), and the time differential of the current steering angle is considered as the regressor of the RLS (ϕ).

$$y = \phi\hat{\theta} \tag{22}$$

where $\hat{\theta}$ represents the estimated value. In this study, $\hat{\theta}$ is defined the estimated control delay of steering system. The defined cost function is used to derive the estimated value by considering the previous data using forgetting factors.

$$J = \sum_{i=0}^{k-1} \lambda^{(k-1)-i} \left| y_i - \phi_i \hat{\theta}_i \right|^2 \tag{23}$$

where λ denotes the forgetting factor and it is determined using the trial and error method. The estimated value is calculated by updating the updating gain (L), the estimated value ($\hat{\theta}$), and the covariance matrix (P) in discrete time with Equations (24)–(26).

$$L_k = P_k \phi_k \left(\lambda + \phi_k^T P_k \phi_k \right)^{-1} \tag{24}$$

$$\hat{\theta}_k = \hat{\theta}_k + L_k (y_k - \phi_k \hat{\theta}_k) \tag{25}$$

$$P_k = \left(I - L_k \phi_k^T \right) P_k \frac{1}{\lambda_k} \tag{26}$$

4. Performance Evaluation

To validate the proposed algorithms, Evaluations 1 and 2 were performed using MATLAB/Simulink, actual DC motor, and CarMaker software 8.1.1 respectively. The RLS-based control delay estimation algorithm was evaluated using the MATLAB/Simulink and actual DC motor platform in real-time in Evaluation 1. Figures 3 and 4 show the used DC motor platform and model schematic for evaluation, respectively.

Figure 3. DC motor platform.

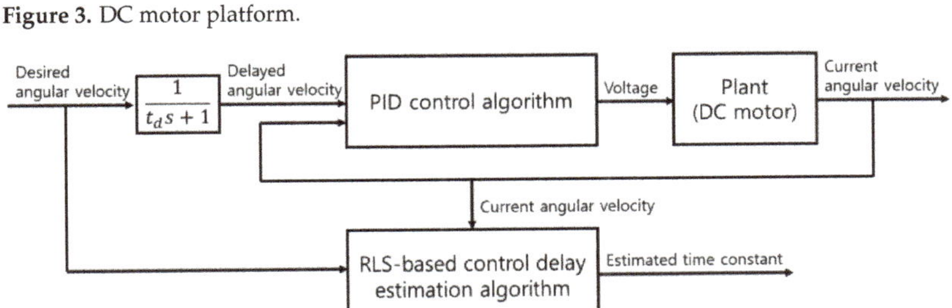

Figure 4. Architecture of the performance evaluation.

The PID controller was applied to track the desired delayed angular velocity by calculating the voltage input. The control delay was applied using the first-order differential equation function in Equation (15) in a Simulink environment. The RLS-based control delay estimation algorithm was used to estimation the control delay of the DC motor in real-time. Table 1 describes the DC motor specification and parameters used for evaluation.

Table 1. Parameters for evaluation 1.

Division		Value
PID controller	P gain	0.15
	I gain	2
	D gain	0.001
RLS	Initial value of covariance matrix	100,000
	Initial value of estimated value	1
	Forgetting factor	0.995

Then, at Evaluation 2, the integrated algorithm with control delay estimation and steering control was evaluated using MATLAB/Simulink and CarMaker software by constructing a waypoint-based path error calculation algorithm and a virtual driving environment. The waypoint-based path error calculation algorithm was designed to calculate the lateral, lateral preview, and yaw angle errors at the preview point. Figure 5 illustrates the architecture of the performance evaluation environment.

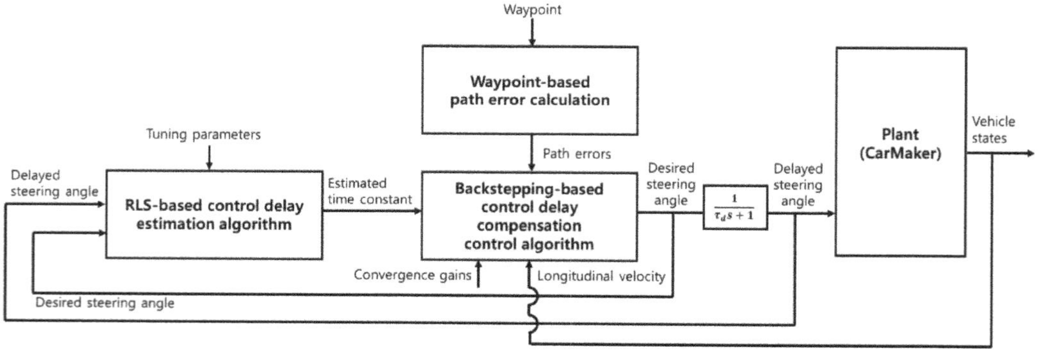

Figure 5. Architecture of the performance evaluation.

In the waypoint-based path error calculation block, path errors, including the lateral error, lateral preview error, and yaw angle error, were calculated based on the current position of the ego vehicle and predefined waypoints. The evaluation scenarios adopted an S-curved path and a four-curved loop path. The longitudinal velocity was maintained constant at 30 km/h and 60 km/h in CarMaker, and the control delay was set to 0.2 s [51] in Scenarios 1 and 2. The control delay is applied by designing the first-order differential equation function in Simulink as shown in Equation (15). The evaluation results were analyzed by comparing the proposed algorithm with a backstepping controller without control delay compensation and with SMC. Table 2 describes the control parameters for Evaluation 2.

Table 2. Control parameters for Evaluation 2 (Scenarios 1, 2).

Division		Value
Backstepping controller	Convergence rate of lateral error	10
	Convergence rate of yaw error	0.5
	Convergence rate of lateral output	10
	Convergence rate of yaw output	0.5
RLS	Initial value of covariance matrix	0.01
	Initial value of estimated value	0.01
	Forgetting factor	0.9993
SMC	Weighting of lateral error	5
	Disturbance boundary	10
	Decay rate of the Lyapunov function	0.5

The SMC was designed to track paths, which are yaw angle error and path error, to zero by defining the sliding surface as shown in Equation (27).

$$\sigma = e_\psi + Ce_y \tag{27}$$

where σ is the sliding surface and C is the weighting of the lateral error. C is set using trial and error method to reasonable result. The cost function was defined based on the Lyapunov function in Equation (28).

$$J = \frac{1}{2}\sigma^2 \tag{28}$$

Then, the control input is derived to satisfy the Lyapunov stability condition.

$$\delta_{d,SMC} = -\left(L_b + \frac{\alpha}{\sqrt{2}}\right) sign(\sigma) \tag{29}$$

where L_b denotes the disturbance boundary and α represents the decay rate of the Lyapunov function. The results of the evaluation are analyzed as follows [52].

4.1. *Evaluation 1 (Using the DC Motor)*

The results of Evaluation 1, which is the control delay estimation algorithm using the RLS with forgetting factor validation, are analyzed through Figures 6–8.

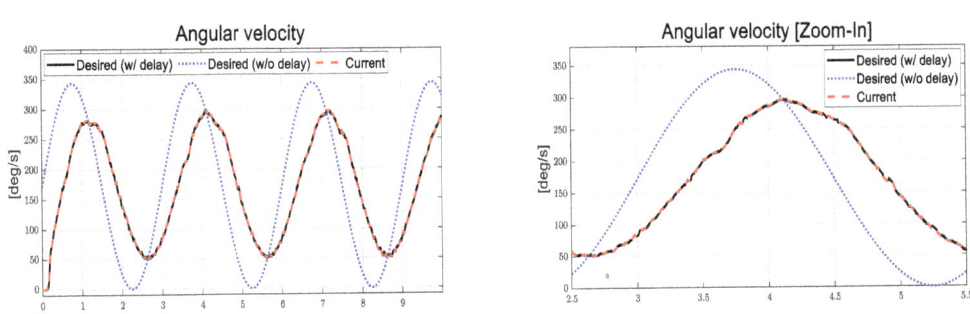

Figure 6. Evaluation 1: angular velocity.

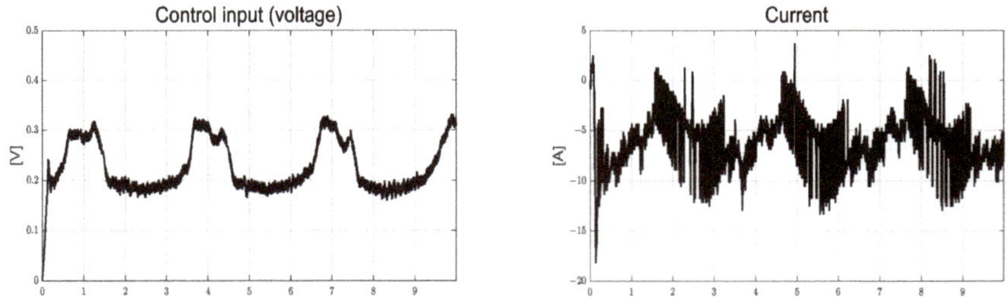

Figure 7. Evaluation 1: control input and current.

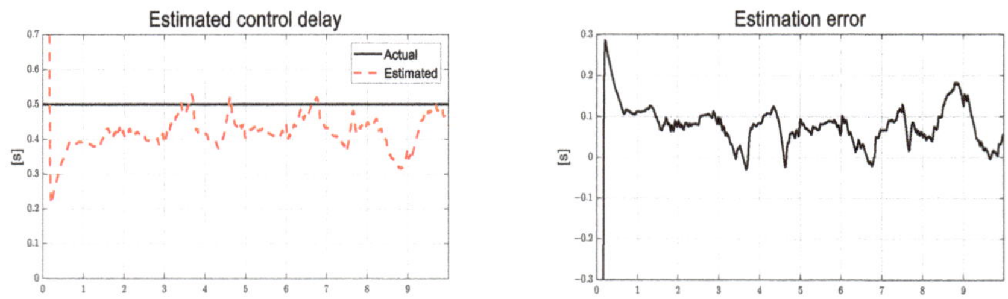

Figure 8. Evaluation 1: estimated control delay and estimation error.

Figure 6 shows the desired and current angular velocities, respectively. The desired angular velocity was applied with the control delay using time constant 0.5 s. Figure 7 shows the current and voltage as control inputs, with the voltage being regulated by a PID controller real-time. As shown in Figure 7, the delayed desired voltage can be tracked reasonably using the PID controller. Figure 8 shows the actual and estimated control delay and estimation error. Although there was an estimation error at the initiation of the evaluation, it converged within 0.5 s. The average estimation error after 0.5 s was approximately 0.065 s. The rate of change of the error was calculated using the linear Kalman filter, and it was confirmed that the estimation error occurred relatively. In the future, when applying the algorithm to a real platform, it is essential to consider uncertainties and disturbances.

4.2. Evaluation 2—Scenario 1 (S-Curved Path)

The backstepping steering control algorithms—with and without control delay compensation—and SMC were compared based on the results of Evaluation 2. Figures 9–15 display the results of the performance evaluation for Evaluation 2—Scenario 1.

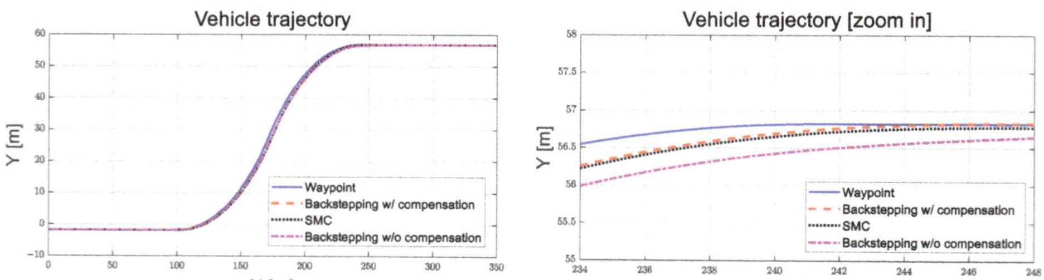

Figure 9. Scenario 1: vehicle trajectory.

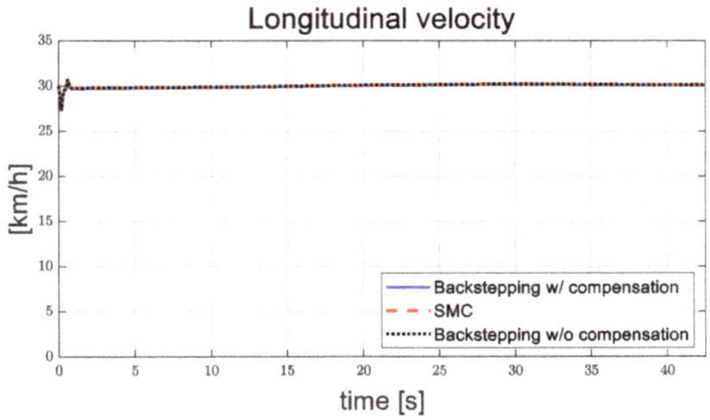

Figure 10. Scenario 1: vehicle longitudinal velocity.

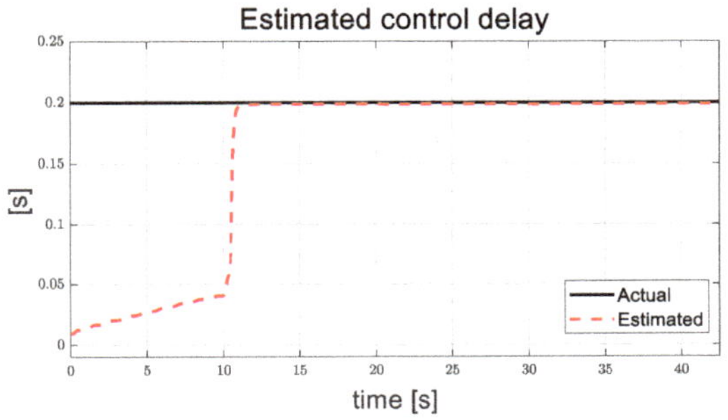

Figure 11. Scenario 1: estimated control delay of the steering system.

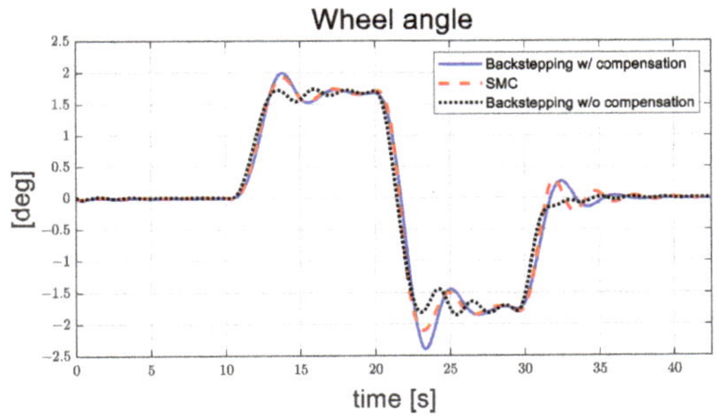

Figure 12. Scenario 1: vehicle wheel angles.

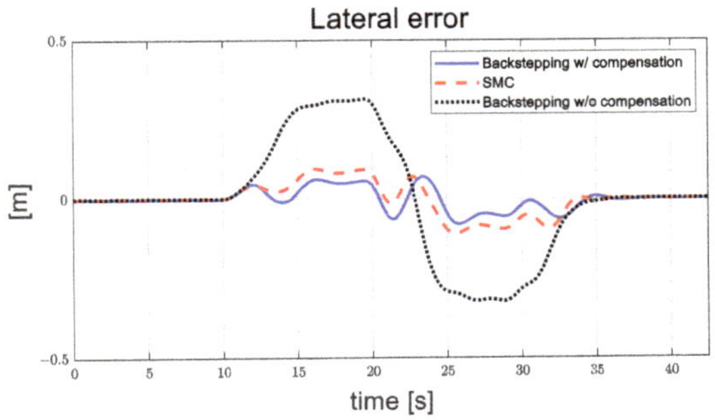

Figure 13. Scenario 1: lateral error.

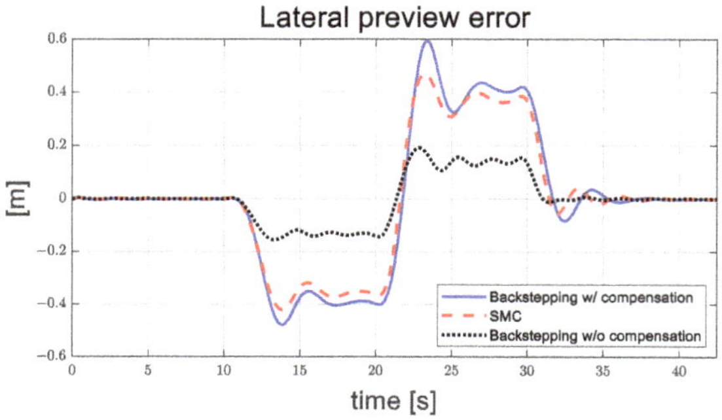

Figure 14. Scenario 1: lateral preview error.

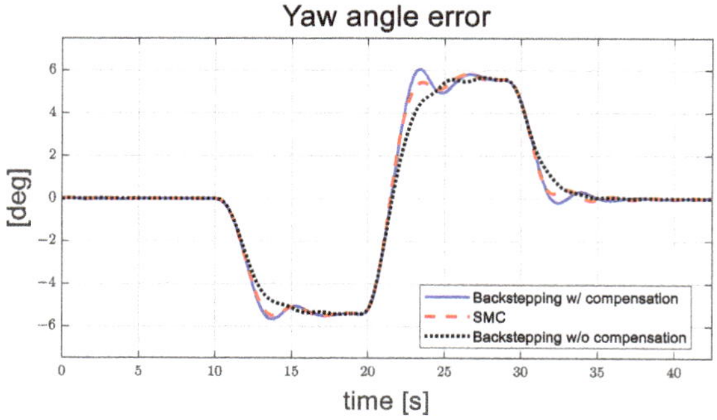

Figure 15. Scenario 1: yaw angle error (preview point).

Figure 9 shows the vehicle trajectory with the proposed algorithm, backstepping steering control without control delay compensation, and SMC. It can be confirmed that the proposed algorithm exhibits a lower lateral error within 0.08 m. However, the other controllers have a lateral error over 0.1 m. The longitudinal velocity was maintained at 30 km/h, which is consistent with the speed initially set for both controllers, as shown in Figure 10. The estimated control delay of the RLS is shown in Figure 11, with reasonable estimation performance after the control input is applied. However, it is difficult to secure estimation performance on a straight road, where less steering input is applied. Therefore, it is necessary to improve the control delay estimation algorithm in the future. The wheel angles are depicted in Figure 12 and show similar magnitudes to the control input. The control errors, namely the lateral error, lateral preview error, and yaw angle error at the preview point, are shown in Figures 13–15. The maximum lateral errors are 0.0782 m, 0.1092 m, and 0.3229 m. The maximum lateral preview errors are 0.5944 m, 0.4690 m, and 0.1908 m, respectively. In addition, the maximum yaw angle errors are 6.0655 deg, 5.7984 deg, and 5.6643 deg. Table 3 presents an analysis of the control errors for both controllers.

Table 3. Control error analysis.

Division		Backstepping w/ Compensation	SMC	Backstepping w/o Compensation
Lateral error	Max	0.0782 [m]	0.1092 [m]	0.3229 [m]
	Average	0.0226 [m]	0.0337 [m]	0.1197 [m]
	Standard deviation	0.0249 [m]	0.0379 [m]	0.1330 [m]

Table 3. *Cont.*

Division		Backstepping w/ Compensation	SMC	Backstepping w/o Compensation
Lateral preview error	Max	0.5944 [m]	0.4690 [m]	0.1908 [m]
	Average	0.1730 [m]	0.1542 [m]	0.0582 [m]
	Standard deviation	0.1944 [m]	0.1733 [m]	0.0664 [m]
Yaw angle error	Max	6.0655 [deg]	5.7984 [deg]	5.6643 [deg]
	Average	2.2209 [deg]	2.2005 [deg]	2.1331 [deg]
	Standard deviation	2.5093 [deg]	2.4772 [deg]	2.3868 [deg]

4.3. Evaluation 2—Scenario 2 (Four-Curved Loop Path)

As in Scenario 1, in Scenario 2, featuring four-curved loop path tracking, the three controllers were compared, and the results of the evaluation are shown in Figures 16–22.

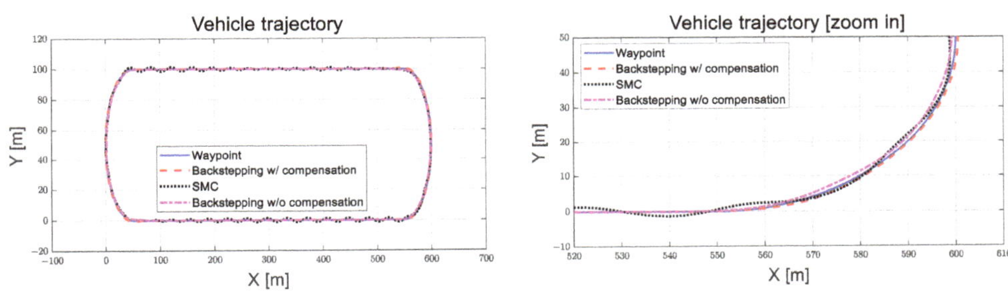

Figure 16. Scenario 2: vehicle trajectory.

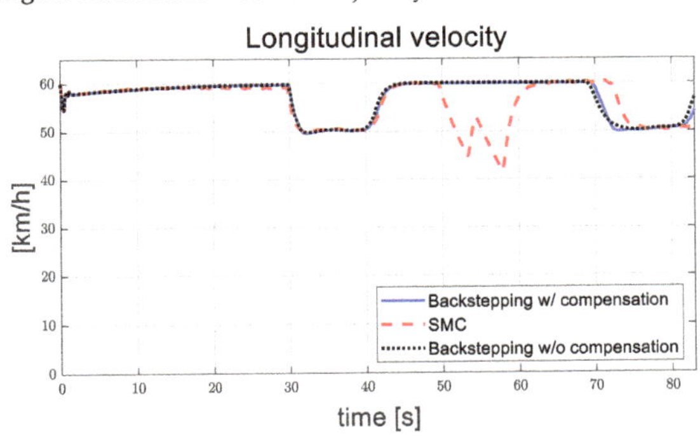

Figure 17. Scenario 2: vehicle longitudinal velocity.

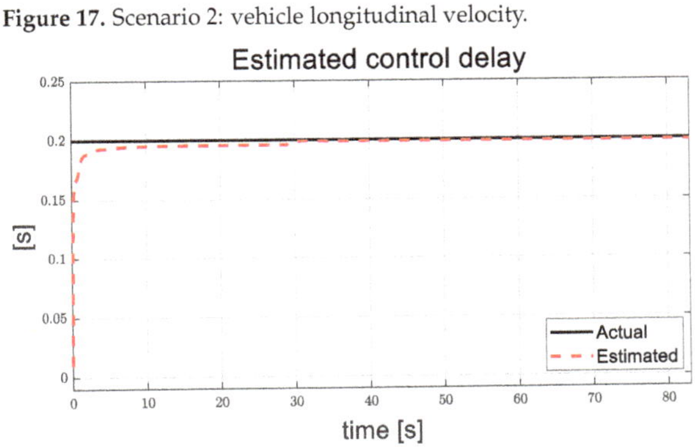

Figure 18. Scenario 2: estimated control delay of the steering system.

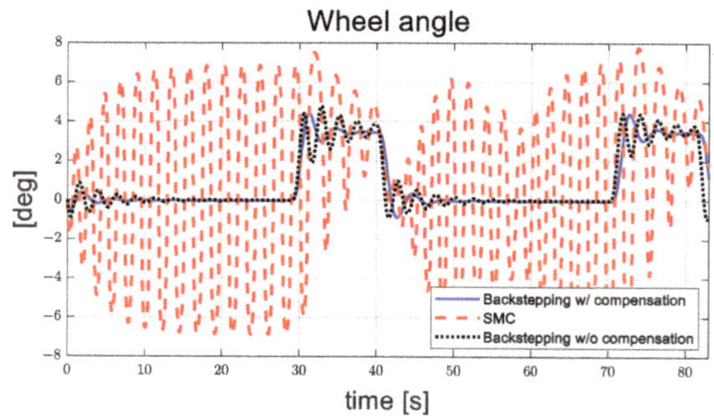

Figure 19. Scenario 2: vehicle wheel angles.

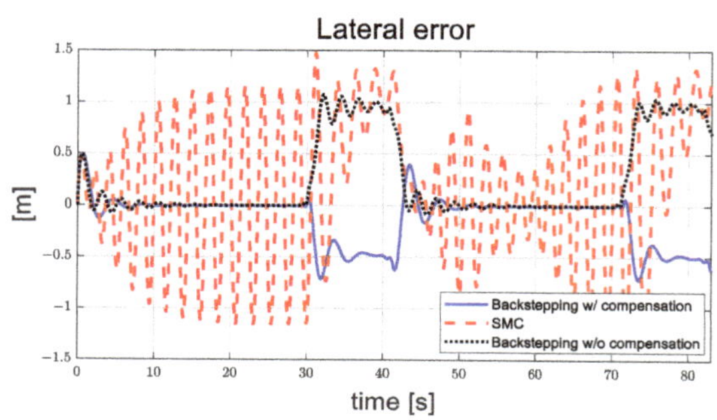

Figure 20. Scenario 2: lateral error.

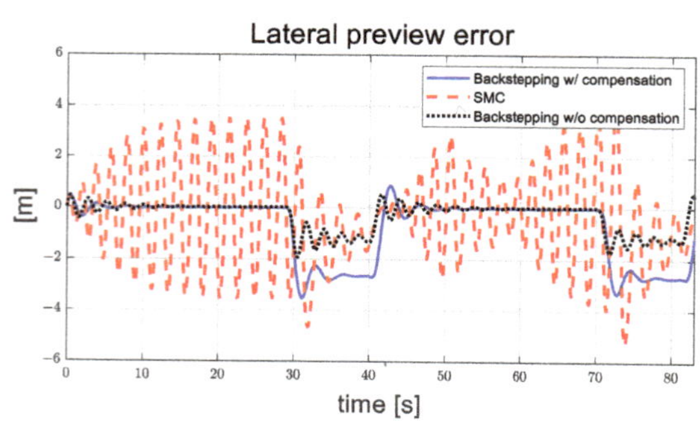

Figure 21. Scenario 2: lateral preview error.

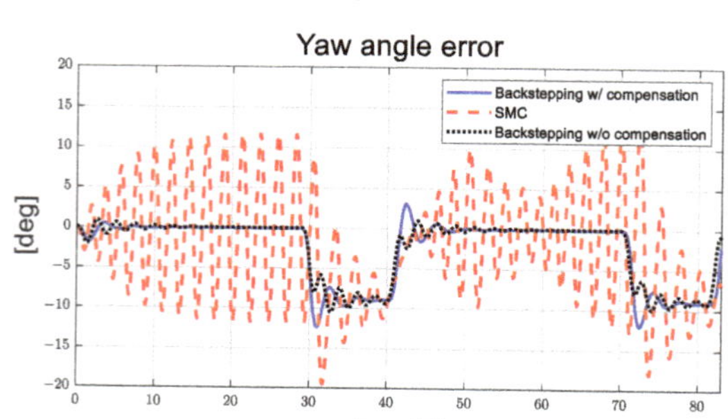

Figure 22. Scenario 2: yaw angle error (preview point).

Figure 16 shows the vehicle trajectory with the three controllers on a four-curved-loop path. Since the same parameters were applied, the SMC exhibits significant chattering phenomena due to improperly determined control parameters, whereas the proposed algorithm shows reasonable performance with a lateral error of about 0.8 m. The longitudinal velocity of the vehicle varied when the control input was applied under backstepping steering control. However, the longitudinal velocity was unstable at 50–60 s under SMC-based steering control because of the chattering phenomenon, as shown in Figure 17. As illustrated in Figure 18, the estimated control delay of the RLS demonstrated reasonable estimation performance after the control input was applied. Although the magnitudes of the control input and wheel angle of the backstepping control were similar, the SMC exhibited chattering phenomena on both straight and cornering roads, as shown in Figure 19. The control errors, including the lateral error, lateral preview error, and yaw angle error at the preview point, are compared and analyzed in Figures 20–22 and Table 4. The proposed steering control algorithm has lower control errors compared to the other controllers; however, the yaw angle error at the preview point is greater than that of the backstepping controller without compensation because the control input has a larger value than the backstepping controller without compensation. The SMC has the largest control errors, measuring 1.4813 m, 5.2869 m, and 19.4741 deg, respectively. It can be confirmed that the proposed algorithm has reasonable performance because the control errors have the lowest values.

Table 4. Control error analysis.

Division		Backstepping w/ Compensation	SMC	Backstepping w/o Compensation
Lateral error	Max	0.7120 [m]	1.4813 [m]	1.0748 [m]
	Average	0.1587 [m]	0.6293 [m]	0.2700 [m]
	Standard deviation	0.2229 [m]	0.3852 [m]	0.3972 [m]
Lateral preview error	Max	3.5123 [m]	5.2869 [m]	1.9212 [m]
	Average	0.7725 [m]	1.6125 [m]	0.3716 [m]
	Standard deviation	1.1605 [m]	1.1369 [m]	0.5230 [m]
Yaw angle error	Max	12.2975 [deg]	19.4741 [deg]	10.5116 [deg]
	Average	2.6112 [deg]	6.2881 [deg]	2.4428 [deg]
	Standard deviation	3.9402 [deg]	4.0023 [deg]	3.6537 [deg]

5. Conclusions

In this study, a Lyapunov approach-based backstepping control has been proposed, considering the estimated control delay in real time. Control delays, such as time delay and actuator lag in the steering system of autonomous mobility, can affect control performance, causing performance and control stability degradation and also compromising operator safety. To address the control delay phenomenon, a backstepping steering control based on the Lyapunov approach has been proposed. The RLS was used to calculate the control delay in real time. The performance of the proposed control algorithm was evaluated using MATLAB/Simulink and CarMaker software under S-curved and four-curved loop path scenarios by comparing SMC, and the backstepping controller without compensation. The reasonableness of the proposed algorithm was validated by achieving the lowest control errors in both scenarios. However, it was necessary to advance the control delay estimation algorithm to ensure better estimation performance on straight roads. In addition, the validation was conducted in a simulation environment, and the uncertainties and disturbances were not considered. Therefore, an actual test platform-based outdoor evaluation is planned for future work. We hope that our method will be helpful not only in the mobile robotics industry, especially in the area of robust actuator control considering control and communication delays in over-actuated systems, such as over-actuated snake robots [53], but also in research focused on resilient robots [54].

Author Contributions: Conceptualization, M.J. and K.O.; methodology, M.J., S.K. and K.O.; software, M.J.; validation, M.J., S.K. and B.Y.; formal analysis, M.J. and K.O.; investigation, M.J.; resources, M.J.; data curation, M.J.; writing—original draft preparation, M.J, S.K. and B.Y.; writing—review and editing, M.J. and K.O.; visualization, M.J.; supervision, K.O.; project administration, K.O.; funding acquisition, K.O. All authors have read and agreed to the published version of the manuscript.

Funding: This study was conducted with the support of the National Research Foundation of Korea (NRF-2022R1F1A1075167) and government funding (Ministry of Science and ICT) for 2022.

Data Availability Statement: The original contributions presented in the study are included in the article. Further inquiries can be directed to the corresponding author/s.

Conflicts of Interest: The authors declare no conflicts of interest.

References

1. Dong, P.; Zhao, J.; Liu, X.; Wu, J.; Xu, X.; Liu, Y.; Guo, W. Practical application of energy management strategy for hybrid electric vehicles based on intelligent and connected technologies: Development stages, challenges, and future trends. *Renew. Sustain. Energy Rev.* **2022**, *170*, 112947. [CrossRef]
2. Urooj, S.; Alrowais, F.; Teekaraman, Y.; Manoharan, H.; Kuppusamy, R. IoT based electric vehicle application using boosting algorithm for smart cities. *Energies* **2021**, *14*, 1072. [CrossRef]
3. Oliveira, L.F.; Moreira, A.P.; Silva, M.F. Advances in forest robotics: A state-of-the-art survey. *Robotics* **2021**, *10*, 53. [CrossRef]
4. Bala, H.; Anowar, S.; Chng, S.; Cheah, L. Review of studies on public acceptability and acceptance of shared autonomous mobility services: Past, present and future. *Transp. Rev.* **2023**, *43*, 970–996. [CrossRef]
5. Lei, J.; Tang, G.Y. Optimal vibration control for active suspension systems with actuator and sensor delays. In Proceedings of the 2008 IEEE International Conference on Systems, Man and Cybernetics, Singapore, 10–12 October 2008; pp. 2828–2833.
6. Gu, B.; Cong, J.; Zhao, J.; Chen, H.; Fatemi Golshan, M. A novel robust finite time control approach for a nonlinear disturbed quarter-vehicle suspension system with time delay actuation. *Automatika* **2022**, *63*, 627–639. [CrossRef]
7. Luo, Q.; Li, C.; Luan, T.H.; Shi, W. Minimizing the delay and cost of computation offloading for vehicular edge computing. *IEEE Trans. Serv. Comput.* **2021**, *15*, 2897–2909. [CrossRef]
8. Zeadally, S.; Guerrero, J.; Contreras, J. A tutorial survey on vehicle-to-vehicle communications. *Telecommun. Syst.* **2020**, *73*, 469–489. [CrossRef]
9. Subramanian, M.; Kokil, P. Delay-dependent stability analysis of network-controlled DC motor with time-invariant delays. In Proceedings of the 2020 IEEE 17th India Council International Conference, New Delhi, India, 10–13 December 2020; pp. 1–7. [CrossRef]
10. Zhang, X.M.; Han, Q.L.; Ge, X.; Ding, D.; Ding, L.; Yue, D.; Peng, C. Networked control systems: A survey of trends and techniques. *IEEE/CAA J. Autom. Sin.* **2019**, *7*, 1–17. [CrossRef]
11. Yu, J.; Zhang, W.; Xiong, W.; Wang, Y. A Sliding Mode Controller with Signal Transmission Delay Compensation for the Parallel DC/DC Converter's Network Control System. *Electronics* **2023**, *13*, 121. [CrossRef]
12. Mishra, P.; Patjoshi, R.K.; Yadav, A.K. A Delay Compensation Approach for IoT-Enabled Networks with Different Control Strategies. In Proceedings of the 2023 Fifth International Conference on Electrical, Computer and Communication Technologies, Erode, India, 22–24 February 2023; pp. 1–5. [CrossRef]
13. Zhang, Y.; Niu, R.; Wang, J.; Liang, H.; Chen, Z.; Huang, Z. Path Tracking Control Algorithm Considering Delay Compensation. In Proceedings of the 2022 7th Asia-Pacific Conference on Intelligent Robot Systems 2022, Tianjin, China, 1–3 July 2022; pp. 64–69. [CrossRef]
14. Lee, J.; Jo, K. Model Predictive Control with Powertrain Delay Consideration for Longitudinal Speed Tracking of Autonomous Electric Vehicles. *World Electr. Veh. J.* **2024**, *15*, 433. [CrossRef]
15. Kennedy, W.T.; Bevly, D.M. *Adaptive Actuator Delay Compensation for a Vehicle Lateral Control System*; SAE Technical Paper; SAE: Warrendale, PA, USA, 2023. [CrossRef]
16. Zhao, Z.; Wang, J. Safe Delay-Compensated Control with Nonlinear Actuator Dynamics. In Proceedings of the 2024 7th International Symposium on Autonomous Systems 2024, Chongqing, China, 7–9 May 2024; pp. 1–6. [CrossRef]
17. Raj, R.; Kos, A. A comprehensive study of mobile robot: History, developments, applications, and future research perspectives. *Appl. Sci.* **2022**, *12*, 6951. [CrossRef]
18. Miskovic, D.; Milic, L.; Cilag, A.; Berisavljevic, T.; Gottscheber, A.; Rakovic, M. Implementation of Robots Integration in Scaled Laboratory Environment for Factory Automation. *Appl. Sci.* **2022**, *12*, 1228. [CrossRef]
19. Hercik, R.; Byrtus, R.; Jaros, R.; Koziorek, J. Implementation of autonomous mobile robot in smartfactory. *Appl. Sci.* **2022**, *12*, 8912. [CrossRef]
20. Thamrongaphichartkul, K.; Worrasittichai, N.; Prayongrak, T.; Vongbunyong, S. framework of IoT platform for autonomous mobile robot in hospital logistics applications. In Proceedings of the 2020 15th International Joint Symposium on Artificial Intelligence and Natural Language Processing (iSAI-NLP), Bangkok, Thailand, 18–20 November 2020; pp. 1–6.

21. Tamantini, C.; di Luzio, F.S.; Cordella, F.; Pascarella, G.; Agrò, F.E.; Zollo, L. A Robotic Health-Care Assistant for the COVID-19 Emergency. *IEEE Robot. Autom. Mag.* **2021**, *28*, 71–81. [CrossRef]
22. Pak, J.; Kim, J.; Park, Y.; Son, H.I. Field evaluation of path-planning algorithms for autonomous mobile robot in smart farms. *IEEE Access* **2022**, *10*, 60253–60266. [CrossRef]
23. Cho, S.; Kim, T.; Jung, D.H.; Park, S.H.; Na, Y.; Ihn, Y.S.; Kim, K. Plant growth information measurement based on object detection and image fusion using a smart farm robot. *Comput. Electron. Agric.* **2023**, *207*, 107703. [CrossRef]
24. Baek, E.T.; Im, D.Y. ROS-based unmanned mobile robot platform for agriculture. *Appl. Sci.* **2022**, *12*, 4335. [CrossRef]
25. Park, M.; Kang, Y. Experimental verification of a drift controller for autonomous vehicle tracking: A circular trajectory using LQR method. *Int. J. Control Autom. Syst.* **2021**, *19*, 404–416. [CrossRef]
26. Amertet, S.; Gebresenbet, G.; Alwan, H.M. Optimizing the performance of a wheeled mobile robots for use in agriculture using a linear-quadratic regulator. *Robot. Auton. Syst.* **2024**, *174*, 104642. [CrossRef]
27. Fnadi, M.; Plumet, F.; Benamar, F. Model predictive control based dynamic path tracking of a four-wheel steering mobile robot. In Proceedings of the 2019 IEEE/RSJ International Conference on Intelligent Robots and Systems, (IROS), Macau, China, 4–8 November 2019; pp. 4518–4523. [CrossRef]
28. Schwartz, M.; Siebenrock, F.; Hohmann, S. Model predictive control allocation of an over-actuated electric vehicle with single wheel actuators. *IFAC-PapersOnLine* **2019**, *52*, 162–169. [CrossRef]
29. Hadian, M.; Zhang, W.J.; Etesami, D. A robust interpolated model predictive control based on recurrent neural networks for a nonholonomic differential-drive mobile robot with quasi-LPV representation: Computational complexity and conservatism. *Int. J. Syst. Sci.* **2024**, *55*, 3257–3271. [CrossRef]
30. Junior, A.D.S.; Birkner, C.; Jazar, R.N.; Marzbani, H. Coupled lateral and longitudinal controller for over-actuated vehicle in evasive maneuvering with sliding mode control strategy. *IEEE Access* **2023**, *11*, 33792–33811. [CrossRef]
31. Hang, P.; Chen, X. Integrated chassis control algorithm design for path tracking based on four-wheel steering and direct yaw-moment control. *Proc. Inst. Mech. Eng. Part I J. Syst. Control Eng.* **2019**, *233*, 625–641. [CrossRef]
32. Lee, J.E.; Kim, B.W. Improving Direct Yaw-Moment Control via Neural-Network-Based Non-Singular Fast Terminal Sliding Mode Control for Electric Vehicles. *Sensors* **2024**, *24*, 4079. [CrossRef]
33. Carlucho, I.; De Paula, M.; Acosta, G.G. An adaptive deep reinforcement learning approach for MIMO PID control of mobile robots. *ISA Trans.* **2020**, *102*, 280–294. [CrossRef]
34. Haddad, J.; Mirkin, B. Adaptive input delays compensation under event-based control signal implementation. *IEEE Trans. Autom. Control.* **2024**, *69*, 4150–4157. [CrossRef]
35. Pérez-Gil, Ó.; Barea, R.; López-Guillén, E.; Bergasa, L.M.; Gómez-Huélamo, C.; Gutiérrez, R.; Díaz-Díaz, A. Deep reinforcement learning based control for Autonomous Vehicles in CARLA. *Multimed. Tools Appl.* **2022**, *81*, 3553–3576. [CrossRef]
36. Kang, N.; Han, Y.; Wang, B.; Guan, T.; Feng, W. Linear quadratic regulator based on extended state observer–based active disturbance rejection control of autonomous vehicle path following control. *Proc. Inst. Mech. Eng. Part I J. Syst. Control Eng.* **2023**, *237*, 102–120. [CrossRef]
37. Liu, K.; Gao, H.; Ji, H.; Hao, Z. Adaptive sliding mode based disturbance attenuation tracking control for wheeled mobile robots. *Int. J. Control Autom. Syst.* **2020**, *18*, 1288–1298. [CrossRef]
38. Jeong, S.; Chwa, D. Sliding-mode-disturbance-observer-based robust tracking control for omnidirectional mobile robots with kinematic and dynamic uncertainties. *IEEE/ASME Trans. Mechatron.* **2020**, *26*, 741–752. [CrossRef]
39. Li, W.; Li, H.; Xu, K.; Huang, Z.; Li, K.; Du, H. Estimation of vehicle dynamic parameters based on the two-stage estimation method. *Sensors* **2021**, *21*, 3711. [CrossRef] [PubMed]
40. Hu, J.; Rakheja, S.; Zhang, Y. Real-time estimation of tire–road friction coefficient based on lateral vehicle dynamics. *Proc. Inst. Mech. Eng. Part D J. Automob. Eng.* **2020**, *234*, 2444–2457. [CrossRef]
41. Xiong, L.; Xia, X.; Lu, Y.; Liu, W.; Gao, L.; Song, S.; Yu, Z. IMU-based automated vehicle body sideslip angle and attitude estimation aided by GNSS using parallel adaptive Kalman filters. *IEEE Trans. Veh. Technol.* **2020**, *69*, 10668–10680. [CrossRef]
42. Henning, K.U.; Speidel, S.; Gottmann, F.; Sawodny, O. Integrated lateral dynamics control concept for over-actuated vehicles with state and parameter estimation and experimental validation. *Control Eng. Pract.* **2021**, *107*, 104704. [CrossRef]
43. Ding, X.; Wang, Z.; Zhang, L.; Wang, C. Longitudinal vehicle speed estimation for four-wheel-independently-actuated electric vehicles based on multi-sensor fusion. *IEEE Trans. Veh. Technol.* **2020**, *69*, 12797–12806. [CrossRef]
44. Wang, J.; Tian, E.; Yan, H. Robust learning control for autonomous vehicle with network delays and disturbances. *J. Intell. Transp. Syst.* **2024**, 1–14. [CrossRef]
45. Luan, Z.; Zhang, J.; Zhao, W.; Wang, C. Trajectory tracking control of autonomous vehicle with random network delay. *IEEE Trans. Veh. Technol.* **2020**, *69*, 8140–8150. [CrossRef]
46. Zhang, Y.; Bai, Y.; Hu, J.; Wang, M. Control design, stability analysis, and traffic flow implications for cooperative adaptive cruise control systems with compensation of communication delay. *Transp. Res. Rec.* **2020**, *2674*, 638–652. [CrossRef]
47. Xu, S.; Peng, H.; Tang, Y. Preview path tracking control with delay compensation for autonomous vehicles. *IEEE Trans. Intell. Transp. Syst.* **2020**, *22*, 2979–2989. [CrossRef]
48. Xu, X.; Liu, Z.; Wang, F.; Xie, J.; Su, P. Trajectory tracking control based on the dual-motor autonomous steering system with time-varying network-induced time delay. *Control Eng. Pract.* **2021**, *116*, 104915. [CrossRef]

49. Chiliveri, V.R.; Kalpana, R.; Subramaniam, U.; Muhibbullah, M.; Padmavathi, L. Novel reaching law based predictive sliding mode control for lateral motion control of in-wheel motor drive electric vehicle with delay estimation. *IET Intell. Transp. Syst.* **2024**, *18*, 872–888. [CrossRef]
50. Tan, Y.; Zhang, K. Real-time distributed cooperative adaptive cruise control model considering time delays and actuator lag. *Transp. Res. Rec.* **2022**, *2676*, 93–111. [CrossRef]
51. Chen, G.; Yao, J.; Hu, H.; Gao, Z.; He, L.; Zheng, X. Design and experimental evaluation of an efficient MPC-based lateral motion controller considering path preview for autonomous vehicles. *Control Eng. Pract.* **2022**, *123*, 105164. [CrossRef]
52. Oh, K.; Seo, J. Recursive least squares based sliding mode approach for position control of DC motors with self-tuning rule. *J. Mech. Sci. Technol.* **2020**, *34*, 5223–5237. [CrossRef]
53. Fukushima, H.; Yanagiya, T.; Ota, Y.; Katsumoto, M.; Matsuno, F. Model predictive path-following control of snake robots using an averaged model. *IEEE Trans. Control Syst. Technol.* **2020**, *29*, 2444–2456. [CrossRef]
54. Zhang, T.; Zhang, W.; Gupta, M.M. Resilient robots: Concept, review, and future directions. *Robotics* **2017**, *6*, 22. [CrossRef]

actuators

MDPI

Article

A Tube-Based Model Predictive Control for Path Tracking of Autonomous Articulated Vehicle

Taeyeon Lee [1] and Yonghwan Jeong [2,*]

[1] Department of Automotive Engineering, Seoul National University of Science and Technology, Seoul 01811, Republic of Korea; cjstkdus6978@seoultech.ac.kr

[2] Department of Mechanical and Automotive Engineering, Seoul National University of Science and Technology, Seoul 01811, Republic of Korea

[*] Correspondence: yh.jeong@seoultech.ac.kr; Tel.: +82-2-970-6322

Abstract: This paper presents tube-based Model Predictive Control (MPC) for the path and velocity tracking of an autonomous articulated vehicle. The target platform of this study is an autonomous articulated vehicle with a non-steerable axle. Consequently, the articulation angle and wheel torque input are determined by the tube-based MPC. The proposed MPC aims to achieve two objectives: minimizing path tracking error and enhancing robustness to disturbances. Furthermore, the lateral stability of the autonomous articulated vehicle is considered to reflect its dynamic characteristics. The vehicle model for the MPC is formulated using local linearization to minimize modeling errors. The reference state is determined using a virtual controller based on the linear quadratic regulator to provide the optimal reference for the MPC solver. The proposed algorithm was evaluated through a simulation study with base algorithms under noise injection into the sensor signal. Simulation results demonstrate that the proposed algorithm achieved the smallest path tracking error, compared to the base algorithms. Additionally, the proposed algorithm demonstrated robustness to external noise for multiple signals.

Keywords: autonomous articulated vehicle; tube-based model predictive control; robust control

1. Introduction

The scope of autonomous driving has expanded beyond passenger vehicles to include commercial vehicles [1]. Trucks and buses, among other commercial vehicles, share similar steering and driving mechanisms with passenger cars, and achieving satisfactory driving performance typically involves adapting supplementary algorithms to meet specific requirements [2,3]. For example, autonomous buses experience notable variations in vehicle mass depending on passenger load, a factor not as pronounced in autonomous cars. Consequently, real-time estimation of mass and center of gravity becomes crucial, with these factors needing incorporation into the controller [4]. Similarly, self-driving trucks exhibit mass and center of gravity variations based on cargo load, necessitating the use of estimators akin to those used for autonomous buses [5]. Therefore, both passenger cars, which have served as primary platforms for autonomous driving research, and commercial vehicles with comparable steering and driving mechanisms can share the basic structure of the vehicle controller. However, significant differences in the actuation mechanism would require the development of a distinct control algorithm, separate from that of passenger autonomous vehicles.

In addition to the conventional front wheel steering mechanism, ground vehicles utilize various steering mechanisms, including four wheel steering [6], skid steering [7], and articulated frame steering [8]. These steering mechanisms are commonly employed alongside front wheel steering. The four-wheel steering mechanism, being an extension of front wheel steering, allows for a similar controller application to determine both front and rear steering angles for path tracking [9,10]. However, skid and articulated frame steering employ significantly different actuation mechanisms, compared to front wheel steering. Skid steering has traditionally been prevalent in construction heavy equipment like wheel loaders, but, in recent years, it has been gradually replaced by articulated steering-based vehicles, due to several advantages offered by articulated steering. Articulated steering vehicles provide benefits such as a wider viewing angle, improved maneuverability, reduced tire wear, and enhanced fuel efficiency, compared to skid vehicles [11]. Consequently, there is an increasing demand for articulated vehicles beyond current usage. As articulated structures are primarily employed in commercial vehicles, there is a growing need for the autonomous driving of articulated vehicles to enhance operational efficiency and minimize manual labor requirements. Therefore, research on a path tracking controller that accommodates the unique steering structure of articulated vehicles is imperative.

An analysis of the articulated vehicle was conducted prior to designing the path tracking controller. The dynamic characteristics of articulated vehicles can be summarized in three key aspects. Firstly, the front and rear bodies exhibit distinct dynamic behavior. As the front body of the vehicle serves as the steering mechanism, it typically experiences greater lateral acceleration and yaw rate, compared to the rear part [12]. Secondly, the articulation structure is more susceptible to lateral instability. Specifically, snaking instability and jackknife phenomena are more likely to occur in articulated vehicles than in front wheel-steering vehicles [13]. This tendency becomes more pronounced at higher driving speeds or when the articulation angle is large, due to traveling on roads with sharp curvature [14]. Additionally, the center of gravity of articulated vehicles is higher than that of passenger cars, as articulated vehicles are primarily used in commercial applications. This elevated center of gravity contributes to reduced lateral stability. Finally, the hydraulic actuator for articulation exhibits a relatively slower response. compared to the electric power steering used in passenger cars [15].

Path tracking controllers for articulated vehicles are classified based on the presence of steering wheels. When a steering wheel is present on the front body of the articulating vehicle, an actuator controlling the articulation of the vehicle body is generally not used. Therefore, the articulation angle becomes a vehicle state determined by front wheel steering and is not used as a control input for path tracking [16–19]. In this scenario, a nonlinear kinematic equation is employed to represent the relationship between front wheel steering and the articulation angle. The kinematic equation is linearized to design a circular path tracking controller [16]. However, since this approach relies on exact linearization, the tracking controller is defined only for straight and circular paths, limiting its applicability to other path geometries. Similarly, a linear vehicle model and a potential field approach are utilized to plan the optimal path for an articulated vehicle with two trailers [17]. However, the potential field implementation as a sigmoid function may struggle to respond effectively in the presence of complex obstacles. Additionally, since the articulation angle is passively determined by front steering in both approaches, snake oscillation or jackknife phenomena can occur if the steering control input is not properly limited based on the vehicle's state. To address the stability problem of articulated vehicles with steerable wheels, phase plane analysis is introduced to maintain lateral stability [18]. However, phase planes depend on the path radius and the vehicle's current state, making them challenging to employ in real-time when following arbitrary paths. This lateral stability issue arises because the articulation angle cannot be actively controlled by front wheel steering. To mitigate this, a steering function is added to the rear body axle of the articulated vehicle to actively control the articulation angle during path tracking [19]. Nevertheless, application to diverse driving situations is challenging, due to the utilization of a fuzzy controller with a complex activation function to determine the appropriate amount of rear wheel steering based on driving conditions. A path tracking controller for articulated vehicles with a front steering axle was proposed using a tube-based MPC [20]. This controller is based on error dynamics using a bicycle model and a planned path to determine the plant model for MPC. Furthermore, the role of tube-based MPC is more about compensating for model error than disturbance to the signal.

In the absence of a steering axle, the articulation angle serves as a steering mechanism, known as Articulated Frame Steering (AFS). Consequently, the goal of the path tracking controller is to determine the articulation angle. Similar to the path tracking problem of front wheel-steering vehicles, error dynamics-based approaches have been proposed for articulated vehicles without steerable wheels. For articulated vehicles, the error state comprises three components: displacement error, heading error, and curvature error. The kinematic model is used to derive error dynamics based on these three errors [21–26]. Typically, a simple PID controller is employed to regulate these errors [21]. The Linear Quadratic Regulator (LQR) is introduced to determine the optimal feedback gain [22]. Additionally, a feedback controller considering lateral and heading errors is proposed to reduce the number of gains [27]. However, since both PID and LQR are feedback controls, there is a disadvantage in that the performance deviation is large, depending on the gain tuning and the disturbance, and the actuator constraints are not explicitly considered in the controller. However, the AFS mechanism has greater inertia than the conventional steering system, so actuator dynamics should be considered to enhance path tracking performance. The Model Predictive Control (MPC) approach is utilized to account for state and input constraints [23–26]. The reference path for MPC is updated to avoid newly detected obstacles [24]. However, in this case, the model of the MPC controller was derived under the assumption that road curvature was constant. However, since the reference path is updated in real time for obstacle avoidance, the constraints of the articulation angle were set tightly to prevent the divergence of the controller. Since the AFS mechanism causes a large slip angle, a switching MPC is designed based on predefined constraint sets to address the varying slip angle caused by articulation steering [25,26]. However, since it is difficult to set the thresholds of the slip angle that distinguish each mode and determine the optimal MPC constraints for each, its use in various conditions is limited.

A geometrical relationship is defined to determine the articulation angle input, similar to front wheel-steering vehicles. The pure pursuit method is introduced to design the path tracking controller for articulated vehicles, akin to its application for front wheel- steering vehicles [28]. To accommodate the characteristics of articulated vehicles, an input map with a predefined global map is utilized to calculate future heading and turning radius. As the pure pursuit method is a feedforward controller, it is susceptible to disturbances. To address this, an integrated controller is designed by incorporating a feedback controller that utilizes lateral and heading errors in conjunction with the pure pursuit controller [29]. The geometrical path tracker may exhibit a large steady-state error as speed increases, potentially leading to performance degradation with changes in speed. When the feedback gain is increased to reduce the steady-state error, control stability is compromised. An MPC for posture tracking is proposed to follow an arbitrary desired path by minimizing errors in longitudinal position, lateral position, the heading angle of the front body, and the heading angle of the rear body [30]. However, the MPC was designed based on the kinematic model, so model errors due to slips were not considered. The virtual terrain field method is suggested to account for the effect of sideslip angles in the kinematic model by generating virtual lateral forces [31]. Since this method is based on the potential field, obtaining a solution in complex road geometries is challenging. To consider the inclined angle of the road, a Udwadia–Kalaba equation-based trajectory tracking controller was proposed to determine the required lateral tire force to track the reference path [32]. In this case, an accurate tire model for converting the lateral force into a steering angle or a refraction angle is required.

Based on a thorough review of previous research, various approaches have been proposed to design path tracking controllers for articulated vehicles without a steerable axle. In this scenario, the articulation angle assumes the same role as the steering angle, leading to the adoption of path tracking algorithms designed for front wheel-steering vehicles. For instance, methods like pure pursuit and Stanley are typical examples of geometry-based controllers. Additionally, error dynamics are derived from the relationship between the kinematic model of the articulated vehicle and the circular reference path.

These defined errors are regulated using feedback controllers, including PID, sliding mode control, and LQR. MPC is introduced to explicitly account for constraints in path tracking. Given the susceptibility of articulated vehicles to rollovers when speed is not properly controlled, an integrated controller specifically designed for articulated vehicles is necessary to address their unique characteristics. Furthermore, previous approaches are susceptible to disturbances and modeling errors. Rapid steering inputs due to disturbances can significantly impact the lateral stability of articulated vehicles. Notably, inaccuracies in the kinematic model and disturbances in sensor signals can lead to a decrease in control performance. Consequently, tube-based MPC for multi-input systems is introduced to develop an integrated controller for the path tracking of articulated vehicles.

The contributions of this paper can be summarized as follows:

1. Integration of controllers for both path tracking and velocity tracking aims to account for the correlation between lateral and longitudinal motion, thereby enhancing lateral stability.
2. The design of a reference tracking-based controller addresses the limitations of point tracking-based controllers and error dynamics defined for a fixed planned path, enabling the proposed controller to be applied to arbitrary desired paths.
3. The utilization of tube-based MPC enhances robustness against external disturbances of sensors and modeling errors.

The remainder of the paper is organized as follows: The overall architecture of the proposed path tracking algorithm is described in Section 2. Section 3 outlines the vehicle model and the process of local linearization. Section 4 consists of two sub-sections: reference state determination and tube-based MPC formulation. Section 5 summarizes the simulation results of the proposed algorithm, compared with base algorithms. The conclusion and avenues for future research are presented in Section 6.

2. Overall Architecture

The overall architecture of the proposed path tracking controller is depicted in Figure 1. The controller utilizes signals from external modules to gather the necessary information for path tracking. Sensors on the Autonomous Articulated Vehicle (AAV) provide vehicle information, including dynamic state and position. In this study, the measured dynamic state includes longitudinal velocity, acceleration, and yaw rate, while position encompasses both global position and heading angle, relative to a fixed coordinate system. Based on the AAV's position, a target path module extracts a segment from a global target path for use in path tracking. The global path is predefined in this study to establish a route towards the destination. The robust path tracking controller employs tube-based MPC to determine hydraulic piston pressure for articulation and traction torque for driving axle operation. The proposed controller consists of two sub-modules. The first module, "reference state decision", determines an appropriate vehicle state to follow the segment from the target path module. To determine the reference state, a virtual controller using LQR is utilized to calculate the virtual input within the prediction horizon. The second module, "tube-based MPC", optimizes control input to track the reference states using a linearized kinematic AAV model and the CVXGEN solver. The proposed MPC determines the hydraulic pressure for the articulation actuator and the required wheel torque for motor and brake actuator. With this structure in place, the controller has the freedom to decide its reference state, regardless of the tracker. Furthermore, since control input optimization is performed by tube-based MPC, our algorithm exhibits robustness against modeling errors and disturbances.

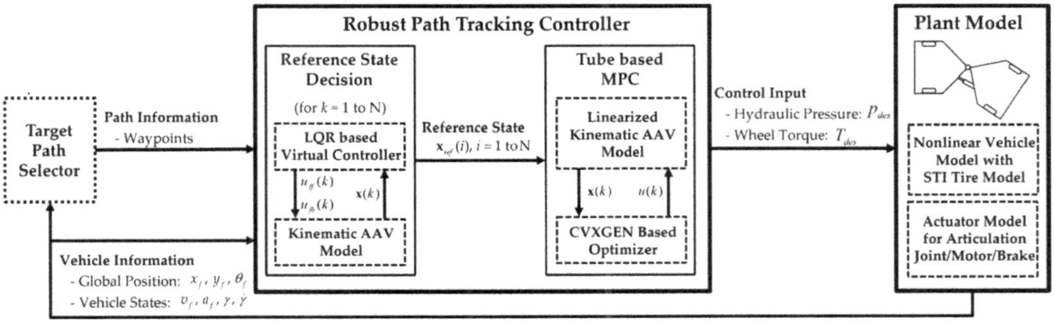

Figure 1. The overall architecture of the proposed algorithm.

3. Vehicle Model

The target platform for this study is an articulated vehicle with non-steerable axles. The transient behavior of such a vehicle mirrors that of a skid-steered wheeled vehicle. However, creating a dynamic model for the skid-steered wheeled vehicle necessitates complex and nonlinear modeling of each wheel's slip dynamics. Furthermore, the nonlinear model increases the computational load on the MPC. Therefore, this study focuses on simplifying the vehicle model for tube-based MPC by concentrating on the characteristics of the AFS mechanism and the Operational Design Domain (ODD) in AAVs. Firstly, the AFS mechanism employs a hydraulic cylinder to amplify small hydraulic pressure into significant articulation torque. Additionally, given that an AAV's body frame inertia is considerably greater than its steering mechanism, its articulation angle rate is slower than that of vehicles with steering capabilities. Secondly, due to structural characteristics susceptible to rollover, an AAV's ODD is set to operate within low-speed ranges. Given that an AAV operates under these low-speed conditions and has a small articulation angle rate, it makes sense to model its behavior as a kinematic model without needing complex tire force models.

The parameters of the AAV's kinematic model are depicted in Figure 2. The AAV is assumed to behave like a bicycle, similar to the commonly used bicycle model for front wheel-steering vehicles. Since each body of the AAV has a non-steerable axle, the steering angle for each axle is set to zero. Instead of the steering angle, the articulation angle γ is defined as a relative heading angle of the front body with respect to the rear body. The global position of the vehicle is represented by the position of the front and rear axles, $P_f(x_f, y_f)$ and $P_r(x_r, y_r)$, respectively. The heading angle of the front and rear bodies are de-noted as θ_f and θ_r, respectively. The distances from the articulation joint to each axle are defined as L_f and L_r. The velocities of each body are represented by v_f and v_r. Since the vehicle model is defined under the assumption of the zero slip, the direction of v_f and v_r is the same, with θ_f and θ_r, respectively.

As previously mentioned, the kinematic model is derived under the assumption of zero slip. Thus, the change rate of the $P_f(x_f, y_f)$ and $P_r(x_r, y_r)$ is calculated, as follows:

$$\begin{cases} \dot{x}_f = v_f \cos\theta_f \\ \dot{y}_f = v_f \sin\theta_f \end{cases} \quad \begin{cases} \dot{x}_r = v_r \cos\theta_r \\ \dot{y}_r = v_r \sin\theta_r \end{cases} \tag{1}$$

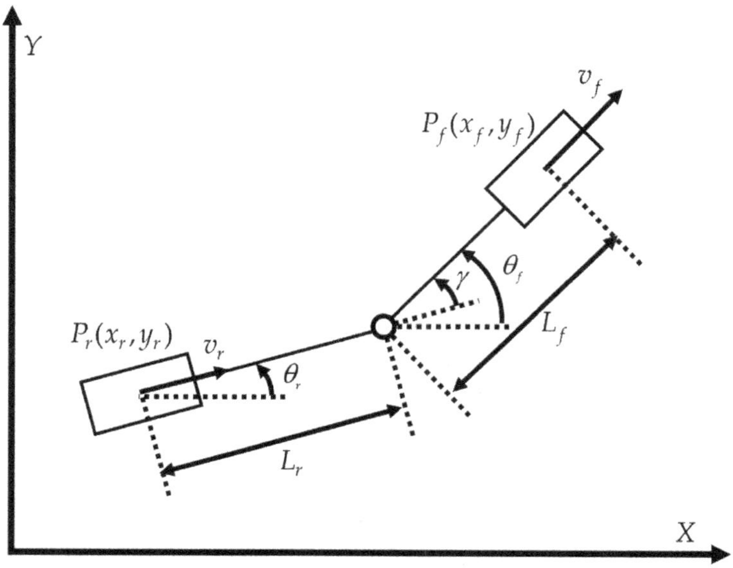

Figure 2. Parameter definition for the vehicle model of the AAV.

Since the articulation angular rate is relatively slower than the controller's sampling time, the change in the heading angle in (1) is negligible. The proposed controller's sampling time is set to 100 ms. By eliminating v_f and v_r, Equation (1) can be transformed into a motion equation for each body, as follows:

$$\begin{aligned} \dot{x}_f \sin\theta_f - \dot{y}_f \cos\theta_f = 0 \\ \dot{x}_r \sin\theta_r - \dot{y}_r \cos\theta_r = 0 \end{aligned} \tag{2}$$

γ is defined by the difference between θ_f and θ_r. Thus, the change rate of γ is calculated, as given in (3).

$$\dot{\gamma} = \dot{\theta}_f - \dot{\theta}_r \tag{3}$$

The change rate of v_f and v_r is the same because the front and rear bodies of the AAV are connected rigidly by the articulation joint. By using these characteristics, (1), and (3), the v_f and v_r, can be represented, as follows:

$$\begin{aligned} v_f &= v_r \cos\gamma + \dot{\theta}_r L_r \sin\gamma \\ v_r &= \dot{\theta}_f L_f \sin^{-1}\gamma + \dot{\theta}_r L_r \tan^{-1}\gamma \end{aligned} \tag{4}$$

The yaw rate of the front body can be represented as a function of γ by substituting (3) into (4). Similarly, the yaw rate of the rear body is also calculated, as follows:

$$\begin{aligned} \dot{\theta}_f &= \frac{v_f \sin\gamma + L_r\dot{\gamma}}{L_f \cos\gamma + L_r} \\ \dot{\theta}_r &= \frac{v_f - v_r \cos\gamma}{L_r \sin\gamma} \end{aligned} \tag{5}$$

The actuator model is defined as a first-order delay model with proportional gain. As mentioned in Section 2, the lateral actuator is a hydraulic piston responsible for generating hydraulic pressure to adjust the articulation angle. For longitudinal motion, the wheel torque on the axle is generated by the AAV's powertrain and brake system. This produced wheel torque modifies the AAV's acceleration. Consequently, the equation for the actuator can be derived, as follows:

$$\begin{aligned}\ddot{\gamma} &= -\tfrac{1}{\tau_\gamma}\dot{\gamma} + \tfrac{k_\gamma}{\tau_\gamma}p_{des} \\ \dot{a}_f &= -\tfrac{1}{\tau_a}a_f + \tfrac{k_a}{\tau_a}\tau_{des}\end{aligned} \tag{6}$$

where, τ_γ and τ_a are the time delay of the hydraulic piston and longitudinal actuator. Generally, the powertrain and brake system have different time delays. However, the larger time delay of the two systems was used as τ_a to prevent the model switching between acceleration and deceleration. k_γ and k_a are the proportional gain for each actuator.

To derive the state equation of the vehicle model, the state and input vectors are defined. The state vector comprises seven states: x position, y position, yaw, velocity, acceleration, articulation angle, and rate. The input vector has a hydraulic pressure, p_{des}, of the articulation cylinder and wheel torque, τ_{des}. The state and input vectors are defined, as follows:

$$\begin{aligned}\mathbf{x} &= \begin{bmatrix} x_f & y_f & \theta_f & v_f & a_f & \gamma & \dot{\gamma} \end{bmatrix}^T \\ u &= \begin{bmatrix} p_{des} & \tau_{des} \end{bmatrix}^T\end{aligned} \tag{7}$$

The state equation of the front body is derived by combining (1), (5), and (6).

$$\dot{\mathbf{x}} = \begin{bmatrix} v_f \cos\theta_f \\ v_f \sin\theta_f \\ \frac{v_f \sin\gamma}{L_f\cos\gamma + L_r} \\ a_f \\ -\frac{1}{\tau_a}a_f \\ \dot{\gamma} \\ -\frac{1}{\tau_\gamma}\dot{\gamma} \end{bmatrix} + \begin{bmatrix} 0 & 0 \\ 0 & 0 \\ 0 & \frac{L_r}{L_f\cos\gamma + L_r} \\ 0 & 0 \\ \frac{k_a}{\tau_a} & 0 \\ 0 & 0 \\ 0 & \frac{k_\gamma}{\tau_\gamma} \end{bmatrix}\mathbf{u} = f(\mathbf{x}, \mathbf{u}) \tag{8}$$

The position and heading angle of the rear body can be derived from the state of the front body, due to their connection through a rigid articulation joint. The position and heading of the rear body are calculated, as follows:

$$\begin{cases} x_r = x_f - L_f\cos\theta_f - L_r\cos\theta_r \\ y_r = y_f - L_f\sin\theta_f - L_r\sin\theta_r \\ \theta_r = \theta_f - \gamma \end{cases} \tag{9}$$

As stated in the introduction, the proposed controller is formulated based on tube-based MPC employing a linear model. Moreover, the designed algorithm is tailored for a digital controller. Thus, in order to configure the tube-based MPC algorithm, Equation (8)'s state equation necessitates both linearization and discretization. Initially, local linearization is applied to linearize (8) in proximity to the current state of the AAV, thereby minimizing linearization errors. Given that local linearization must be performed for each control sampling interval, the resultant linearized model becomes time-varying, as described below:

$$\begin{aligned}\dot{\mathrm{x}}(t) &= \mathrm{A}(t)\mathrm{x}(t) + \mathrm{B}(t)\mathrm{u}(t) \\ &= \left(\left.\frac{\partial f(\mathrm{x},\mathrm{u})}{\partial \mathrm{x}}\right|_{\mathrm{x}(t),\mathrm{u}(t-1)}\right)\mathrm{x}(t) + \left(\left.\frac{\partial f(\mathrm{x},\mathrm{u})}{\partial \mathrm{u}}\right|_{\mathrm{x}(t),\mathrm{u}(t-1)}\right)\mathrm{u}(t)\end{aligned} \tag{10}$$

The first order Taylor polynomial is used to approximate the (10) to discrete time model.

$$\begin{aligned}\mathrm{x}(k+1) &= \mathrm{A}_d(k)\mathrm{x}(k) + \mathrm{B}_d(k)\mathrm{u}(k) \\ \textit{where}\ \ \mathrm{A}_d(k) &= \mathrm{I} + \mathrm{A}(t)\Delta T \\ \mathrm{B}_d(k) &= \mathrm{B}(t)\Delta T\end{aligned} \tag{11}$$

where ΔT is the sampling time of the controller. In this study, the MPC uses the ΔT of 100 ms.

4. Robust Path Tracking Controller

The tube-based MPC framework is employed to configure both the path tracking and velocity controllers for the AAV. As illustrated in Figure 1, the proposed controller comprises two distinct sub-modules: the reference state decision and the tube-based MPC. The reference state decision module is specifically designed to provide the reference state vector, guiding the AAV along the desired path. Given that constraints are explicitly accounted for within the MPC framework, the reference state decision module primarily focuses on determining the theoretical optimal states only considering the constraints for the maximum control input. On the other hand, the tube-based MPC module is responsible for optimizing the vehicle's trajectory to adhere closely to this reference state, while concurrently computing control inputs that satisfy both state and input constraints. Further elaboration of this proposed robust path tracking controller for AAV is provided in the subsequent sub-sections.

4.1. Reference State Decision

The reference state serves as a replacement for the error dynamics typically utilized in defining the path tracking controller. To mitigate computational overhead, the proposed reference state decision module relies on a kinematic vehicle model augmented with a virtual controller. This virtual controller computes virtual lateral and longitudinal inputs aimed at tracking the path segment provided by the target path module. The kinematic vehicle model, expounded in Section 3, integrates these virtual inputs to forecast the AAV's future state at the subsequent sampling instance. Subsequently, this virtual input is re-evaluated concerning the AAV's most recent state and utilized to derive its forthcoming state. This iterative procedure persists within the prediction horizon of the MPC. Consequently, any form of control algorithm can be employed to ascertain virtual inputs for the reference state decision. Specifically, as optimal inputs, accounting for various constraints, are determined using MPC; employing a straightforward virtual controller that minimizes path tracking error to the utmost extent is deemed more efficacious.

The virtual controller is formulated based on the LQR controller to compute the optimal control input for regulating path tracking error. In contrast to the design of actual controllers, the reference state decision module ideally accesses state variables and accurately predicts vehicle behavior through the linear kinematic model. Hence, employing linear optimal controllers proves more efficient in minimizing computational burden and determining optimal inputs, compared to other robust control techniques. To delineate the LQR controller for path tracking, the error dynamics of the vehicle model are derived from the disparity between the reference state and the vehicle state. Initially, the error $\mathbf{e}(k)$ is defined for the reference state decision, as follows:

$$\mathbf{e}(k) = \mathbf{x}_{ref}(k) - \mathbf{x}(k) \tag{12}$$

where $\mathbf{x}_{ref}(k)$ is the reference state. Under the assumption of the constant reference, the error dynamics is derived, as follows:

$$\begin{aligned} \dot{\mathbf{e}}(k) &= \dot{\mathbf{x}}_{ref}(k) - \dot{\mathbf{x}}(k) \\ &= -\dot{\mathbf{x}}(k) \\ &= -\mathbf{A}(k)\mathbf{x}(k) - \mathbf{B}(k)\mathbf{u}(k) \\ &= -\mathbf{A}(k)\left(\mathbf{x}_{ref}(k) - \mathbf{e}(k)\right) - \mathbf{B}(k)\mathbf{u}(k) \\ &= \mathbf{A}(k)\mathbf{e}(k) - \mathbf{B}(k)\mathbf{u}(k) - \mathbf{A}(k)\mathbf{x}_{ref}(k) \end{aligned} \tag{13}$$

The control input $\mathbf{u}(k)$ in (13) is divided into two parts, feedforward input $\mathbf{u}_{ff}(k)$ and feedback input $\mathbf{u}_{fb}(k)$. The $\mathbf{u}_{ff}(k)$ is defined, as follows:

$$\mathbf{u}_{ff}(k) = \mathbf{B}^{-1}(k)\mathbf{A}(k)\mathbf{x}_{ref}(k) \tag{14}$$

After applying the feedforward input, the error dynamics is summarized, as follows:

$$\begin{aligned} \dot{\mathbf{e}}(k) &= \mathbf{A}(k)\mathbf{e}(k) - \mathbf{B}(k)\left(\mathbf{u}_{ff}(k) + \mathbf{u}_{fb}(k)\right) - \mathbf{A}(k)\mathbf{x}_{ref}(k) \\ &= \mathbf{A}(k)\mathbf{e}(k) - \mathbf{B}(k)\mathbf{u}_{fb}(k) \end{aligned} \tag{15}$$

To apply the (15) into the simulation environment, the error dynamics is discretized based on the Euler approximation, as follows:

$$\begin{aligned} \mathbf{e}(k+1) &= \mathbf{e}(k) + \dot{\mathbf{e}}(k) \cdot \Delta T \\ &= \mathbf{e}(k) + \left(\mathbf{A}(k)\mathbf{e}(k) - \mathbf{B}(k)\mathbf{u}_{fb}(k)\right) \cdot \Delta T \\ &= (\mathbf{I} + \mathbf{A}(k) \cdot \Delta T)\mathbf{e}(k) - \mathbf{B}(k) \cdot \Delta T \cdot \mathbf{u}_{fb}(k) \\ &= (\mathbf{I} + \mathbf{A}(k) \cdot \Delta T + \mathbf{B}(k) \cdot \Delta T \cdot \mathbf{K})\mathbf{e}(k) \\ \mathbf{u}_{fb}(k) = & \ -\mathbf{K}\mathbf{e}(k) \end{aligned} \tag{16}$$

As the dynamics provided in (16) represent a time-varying system, the finite horizon LQR approach is employed to compute the feedback gain **K** [33]. Given that the system is discrete, the reference state $\mathbf{x}_{ref}(k+1)$ is determined by the $\mathbf{u}_{fb}(k)$ and $\mathbf{u}_{ff}(k)$. In other words, the control input relied on the $\mathbf{x}_{ref}(k)$. Consequently, it becomes feasible to compute the reference state by advancing the system and controller within the prediction horizon of the MPC.

4.2. Tube-Based MPC

The core concept of tube-based MPC revolves around partitioning the control authority into two components: the initial segment steers the noise-free system towards the origin, while the subsequent segment offsets deviations from the ideal system [34]. The representation of the actual system incorporating noise is formulated, as follows:

$$\begin{aligned} \mathbf{x}(k+1) = \mathbf{A}(k)\mathbf{x}(k) + \mathbf{B}(k)\mathbf{u}(k) + \mathbf{w}(k) \\ \text{s.t. } \mathbf{x}(k) \in \mathcal{X} \\ \mathbf{u}(k) \in \mathcal{U} \\ \mathbf{w}(k) \in \mathcal{W} \end{aligned} \tag{17}$$

where $\mathbf{x}(k)$ is the actual state vector and $\mathbf{w}(k)$ is a bounded disturbance vector. Following the core concept of tube-based MPC, the control input comprises two constituent parts, delineated as follows:

$$\mathbf{u}(k) = \bar{\mathbf{u}}(k) + \mathbf{K}(\mathbf{x}(k) - \mathbf{z}(k)) \tag{18}$$

where $\bar{\mathbf{u}}(k)$ is the control input determined by the MPC based on the noise-free system. $\mathbf{z}(k)$ denotes the state vector of the noise-free system, and **K** signifies the state feedback gain utilized to manage the discrepancies between the actual system and the ideal one. In other words, the second term of (18) denotes the control input aimed at addressing the disparity between the actual and ideal systems. When defining the error $\mathbf{e}(k) = \mathbf{x}(k) - \mathbf{z}(k)$, the error dynamics of the system can be expressed, as follows:

$$\begin{aligned} \mathbf{e}(k+1) &= \mathbf{x}(k+1) - \mathbf{z}(k+1) \\ &= \mathbf{A}(k)x(k) + \mathbf{B}(k)u(k) + \mathbf{w}(k) - \mathbf{A}(k)z(k) - \mathbf{B}(k)\bar{u}(k) \\ &= \mathbf{A}(k)x(k) + \mathbf{B}(k)\bar{u}(k) + \mathbf{B}(k)\mathbf{K}(\mathbf{x}(k) - \mathbf{z}(k)) \\ &\quad + \mathbf{w}(k) - \mathbf{A}(k)z(k) - \mathbf{B}(k)\bar{u}(k) \\ &= (\mathbf{A}(k) + \mathbf{B}(k)\mathbf{K})(k)(\mathbf{x}(k) - \mathbf{z}(k)) + \mathbf{w}(k) \\ &= (\mathbf{A}(k) + \mathbf{B}(k)\mathbf{K})(k)\mathbf{e}(k) + \mathbf{w}(k) \end{aligned} \tag{19}$$

If the dynamics ($\mathbf{A}(k)$ + $\mathbf{B}(k)$ **K**) are stable and the set $\mathcal{W}$ is bounded, there is some set $\mathcal{E}$ that $\mathbf{e}(k)$ remains inside for all time. In other words, if a bounded disturbance is applied to a stable closed loop system, $\mathbf{e}(k)$ does not diverge. In this case, the constraints for the MPC based on the noise free system are tightened, as follows:

$$\begin{aligned} \mathbf{z}(k) &\in \mathcal{X} \ominus \mathcal{E} \\ \bar{\mathbf{u}}(k) &\in \mathcal{U} \ominus \mathbf{K}\mathcal{E} \end{aligned} \tag{20}$$

The state feedback control is designed to regulate the $\mathbf{e}(k)$, so the MPC based on the actual system can satisfy the tightened constraints. Initially, the invariant set $\mathcal{E}$ should be defined to derive these tightened constraints. In this study, the uncertainties of the state variables are modeled as a combination of modeling errors and sensor noise, which are employed to measure the position and dynamic states of the AAV. Specifically, the noise characteristics of the Differential Global Positioning System (DGPS) determine the standard deviation of the position and heading angle errors in the state vector, as presented in Equation (7). For velocity, the characteristics from a commercial vehicle's wheel speed sensor are utilized. Gaussian white noise is considered for acceleration measurements, based on the noise characteristics of the Inertial Measurement Unit (IMU). Additionally, noise levels for articulation angle and rate measurements are determined by magnetic encoder characteristics. Table 1 provides a summary of the standard deviations for Gaussian white noise for each component of the state vector.

Table 1. Noise specification.

Variable	Standard Deviation	Variable	Standard Deviation
x_f	0.5 m	v_f	1 m/s
y_f	0.5 m	a_f	0.2 m/s^2
θ_f	5 deg	γ	0.5 deg

The MPC problem for the ideal system is formulated to calculate the $\bar{\mathbf{u}}(k)$. Considering that the primary objective of the proposed algorithm is to follow the reference state, the cost function is defined to minimize the tracking error of the reference state, presented as follows:

$$\begin{aligned} J &= \sum_{k=1}^{N_p} \left[\begin{array}{c} \left(\mathbf{x}(k) - \tilde{\mathbf{x}}_{ref}(k)\right)^T \mathbf{Q}\left(\mathbf{x}(k) - \tilde{\mathbf{x}}_{ref}(k)\right) \\ +\bar{\mathbf{u}}(k)^T \mathbf{R}\bar{\mathbf{u}}(k) \end{array} \right] + \rho\varepsilon, \\ N_p &= T_p/\Delta T \end{aligned} \tag{21}$$

where T_p represents the prediction horizon of the MPC, while ΔT denotes the MPC's sampling time, as utilized in (11). Thus, the maximum prediction step N_p is derived by dividing T_p by ΔT. ε stands for a slack variable, introduced to prevent optimization failure. The slack variable is instrumental in reconciling constraints and the cost function. In cases where optimization without a slack variable fails to converge, the slack variable alleviates one of the constraints, facilitating the MPC in finding an optimal solution. A weight ρ adjusts the extent of constraint relaxation. To mitigate risks associated with the slack variable, a modification is implemented to utilize it in establishing a lower bound for longitudinal acceleration constraints. The weight matrices for reference state tracking and input minimization are denoted as $\mathbf{Q}$ and $\mathbf{R}$, respectively.

$$\mathbf{Q} = \begin{bmatrix} Q_x & & & & & & \\ & Q_y & & & & & \\ & & Q_\theta & & & & \\ & & & Q_v & & & \\ & & & & Q_a & & \\ & & & & & Q_\gamma & \\ & & & & & & Q_{\dot{\gamma}} \end{bmatrix}, \mathbf{R} = \begin{bmatrix} R_p & \\ & R_\tau \end{bmatrix} \tag{22}$$

The off-diagonal term of the **Q** and **R** is zero. The diagonal terms of **Q** are tuned to reflect the scale and importance of each variable of the state vector. Similarly, R_p and R_τ are the weights for each control input and adjusted to account for the difference scales of the inputs.

The proposed MPC integrates constraints categorized into three groups: state, input, and dynamic constraints. State constraints are instituted to uphold the vehicle's state within predetermined bounds, thereby averting excessive motion of the AAV. These state constraints entail defining a margin around the reference state for the position and heading angle of the AAV. Additionally, minimum and maximum values are stipulated as constraints for velocity, acceleration, articulation angle, and its rate. The state constraints can be summarized, as follows:

$$\begin{gathered} x_{f,ref}(k) - x_{mgn} \leq x_f(k) \leq x_{f,ref}(k) + x_{mgn} \\ y_{f,ref}(k) - y_{mgn} \leq y_f(k) \leq y_{f,ref}(k) + y_{mgn} \\ \theta_{f,ref}(k) - \theta_{mgn} \leq \theta_f(k) \leq \theta_{f,ref}(k) + \theta_{mgn} \\ v_{f,\min} \leq v_f(k) \leq v_{f,\max} \\ a_{f,\min} \leq a_f(k) \leq a_{f,\max} \\ -\gamma_{\max} \leq \gamma(k) \leq \gamma_{\max} \\ -\dot{\gamma}_{\max} \leq \dot{\gamma}(k) \leq \dot{\gamma}_{\max} \end{gathered} \tag{23}$$

The input constraints are defined with the maximum values of each, as follows:

$$\begin{gathered} -p_{\max} \leq p_{des}(k) \leq p_{\max} \\ -\tau_{\max} \leq \tau_{des}(k) \leq \tau_{\max} \end{gathered} \tag{24}$$

The vehicle model derived in Section 3 serves as a dynamic constraint for the MPC. The discretized linear model is employed within the prediction horizon to delineate the dynamic constraint, presented as follows:

$$\begin{aligned} \mathrm{x}(2) &= \mathrm{A}_d(1)\mathrm{x}(1) + \mathrm{B}_d(1)\bar{\mathrm{u}}(1) \\ &\vdots \\ \mathrm{x}(N_p) &= \mathrm{A}_d(N_p-1)\mathrm{x}(N_p-1) + \mathrm{B}_d(N_p-1)\bar{\mathrm{u}}(N_p-1) \end{aligned} \tag{25}$$

where N_p is the maximum prediction step. Table 2 provides a summary of the parameters for the MPC. Among the parameters listed in Table 2, there are a total of nine elements comprising **Q** and **R**, making manual tuning challenging. Therefore, an automated parameter search method becomes necessary. Given that the number of variables exceeds that of typical optimization problems, a grid search approach was employed [35]. The search range for each element was determined considering the dimensions of each state variable and input, with lateral error serving as the criterion for optimal judgment.

Table 2. Parameters for AAV and the proposed algorithm.

Parameter	Value	Parameter	Value
L_f	0.605 m	ΔT	0.1 s
L_r	0.895 m	N_p	20
τ_f	0.2 s	τ_a	0.05 s
Q_x	1	x_{mgn}	0.5 m
Q_y	75	y_{mgn}	0.5 m
Q_θ	100	θ_{mgn}	3 deg
Q_v	10	$v_{f,\min}$	0 m/s
Q_a	20	$v_{f,\max}$	5 m/s
Q_γ	100	$a_{f,\min}$	−3 m/s^2
$Q_{\dot{\gamma}}$	150	$a_{f,man}$	1 m/s^2
R_p	0.1	γ_{man}	50 deg
R_τ	0.01	$\dot{\gamma}_{\max}$	90 deg/s
p_{max}	50 bar	$\tau_{\max}$	1500 Nm

After determining the $\bar{\mathbf{u}}(k)$, the feedback gain K regulating the model deviation is determined using the LQR, designed to regulate the dynamics $\mathbf{A}(k) + \mathbf{B}(k)\mathbf{K}$ in (19). Finite horizon LQR is employed to compute the feedback gain K for the time-varying dynamics [33]. Given that the purpose of state feedback control is to regulate model error, the error dynamics are defined by utilizing the difference between the predicted states from the MPC and the measured states of the AAV. According to the error defined in (19), the measured state at k step is used as $\mathbf{x}(k)$, while the predicted states of the MPC are represented as $\mathbf{z}(k)$. Since the predicted states of the MPC are determined at step $k-1$, a coordinate transformation is necessary to align the local coordinate system between steps $k-1$ and k. The change in position and heading angle between adjacent timestamps is computed under the assumption of constant motion, utilizing Euler approximation. The position and heading relationships between steps $k-1$ and k are calculated, as follows:

$$\begin{bmatrix} x_f(k) \\ y_f(k) \\ \theta_f(k) \end{bmatrix} = \begin{bmatrix} x_f(k-1) + d(k-1)\cos\begin{pmatrix} \theta_f(k-1) + \Delta T\dot{\theta}_f(k-1) \\ +\frac{1}{2}\Delta T^2\ddot{\theta}_f(k-1) \end{pmatrix} \\ y_f(k-1) + d(k-1)\sin\begin{pmatrix} \theta_f(k-1) + \Delta T\dot{\theta}_f(k-1) \\ +\frac{1}{2}\Delta T^2\ddot{\theta}_f(k-1) \end{pmatrix} \\ \theta_f(k-1) + \Delta T\dot{\theta}_f(k-1) + \frac{1}{2}\Delta T^2\ddot{\theta}_f(k-1) \end{bmatrix}$$
$$d(k-1) = v_f(k-1)\Delta T + \tfrac{1}{2}\Delta T^2 a_f(k-1) \tag{26}$$

where ΔT is the sampling time of the MPC given in Table 2. A coordinate transformation matrix $\mathbf{T}(k)$ between $k-1$ and k step are defined, as follows:

$$\begin{aligned} &\begin{bmatrix} dx_f(k) & dy_f(k) & d\theta_f(k) \end{bmatrix} \\ &= \begin{bmatrix} x_f(k) - x_f(k-1) y_f(k) - y_f(k-1) x_f(k) - x_f(k-1) \end{bmatrix} \\ &\mathbf{T}(k) = \begin{bmatrix} \cos\left(d\theta_f(k)\right) & -\sin\left(d\theta_f(k)\right) & 0 & dx_f(k) \\ \sin\left(d\theta_f(k)\right) & \cos\left(d\theta_f(k)\right) & 0 & dy_f(k) \\ 0 & 0 & 1 & 0 \\ 0 & 0 & 0 & 1 \end{bmatrix} \end{aligned} \tag{27}$$

To define the error $\mathbf{e}(k)$, the $\mathbf{z}(k)$ is defined, by using the $\mathbf{T}(k)$ and the predicted states of the MPC, which is determined at step $k-1$. To prevent confusion between the measured and predicted states, a lower subscript p is employed to denote the predicted states of the MPC.

$$\begin{aligned} \mathbf{z}(k) &= \begin{bmatrix} \mathbf{p}_1(k) \\ \mathbf{p}_2(k) \\ \theta_f(k|k-1) - d\theta_f(k) \\ v_{f,p}(k|k-1) \\ a_{f,p}(k|k-1) \\ \gamma_p(k|k-1) \\ \dot{\gamma}_p(k|k-1) \end{bmatrix} \\ \mathbf{p}(k) &= \mathbf{T}(k) \cdot \begin{bmatrix} x_f(k|k-1) \\ y_f(k|k-1) \\ 0 \\ 1 \end{bmatrix} \end{aligned} \tag{28}$$

where $\mathbf{p}_1(k)$ and $\mathbf{p}_2(k)$ mean the first and second element of $\mathbf{p}$(k), which are the x and y position.

4.3. Vehicle Model Validation

A validation of the vehicle model for MPC should precede any simulation study. Experimental results from a right-turn scenario are used to assess the modeling error of the vehicle model. Experimental data were obtained using an autonomous electric road sweeper developed jointly by the Korea Institute of Industrial Technologies (KIIT) and AM Special Vehicle (AMSV) in 2019 [36]. This autonomous electric road sweeper utilized the AFS mechanism for vehicle steering, while IMU and DGPS systems were employed to measure the vehicle's accurate states. Additionally, an encoder was incorporated into the articulation joint to measure the articulation angle. Figure 3 illustrates both the experimental and simulation results for this right-turn scenario. As depicted in Figure 3, the simulation results obtained from the kinematic vehicle model closely match the experimental results. However, there are phase delays in speed and articulation angle between the experiment and simulation, as shown in Figure 3b,c. Yaw rate and lateral acceleration, illustrated in Figure 3d,e, exhibit errors resembling white noise. Therefore, we model the yaw rate and acceleration modeling errors as Gaussian white noise.

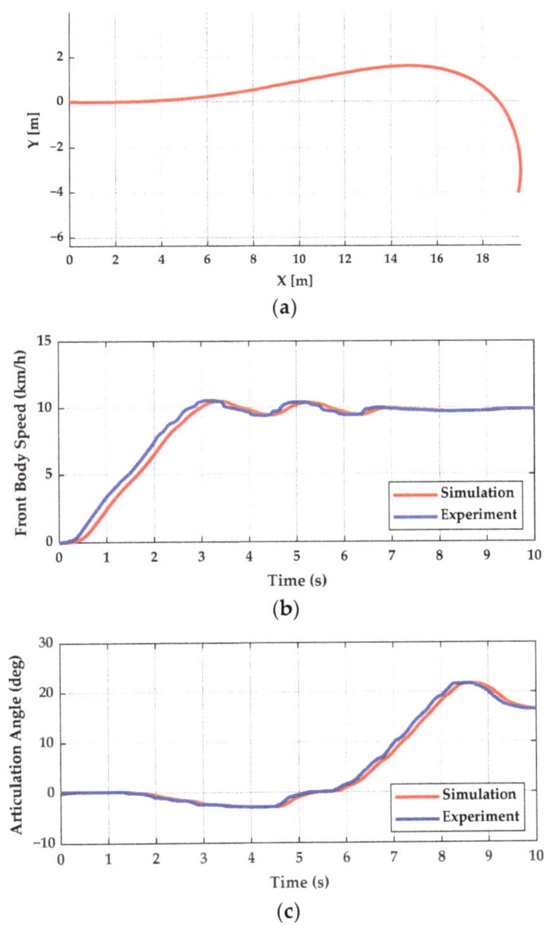

Figure 3. *Cont.*

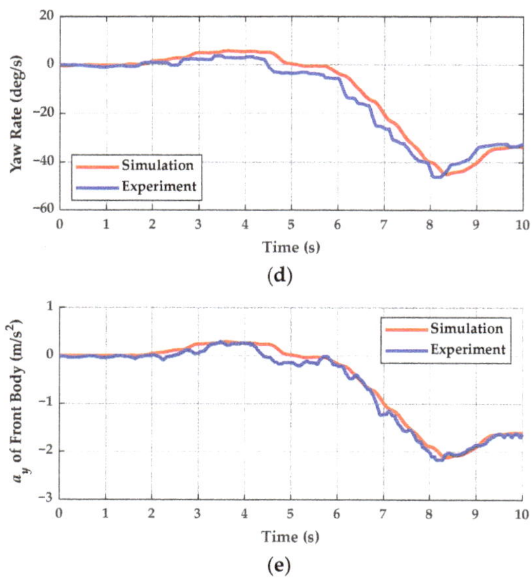

Figure 3. Comparison between the simulation results of the kinematic vehicle model and experimental results: (**a**) trajectory of the simulation result; (**b**) speed of the front body; (**c**) articulation angle; (**d**) yaw rates of the front body; (**e**) lateral accelerations of the front body.

4.4. Base Algorithms

In this study, base algorithms were employed. Given that the AFS mechanism can be approximated as front wheel steering, both the pure pursuit and Stanley methods were selected as base algorithms. The outputs from these two methods are utilized as desired articulation angles for the plant model. However, conventional path tracking algorithms are typically designed under the assumption of front wheel steering. Consequently, parameters for these base algorithms had to be redefined to align with the characteristics of an AAV. As illustrated in Figure 4, the AFS mechanism can be transformed into a front wheel-steering vehicle with a variable wheelbase. $P_{f,v}(x_{f,v}, y_{f,v})$ represents the position of the virtual front axle.

Thus, the instance wheelbase L_{inst} with articulation angle γ, is calculated, as follows:

$$L_{inst} = \frac{L_f}{\cos\gamma} + L_r \tag{29}$$

When applying the (29) in pure pursuit controller, the articulation angle of the previous sampling time is used to approximate L_{inst} to avoid the recursive formula. The Stanley method for the AAV employs the modified cross-track error, defined as the minimum distance between the target path and $P_{f,v}(x_{f,v}, y_{f,v})$. Further details of the pure pursuit and Stanley methods are provided in [37,38].

The distinctive mechanism and structure of the AAV could potentially lead to performance degradation in path tracking algorithms originally designed for front wheel steering. Consequently, an adaptive controller is chosen as another base algorithm to address the limitations associated with approximating front wheel steering. This adaptive controller, operating independently of a vehicle model, is referred to as a model-free controller. The design of this model-free controller is based on the assumption that changes in yaw rate in response to the vehicle's steering angle can be modeled as a first-order delay system. Essentially, lateral dynamics are assumed to follow a first-order delay model, as follows:

$$\dot{\theta}_r = \frac{\lambda}{\tau s + 1}\gamma \tag{30}$$

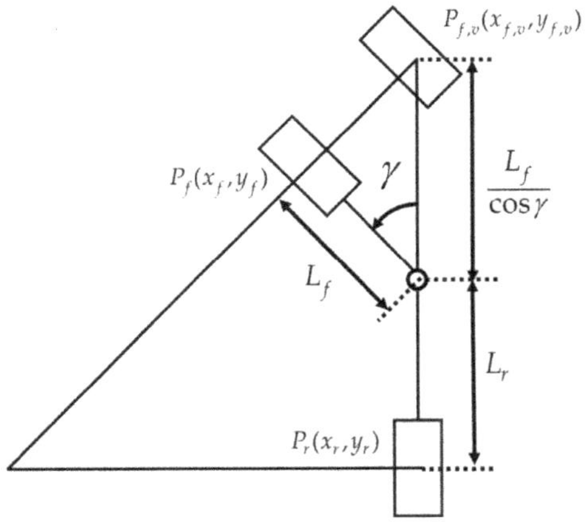

Figure 4. The virtual front axle.

To consider the structure of the AAV, the $\dot{\theta}_r$ and γ are used as the yaw rate and steering angle of the vehicle, respectively. λ and τ are the adaptive yaw rate gain and time constant of the yaw rate generation delay. As shown in (30), the simplified lateral dynamics does not take the vehicle model into account. Instead, λ is adapted to minimize the $\dot{\theta}_r$ error, as follows:

$$\dot{\hat{\lambda}} = -k \cdot \gamma \cdot \left(\dot{\theta}_{r,des} - \dot{\theta}_r\right) \quad (31)$$

where $\dot{\theta}_{r,des}$ is the desired yaw rate of the AAV's rear body. Based on (31), the articulated angle command is calculated, as follows:

$$\gamma = \frac{1}{\hat{\lambda}}\left(\dot{\theta}_{r,des} + \tau \cdot \ddot{\theta}_{r,des}\right) \quad (32)$$

The $\dot{\theta}_{r,des}$ should be defined first to apply the model-free path tracking controller. In this study, the results of the reference state decision module were utilized to define the $\dot{\theta}_{r,des}$. Since the reference state decision module determined the reference state of the front body, the relationship between front and rear bodies given in (9) was used to derive the $\dot{\theta}_{r,des}$. The final base algorithm, denoted as 'MPC' in subsequent results, is designed using the same linear MPC framework as the proposed algorithm. However, this MPC controller is configured without the robust control method. In other words, no uncertainty tube is defined for the state vector. Thus, the base MPC only used $\bar{\mathbf{u}}(k)$ to track the same reference state of the tube-based MPC.

The pure pursuit, Stanley method, and model-free controller all calculate the desired articulation angle. As discussed earlier in Section 5.1, the actuator for the articulation angle is modeled as a hydraulic cylinder. Consequently, an additional lower-level controller is introduced to determine the pressure command for this hydraulic cylinder. The pressure command is computed using a PID controller to minimize the articulation angle error. For longitudinal motion, the base algorithms require an additional controller to track the desired velocity. This velocity error is also regulated by a PID controller. Therefore, the actuator input required to track the output of these base algorithms can be determined, as follows:

$$\begin{aligned} e_\gamma &= \dot{\gamma}_{des} - \dot{\gamma} \\ p_{des} &= K_{P,\gamma} e_\gamma + K_{I,\gamma} \int e_\gamma dt + K_{D,\gamma} \dot{e}_\gamma \\ e_v &= v_{des} - v_f \\ \tau_{des} &= K_{P,v} e_v + K_{I,v} \int e_v dt + K_{D,v} \dot{e}_v \end{aligned} \quad (33)$$

5. Results

5.1. Simulation Model and Environment

The effectiveness of the proposed path tracking controller has been assessed through a simulation study. Simulations were conducted in two scenarios: one with sensor measurement noise and one without, to evaluate the robustness of the proposed method. To collectively consider the impact of model errors, a different model, distinct from the vehicle model used in MPC, was employed as the plant model for simulation. This plant model is designed based on the nonlinear dynamic bicycle model, incorporating the STI tire model. The actuator for the articulation joint is modeled with second-order dynamics to account for the stiffness and damping of the hydraulic cylinder. For longitudinal motion, the longitudinal tire force from the STI model served as an input for wheel dynamics at each axle [39]. In this case, first-order dynamics were added to the wheel dynamics input to account for the actuation delay of the powertrain and brake system. Additional details about this simulation model can be found in [40]. The simulations of the plant model and controller were implemented using MATLAB/Simulink. Since the proposed MPC is designed using linearized discrete vehicle models, it formulates an MPC problem as a linear quadratic problem. For this study, the CVXGEN solver was utilized to implement this MPC problem within the MATLAB/Simulink environment [41].

5.2. Simulation Results

A simulation study was conducted to assess the robustness of the proposed algorithm. The target path for this study was designed as an S-shaped path, comprising straight lines and arcs. Path tracking algorithms typically face challenges when encountering sudden changes in road curvature. To examine the algorithm's response to such scenarios, segments of the target path were directly connected, creating discontinuous curvature changes. Specifically, two arcs with a radius of 4 m were linked to form the S-shaped path, with a straight road connected before and after these arc segments. The point where these two arcs meet exhibits the most abrupt curvature change. This target path is represented as a solid black line in Figure 5a.

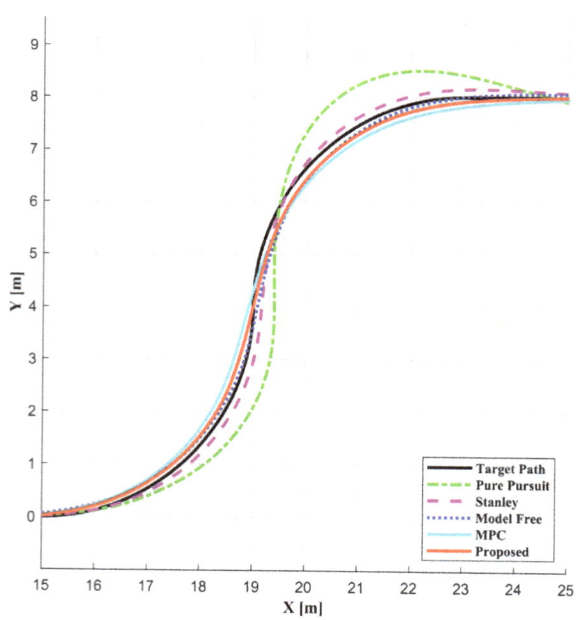

(a)

Figure 5. *Cont.*

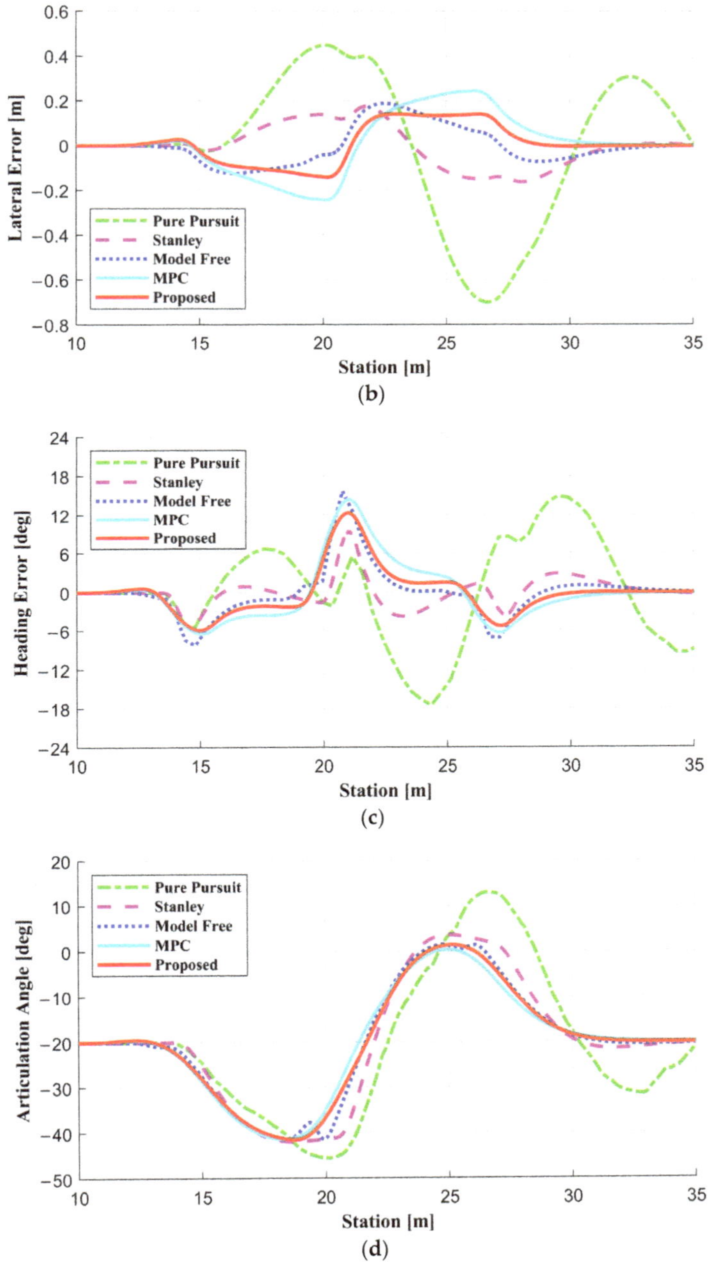

Figure 5. *Cont.*

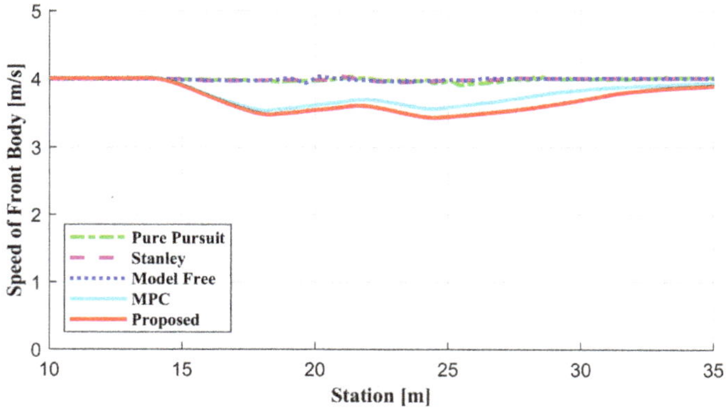

(e)

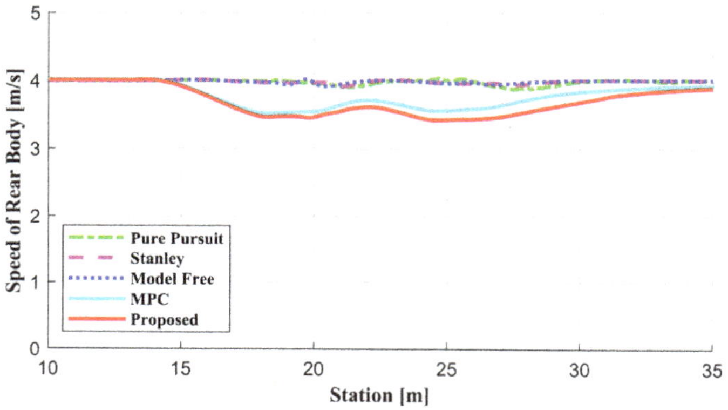

(f)

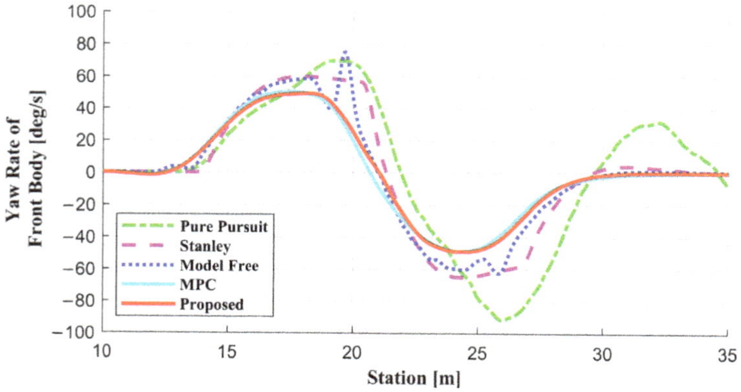

(g)

Figure 5. *Cont.*

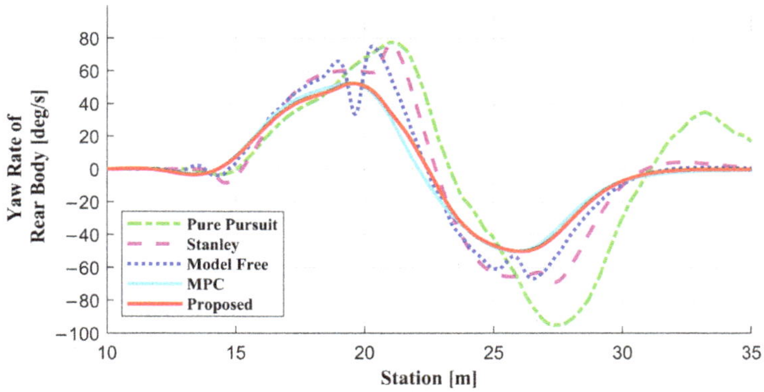

(h)

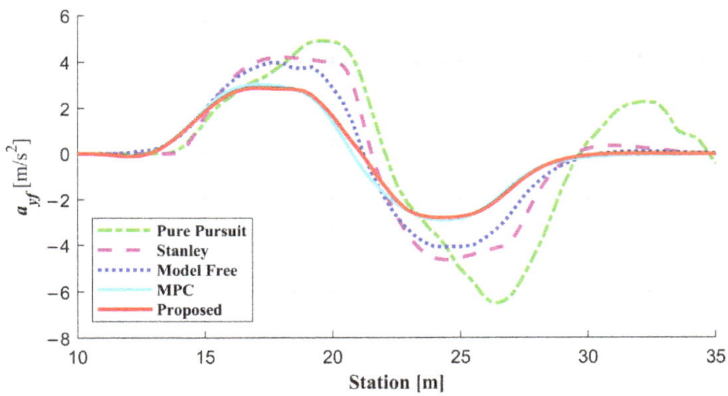

(i)

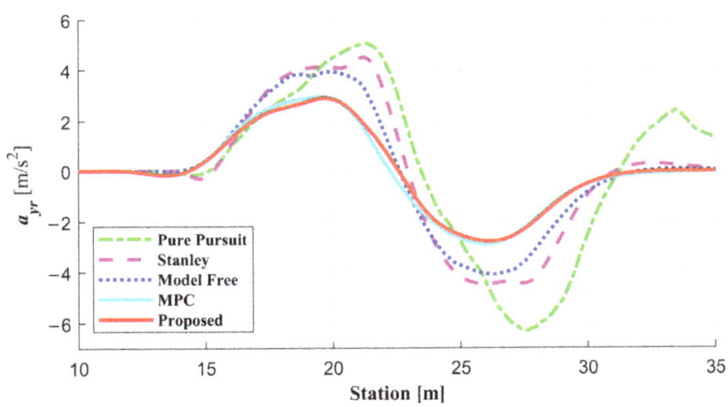

(j)

Figure 5. *Cont.*

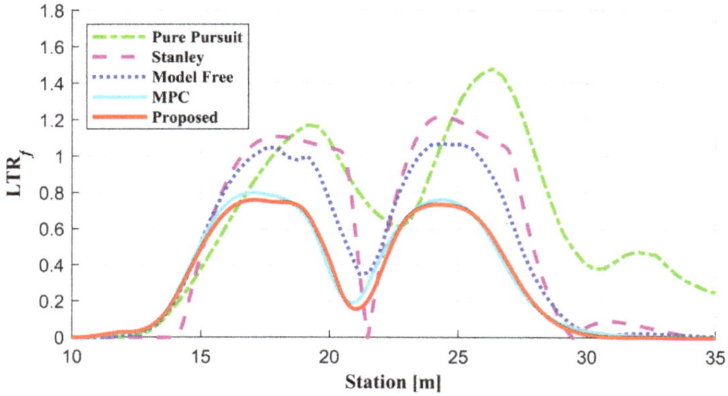

(**k**)

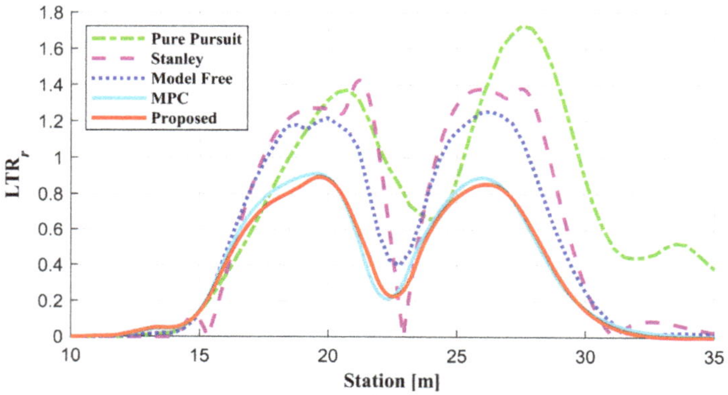

(**l**)

Figure 5. Simulation results of noise-free case: (**a**) driving trajectory; (**b**) lateral error; (**c**) heading angle error; (**d**) articulation angle; (**e**) speed of the front body; (**f**) speed of the rear body; (**g**) yaw rate of the front body; (**h**) yaw rate of the rear body; (**i**) lateral acceleration of the front body; (**j**) lateral acceleration of the rear body; (**k**) LTR of the front body; (**l**) LTR of the rear body.

The desired velocity for longitudinal motion was set at 4 m/s, a speed at which an AAV is prone to rollovers. To assess the potential for rollovers in AAVs, a Load Transfer Ratio (LTR) is employed as an index. LTR is computed from each tire's vertical force, making it applicable to various vehicle types. Rollovers can occur if either of the two bodies of the vehicle rolls over; therefore, LTR was calculated for both bodies, as follows:

$$\mathrm{LTR}_i = \frac{|F_{zi,L} - F_{zi,R}|}{F_{zi,L} + F_{zi,R}}, \quad i = f, r \tag{34}$$

where $F_{zi,L}$ and $F_{zi,R}$ means the vertical tire force of the left and right tires, respectively. The subscribe f and r indicate the front and rear bodies. A total of eight Key Performance Indices (KPIs) are defined to evaluate the path tracking performance. These include the mean, standard deviation, and maximum value of the lateral and heading errors, as well as the maximum of the lateral acceleration and LTR.

Figure 5 illustrates the simulation results for a noise-free scenario. The road friction coefficient is set to 0.85, simulating conditions where a rollover could potentially occur. In cases of low road friction coefficients, sufficient tire force to cause a rollover cannot be generated. Figure 5a describes the trajectory of each algorithm, showing the history of $P_f(x_f, y_f)$. All algorithms used in this study successfully followed the target path, with trajectories closely matching the target path, except for the pure pursuit method. The pure pursuit controller exhibited significant overshooting, deviating outside the target path due to its inherent vulnerability to different AAV structures. As mentioned in Section 4.4, an AAV's effective wheelbase depends on its articulation angle, making pure pursuit controllers less effective on high-curvature roads that require large articulation angles for the desired yaw motion. In contrast, both the Stanley method and model-free controllers do not consider wheelbase when determining the articulation angle, and they outperformed pure pursuit in this scenario. Both pure pursuit and Stanley methods, relying on single preview points, often resulted in vehicles driving along the inner corners of their target paths. However, actuation delays resulted in slow responses in articulation, causing vehicle trajectories to deviate outward from their paths when following second arcs. In contrast, by adjusting the yaw rate gain, model-free controllers significantly improved path tracking performance, compared to both pure pursuit and Stanley methods.

The simulation results for both the MPC and the proposed algorithm were quite similar. This similarity arises because no noise was introduced into the simulation, making it challenging for these two MPC-based controllers to exhibit performance differences related to robustness. As a result, as depicted in Figure 5b–d, the lateral and heading errors, as well as the articulation angle, displayed similar values, except for those from the pure pursuit method. However, both the MPC and the proposed algorithm experienced slight deceleration when navigating curvy roads, as illustrated in Figure 5e,f. This occurred because the lateral acceleration reached the MPC constraints, as shown in Figure 5i,j. An examination of the LTR values in Figure 5k,l reveals that all controllers, except those based on MPC, had an LTR value exceeding 1, indicating a potential rollover. This is because the pure pursuit and Stanley methods are geometry-based path trackers. In other words, the dynamic behavior of the AAV, such as rollover, is not considered in the pure pursuit and Stanley methods. Even if the steering input is applied to the level that the vehicle will rollover, the steering input does not decrease. Similarly, the model-free controller only considers the relationship between yaw rate and steering input as a first-order delay model. Therefore, dynamic behaviors are not considered the same as geometry-based controllers in model-free controllers. In addition, an articulated vehicle without a steering axle cannot be restored to a stable condition without proper steering or velocity control, unlike a front wheel-steering vehicle. This also explains a situation in which the base algorithm cannot recover in a situation where a rollover is about to occur. Therefore, addressing AAVs' vulnerability to rollovers becomes challenging without appropriate longitudinal controllers working in tandem. Nevertheless, under ideal simulation conditions, both the MPC and the proposed algorithm demonstrate similar performance, with constraints applied within the MPC to prevent rollovers.

The simulation results of the noise injection case are summarized in Figure 6. In contrast to the simulation results without noise, the results of the base algorithms, except for MPC, were omitted due to their lack of robustness against disturbances, leading to controller divergence. Consequently, Figure 6 displays the results of MPC and the proposed algorithm for cases where noise was applied to the sensor measurement and the case where it was not. In Figure 6a, both MPC and the proposed algorithm can be observed to follow the target path without divergence, even with disturbances injected. Notably, the proposed algorithm exhibited a nearly identical driving trajectory to the noise-free case. However, MPC, which demonstrated similar performance to the proposed algorithm in the absence of noise, proved vulnerable to disturbances, resulting in significantly degraded performance in the noise case. This phenomenon becomes more pronounced in Figure 6b,c, which depict lateral and heading errors. The injected noise caused both algorithms to exhibit some level of vibration, but the proposed algorithm managed to keep the error within a range similar to that in the noise-free case. Similar trends were observed in parameters such as speed, yaw rate, and lateral acceleration. Nevertheless, when examining the LTR in Figure 6k,l, it becomes apparent that MPC, used as the base algorithm, exceeded a value of 1 in both LTR_f and LTR_r at the first corner. In contrast, the proposed algorithm effectively controlled the input to prevent rollover, even when exposed to high disturbances, confirming its robust control performance.

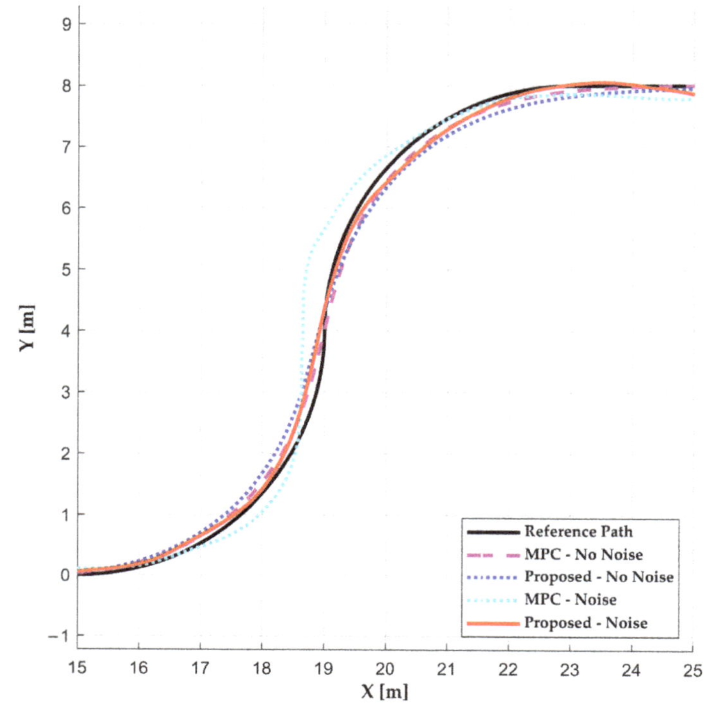

(a)

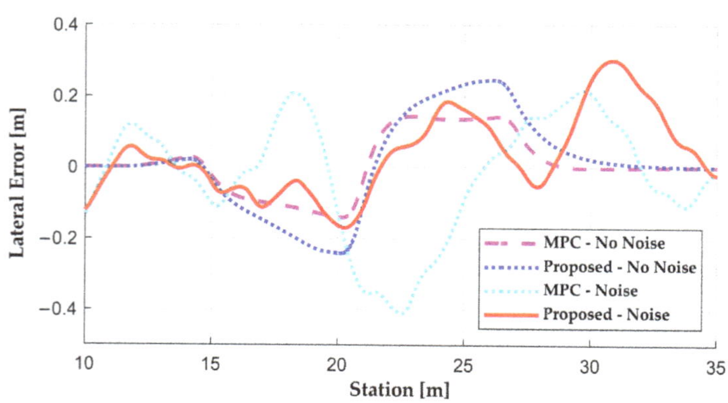

(b)

Figure 6. *Cont.*

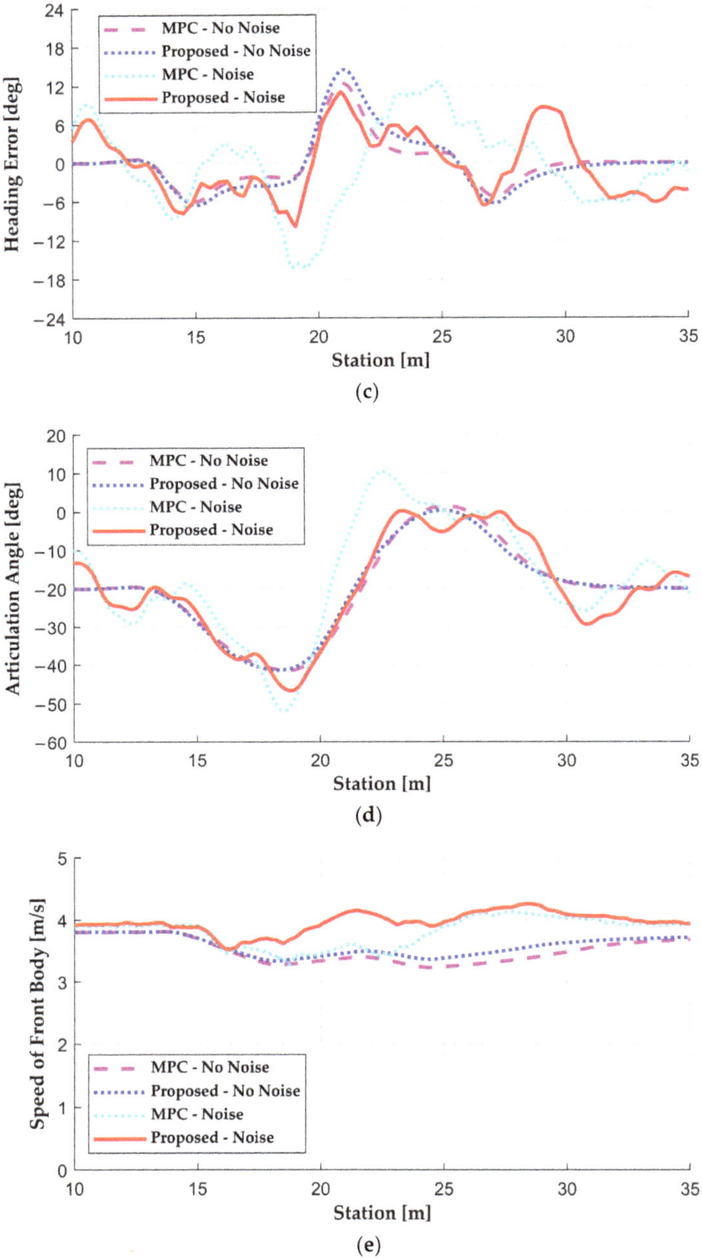

Figure 6. *Cont.*

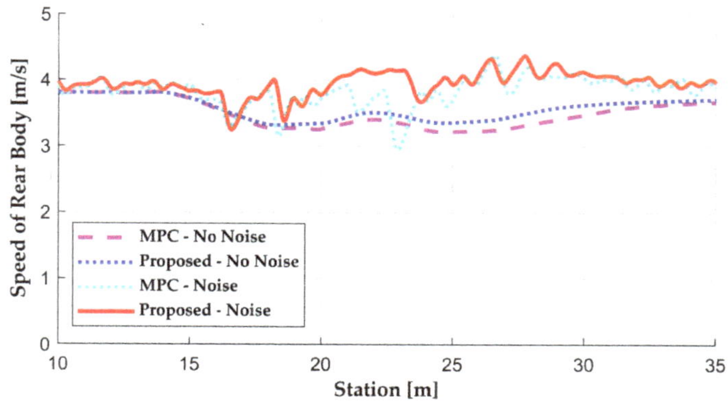

(f)

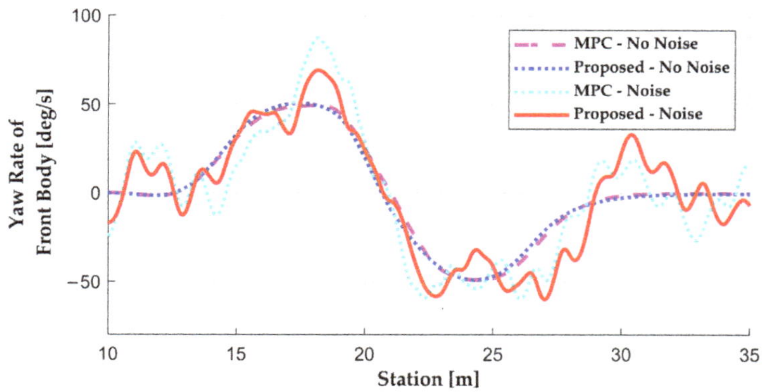

(g)

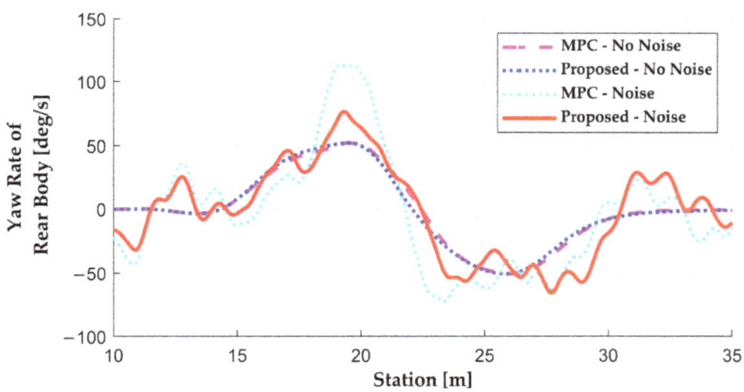

(h)

Figure 6. *Cont.*

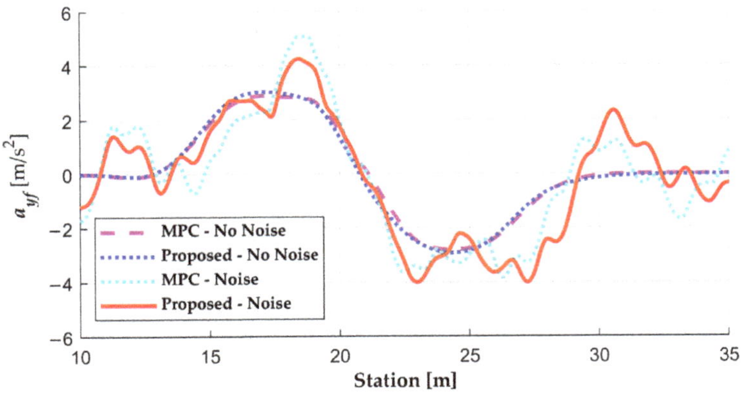

(**i**)

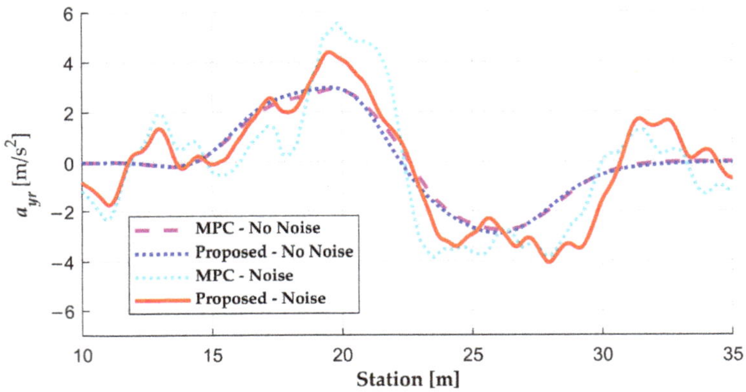

(**j**)

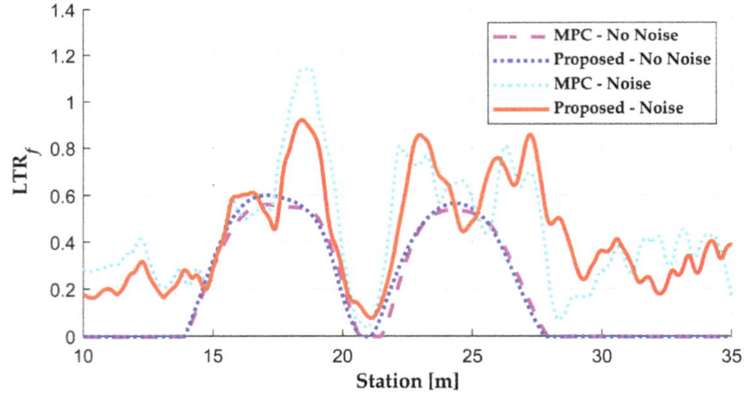

(**k**)

Figure 6. *Cont.*

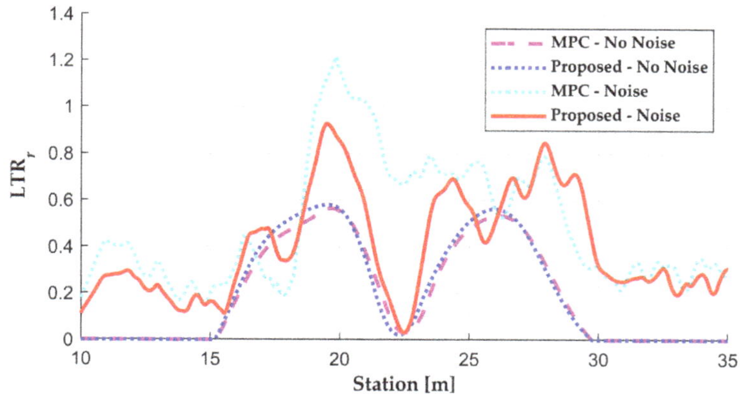

(l)

Figure 6. Simulation results of noise case: (**a**) driving trajectory; (**b**) lateral error; (**c**) heading angle error; (**d**) articulation angle; (**e**) speed of the front body; (**f**) speed of the rear body; (**g**) yaw rate of the front body; (**h**) yaw rate of the rear body; (**i**) lateral acceleration of the front body; (**j**) lateral acceleration of the rear body; (**k**) LTR of the front body; (**l**) LTR of the rear body.

KPIs from the simulation results are summarized in Table 3. When considering scenarios without noise injection, the proposed algorithm showed KPIs related to lateral error compared to other base algorithms, except for pure pursuit. Notably, pure pursuit exhibited a mean and standard deviation of the lateral error nearly three times larger than the other algorithms. Regarding heading errors, the Stanley method showed the smallest mean and standard deviation, while the model-free controller and proposed controller exhibited similar values. Similar to lateral error, pure pursuit also showed the largest heading error, compared to other algorithms. Only considering the lateral and heading errors, the Stanley method provided the best performance among the base and proposed algorithms. However, when considering all KPIs, the proposed algorithm emerged as the top performer. For lateral stability and pure pursuit, the Stanley method and the model-free controller exhibited larger lateral acceleration, compared to MPC and the proposed algorithm. In comparison with MPC-based algorithms, the maximum lateral acceleration of other methods reached approximately 1.3 to 2 times that of MPC. This phenomenon stems from the limitation of determining control input by considering only a single position on the target path, unlike MPC, which factors in future vehicle motions. As indicated in Table 3, the increase in lateral acceleration corresponded to an increase in LTR. Even with no noise injection, only MPC and the proposed algorithm successfully kept the maximum LTR within 1, preventing rollovers.

Table 3. KPIs of the simulation results.

Criteria	Noise	Pure Pursuit	Stanley	Model-	MPC	Proposed
Mean (SD) of lateral error [m]	Off	0.1660 (0.1975)	0.0422 (0.0585)	0.0389 (0.0530)	0.0696 (0.0893)	0.0447 (0.0568)
	On	-	-	-	0.1286 (0.0983)	0.0863 (0.0702)
Max of lateral error [m]	Off	0.6983	0.1770	0.1901	0.2470	0.1429
	On	-	-	-	0.4093	0.3022

Table 3. *Cont.*

Criteria	Noise	Pure Pursuit	Stanley	Model-	MPC	Proposed
Mean (SD) of heading error [deg]	Off	4.3795 (4.8417)	0.9832 (1.6004)	1.3902 (2.7911)	2.2949 (3.1920)	1.7151 (2.6232)
	On	-	-	-	4.8468 (3.8542)	3.7865 (2.4860)
Max of heading error [deg]	Off	17.4951	9.3390	15.7492	14.4686	12.3413
	On	-	-	-	16.2658	11.0544
Max of ay [m/s^2]	Off	5.8621	4.6436	4.2733	3.0435	2.9409
	On	-	-	-	5.5699	4.3923
Max of LTR	Off	1.7861	1.4252	1.2565	0.9139	0.8942
	On	-	-	-	1.2175	0.9284

KPIs for simulations with noise injection are also summarized in Table 3. Due to the divergence of the pure pursuit, Stanley, and model-free controllers when signal corruption occurred, their results are not included in Table 3. Comparing MPC and the proposed algorithm, the KPIs of the proposed algorithm delivered superior results. Hence, the proposed algorithm demonstrates robust path tracking performance while adhering to constraints.

6. Conclusions

An integrated controller for path tracking and velocity control is proposed to address the lateral stability of the AAV and enhance its resilience to external disturbances. This robust path tracking controller comprises two key modules: the reference state decision and tube-based MPC. In contrast to conventional approaches, the reference state decision module replaces error dynamics. It computes the reference state within the prediction horizon by integrating the kinematic vehicle model with the control input from the virtual controller. An LQR controller is employed to provide the optimal reference state. Tube-based MPC, the second component, calculates the optimal control input to track the reference state while adhering to state, input, and dynamic constraints. These constraints take into account factors like lateral stability and actuation limits, which can significantly affect performance. To evaluate the performance of the proposed robust path tracking controller, a simulation study was conducted. White noise was introduced to the measured signals from the plant model to assess robustness. The results demonstrated improved robustness and path tracking performance for the proposed algorithm, even under scenarios with measurement noise for multiple signals.

The motion control algorithm for the AAV can be improved in three ways. Firstly, replacing the kinematic vehicle model with a nonlinear dynamic model can significantly improve performance. Given that the AAV has non-steerable wheels, accounting for the non-negligible slip angle becomes crucial, especially as driving speeds increase. Therefore, designing a robust nonlinear controller is essential to extend the operational design domain of the AAV. Secondly, expanding the system to include more bodies is another potential improvement. For instance, adding an additional trailer to the rear body of the AAV requires the inclusion of a passive articulation joint. As a result, it becomes imperative to extend the applicability of the proposed algorithm to multi-trailer scenarios. Lastly, the implementation of the MPC and its testing on an actual AAV is a crucial step. This practical implementation allows for the validation and fine-tuning of the algorithm, ensuring that it works effectively in real-world AAVs.

Author Contributions: Conceptualization, T.L. and Y.J.; methodology, Y.J.; software, T.L. and Y.J.; validation, T.L. and Y.J.; formal analysis, T.L.; investigation, T.L.; resources, Y.J.; data curation, T.L.; writing—original draft preparation, Y.J.; writing—review and editing, Y.J.; visualization, T.L.; supervision, Y.J.; project administration, Y.J.; funding acquisition, Y.J. All authors have read and agreed to the published version of the manuscript.

Funding: This study was supported by the Research Program funded by the SeoulTech (Seoul National University of Science and Technology).

Data Availability Statement: Data are contained within the article.

Conflicts of Interest: The authors declare no conflicts of interest.

References

1. Viscelli, S. *Driverless? Autonomous Trucks and the Future of the American Trucker, Center for Labor Research and Education*; University of California, Berkely and Working Partnerships USA: San Francisco, CA, USA, 2018.
2. Rana, M.M.; Hossain, K. Connected and autonomous vehicles and infrastructures: A literature review. *Int. J. Pavement Res. Technol.* **2021**, *16*, 264–284. [CrossRef]
3. Iclodean, C.; Cordos, N.; Varga, B.O. Autonomous shuttle bus for public transportation: A review. *Energies* **2020**, *13*, 2917. [CrossRef]
4. Jo, A.; Lee, H.; Seo, D.; Yi, K. Model-reference adaptive sliding mode control of longitudinal speed tracking for autonomous vehicles. *Proc. Inst. Mech. Eng. D J. Automob. Eng.* **2023**, *237*, 493–515. [CrossRef]
5. Zhang, Y.; Zhang, Y.; Ai, Z.; Feng, Y.; Zhang, J.; Murphey, Y.L. Estimation of electric mining haul trucks' mass and road slope using dual level reinforcement estimator. *IEEE Trans. Veh. Technol.* **2019**, *68*, 10627–10638. [CrossRef]
6. Hiraoka, T.; Nishihara, O.; Kumamoto, H. Automatic path-tracking controller of a four-wheel steering vehicle. *Veh. Syst. Dyn.* **2009**, *47*, 1205–1227. [CrossRef]
7. Kang, J.; Kim, W.; Lee, J.; Yi, K. Design, implementation, and test of skid steering-based autonomous driving controller for a robotic vehicle with articulated suspension. *J. Mech. Sci. Technol.* **2010**, *24*, 793–800. [CrossRef]
8. Yin, Y.; Rakheja, S.; Yang, J.; Boileau, P.E. Design optimization of an articulated frame steering system. *Proc. Inst. Mech. Eng. D J. Automob. Eng.* **2018**, *232*, 1339–1352. [CrossRef]
9. Hang, P.; Chen, X.; Luo, F. LPV/H∞ controller design for path tracking of autonomous ground vehicles through four-wheel steering and direct yaw-moment control. *Int. J. Automot. Technol.* **2019**, *20*, 679–691. [CrossRef]
10. Hiraoka, T.; Nishihara, O.; Kumamoto, H. Model-following sliding mode control for active four-wheel steering vehicle. *Rev. Automot. Eng.* **2004**, *25*, 305.
11. Shamah, B.; Wagner, M.D.; Moorehead, S.; Teza, J.; Wettergreen, D.; Whittaker, W.L. Steering and control of a passively articulated robot. In Proceedings of the SPIE 4571, Sensor Fusion and Decentralized Control in Robotic Systems IV, Boston, MA, USA, 28–31 October 2001.
12. Sunday, B. Stability analysis of a semi-trailer articulated vehicle: A review. *Int. J. Automot. Sci. Technol.* **2021**, *5*, 131–140.
13. He, Y.; Khajepour, A.; McPhee, J.; Wang, X. Dynamic modelling and stability analysis of articulated frame steer vehicles. *Int. J. Heavy Veh. Syst.* **2005**, *12*, 28–59. [CrossRef]
14. Lei, T.; Wang, J.; Yao, Z. Modelling and stability analysis of articulated vehicles. *Appl. Sci.* **2021**, *11*, 3663. [CrossRef]
15. Shen, Y.; Xie, J.; Zhang, W. Study of hydraulic steering process for intelligent autonomous articulated vehicle. *SAE Tech. Pap.* **2018**. [CrossRef]
16. Sampei, M.; Tamura, T.; Kobayashi, T.; Shibui, N. Arbitrary path tracking control of articulated vehicles using nonlinear control theory. *IEEE Trans. Control Syst. Technol.* **1995**, *3*, 125–131. [CrossRef]
17. Mohamed, A.; Ren, J.; Lang, H.; El-Gindy, M. Optimal path planning for an autonomous articulated vehicle with two trailers. *Int. J. Autom. Control* **2018**, *12*, 449–465. [CrossRef]
18. Bolzern, P.; DeSantis, R.M.; Locatelli, A.; Masciocchi, D. Path-tracking for articulated vehicles with off-axle hitching. *IEEE Trans. Control Syst. Technol.* **1998**, *6*, 515–523. [CrossRef]
19. Oreh, S.T.; Kazemim, R.; Azadi, S. A new desired articulation angle for directional control of articulated vehicles. *Proc. Inst. Mech. Eng. K J. Multi-body Dyn.* **2012**, *226*, 298–314. [CrossRef]
20. Jeong, D.; Choi, S.B. Tube-based Robust Model Predictive Control for Tracking Control of Autonomous Articulated Vehicles. *IEEE Trans. Intell. Veh.* **2024**, *9*, 2184–2196. [CrossRef]
21. Alshaer, B.J.; Darabseh, T.T.; Alhanouti, M.A. Path planning, modeling and simulation of an autonomous articulated heavy construction machine performing a loading cycle. *Appl. Math. Model.* **2013**, *37*, 5315–5325. [CrossRef]
22. Meng, Y.; Gan, X.; Wang, Y.; Gu, Q. LQR-GA controller for articulated dump truck path tracking system. *J. Shanghai Jiaotong Univ. (Sci.)* **2019**, *24*, 78–85. [CrossRef]
23. Nayl, T.; Nikolakopoulos, G.; Gustafsson, T. Effect of kinematic parameters on MPC based on-line motion planning for an articulated vehicle. *Robot. Auton. Syst.* **2015**, *70*, 16–24. [CrossRef]

24. Nayl, T.; Nikolakopoulos, G.; Gustafsson, T. On-line path planning for an articulated vehicle based on model predictive control. In Proceedings of the 2013 IEEE international conference on control applications (CCA), Hyderabad, India, 28–30 August 2013.
25. Nayl, T.; Nikolakopoulos, G.; Gustafsson, T. A full error dynamics switching modeling and control scheme for an articulated vehicle. *Int. J. Control Autom. Syst.* **2015**, *13*, 1221–1232. [CrossRef]
26. Nayl, T.; Nikolakopoulos, G.; Gustafsson, T. Switching model predictive control for an articulated vehicle under varying slip angle. In Proceedings of the 2012 20th Mediterranean Conference on Control & Automation (MED), Barcelona, Spain, 3–6 July 2012.
27. Marshall, J.; Barfoot, T.; Larsson, J. Autonomous underground tramming for center-articulated vehicles. *J. Field Robot.* **2008**, *25*, 400–421. [CrossRef]
28. Rains, G.C.; Faircloth, A.G.; Thai, C.; Raper, R.L. Evaluation of a simple pure pursuit path-following algorithm for an autonomous, articulated-steer vehicle. *Appl. Eng. Agric.* **2014**, *30*, 367–374.
29. Yao, D.; Xie, H.; Qiang, W.; Liu, Y.; Xiong, S. Accurate trajectory tracking with disturbance-resistant and heading estimation method for self-driving vibratory roller. *IFAC-PapersOnLine* **2018**, *51*, 754–758. [CrossRef]
30. Liu, Z.; Yue, M.; Guo, L.; Zhang, Y. Trajectory planning and robust tracking control for a class of active articulated tractor-trailer vehicle with on-axle structure. *Eur. J. Control.* **2020**, *54*, 87–98. [CrossRef]
31. Gao, Y.; Cao, D.; Shen, Y. Path-following control by dynamic virtual terrain field for articulated steer vehicles. *Veh. Syst. Dyn.* **2020**, *58*, 1528–1552. [CrossRef]
32. Li, X.; Gong, X.; Huang, J.; Chen, Y.H. Trajectory-tracking controller for vehicles on inclined road based on Udwadia–Kalaba equation. *Green Energy Intell. Transp.* **2022**, *1*, 100021. [CrossRef]
33. Fong, J.; Tan, Y.; Crocher, V.; Oetomo, D.; Mareels, I. Dual-loop iterative optimal control for the finite horizon LQR problem with unknown dynamics. *Syst. Control Lett.* **2018**, *111*, 49–57. [CrossRef]
34. Lee, C.; Van Tran, Q.; Kim, J. Robust Path Tracking and Obstacle Avoidance using Tube-based Model Predictive Control for Surface Vehicles. *IFAC-Pap.* **2022**, *55*, 301–306. [CrossRef]
35. Youness, G.; Phan, N.U.T.; Boulakia, B.C. BootBOGS: Hands-on optimizing Grid Search in hyperparameter tuning of MLP. In Proceedings of the AICCSA 2023: 20th ACS/IEEE International Conference on Computer Systems and Applications, Giza, Egypt, 4–7 December 2023.
36. Korea IT News. Gwangju-Si and Korea Institute of Industrial Technology to Introduce Two Unmanned Special Vehicles Next Month. Available online: https://english.etnews.com/20201118200004 (accessed on 20 November 2020).
37. Coulter, R. *Implementation of the Pure Pursuit Path Tracking Algorithm*; Carnegie Mellon University: Pittsburgh, PA, USA, 1990; pp. 1–9.
38. Thrun, S.; Montemerlo, M.; Dahlkamp, H.; Stavens, D.; Aron, A.; Diebel, J.; Fong, P.; Gale, J.; Halpenny, G.; Lau, K.; et al. Stanley: The robot that won the DARPA Grand Challenge. *J. Field Robot.* **2006**, *23*, 661–692. [CrossRef]
39. Liu, Q.; Solis, D.; Pan, W. Analysis of the STI tire Model. *SAE Tech. Pap.* **2002**. [CrossRef]
40. Jeong, Y. Integrated Vehicle Controller for Path Tracking with Rollover Prevention of Autonomous Articulated Electric Vehicle Based on Model Predictive Control. *Actuators* **2023**, *12*, 41. [CrossRef]
41. Mattingley, J.; Boyd, S. CVXGEN: A code generator for embedded convex optimization. *Optim. Eng.* **2012**, *13*, 1–27. [CrossRef]

actuators

Article

Yaw Stability Control of Unmanned Emergency Supplies Transportation Vehicle Considering Two-Layer Model Predictive Control

Minan Tang [1,2,*], Yaqi Zhang [1], Wenjuan Wang [2], Bo An [1] and Yaguang Yan [1]

[1] School of Automation and Electrical Engineering, Lanzhou Jiaotong University, Lanzhou 730070, China; 12211534@stu.lzjtu.edu.cn (Y.Z.); 12211434@stu.lzjtu.edu.cn (B.A.); 12211440@stu.lzjtu.edu.cn (Y.Y.)

[2] School of New Energy and Power Engineering, Lanzhou Jiaotong University, Lanzhou 730070, China; 11210950@stu.lzjtu.edu.cn

* Correspondence: tangminan@mail.lzjtu.cn; Tel.: +86-1389-368-8178

Abstract: The transportation of emergency supplies is characterized by real-time, urgent, and non-contact, which constitute the basic guarantee for emergency rescue and disposal. To improve the yaw stability of the four-wheel-drive unmanned emergency supplies transportation vehicle (ESTV) during operation, a two-layer model predictive controller (MPC) method based on a Kalman filter is proposed in this paper. Firstly, the dynamics model of the ESTV is established. Secondly, the improved Sage–Husa adaptive extended Kalman filter (SHAEKF) is used to decrease the impact of noise on the ESTV system. Thirdly, a two-layer MPC is designed for the yaw stability control of the ESTV. The upper-layer controller solves the yaw moment and the front wheel steering angle of the ESTV. The lower-layer controller optimizes the torque distribution of the four tires of the ESTV to ensure the self-stabilization of the ESTV operation. Finally, analysis and verification are carried out. The simulation results have verified that the improved SHAEKF can decrease the state estimation error by more than 78% and achieve the noise reduction of the ESTV state. Under extreme conditions of high velocity and low adhesion, the average relative error is within 6.77%. The proposed control method can effectively prevent the instability of the ESTV and maintain good yaw stability.

Keywords: emergency supplies transportation vehicle; yaw stability; two-layer model predictive control; improved Sage–Husa adaptive extended Kalman filter; dynamics model

1. Introduction

Emergency supplies transportation plays an important role in production disaster relief, medical assistance, and dangerous situation disposal, and it has important application value for protecting social property and people's safety [1]. The unmanned emergency supplies transportation vehicle (ESTV) has a high energy utilization rate, which can assist or replace human work and reduce labor input. Therefore, the attention gradually increased [2]. In the context of artificial intelligence, ESTV has been widely used in all aspects of society. With the continuous improvement and development of the performance of the ESTV, stability control under the regular operating conditions of good roads is easy to achieve, but due to uncertain factors such as the ground adhesion coefficient, especially the stability under complex road conditions such as low adhesion after snow and ice, still faces great challenges [3]. Therefore, ensuring the safety and stability of ESTV under various complex road conditions is an important research direction for vehicle dynamics control [4].

Yaw stability control is of important significance for ensuring vehicle safety [5]. To promote the rapid development of four-wheel-drive vehicles, many scholars have carried out related research on improving their yaw stability, such as PID control [6,7], sliding mode control [8,9], adaptive control [10,11], neural network control [12,13], model predictive control (MPC) [14,15], and other methods. Among them, the MPC can make local optimization adjustments at each time step and obtain optimal control input. It is an effective method to deal with multivariable and constrained problems and can compensate for the uncertainties caused by model error, disturbances, and other factors in real time, with good dynamic control performance [16].

The yaw stability control system can be divided into integral stability controllers and hierarchical stability controllers according to the system structure. The integral structure control strategy combines the yaw stability controller design with the drive torque distribution, solves the dynamic coupling between the vehicle and the actuator, and directly optimizes the control input of the system with good control performance [17]. Qu et al. [18] used an extended state observer to estimate the disturbance after linearization, and the LMPC controller was used for yaw stability control to provide the front wheel steering angle and drive torque for four wheels. Wang et al. [19] proposed a coordination control strategy with a two-layer controller. The upper controller used adaptive backstepping technology to obtain the expected additional yaw moment, and the lower controller used the quadratic programming method to calculate the torque distribution between the four wheels so that the tire adhesion could be fully utilized. Chowdhri et al. [20] calculates the control action of four wheels considering reaction time and vehicle limitations with a g–g diagram and Kamm's circle constraints. Since the integrated control of a vehicle is a highly nonlinear control problem, the computational burden of stability control based on a nonlinear predictive controller is large, and meeting the real-time requirements of a fast dynamic system of a vehicle becomes a control challenge. The hierarchical stability controller treats the upper-layer yaw stability control separately from the lower-layer optimal distribution of the drive torque. Liu et al. [21] constructed a coordinated yaw and roll motion controller based on data-driven multi-MPC in the upper-layer controller, and the lower-layer controller combined the tire slip rate and the vertical load transfer amount and used the quadratic programming algorithm to transform the fusion expected yaw moment into the optimal drive torque of each wheel. Zhang et al. [15] designed the upper supervisor controller to calculate the compensation torque required when the vehicle turns and used Gaussian process regression to reduce the error. The lower-layer controller distributes the torque to the wheels in the form of braking torque. This hierarchical structure reduces the system dimensionality of the centralized optimization strategy and computational burden [9].

The ESTV is inevitably affected by environmental factors in actual work. To obtain the required signal, suppress noise, and improve the accuracy of measurement information, it is particularly important to use filtering methods to remove noise [22]. Kalman filtering (KF) is a filtering algorithm that filters out various mixed noise disturbances from signal observation and estimates the required signal, which is widely used in intelligent vehicle fields, such as positioning [23], parameter identification [24,25], target tracking [26,27], etc. Both the KF [28,29] and the extended Kalman filter (EKF) [30,31] are based on the premise that the statistical characteristics of the process noise and measurement noise of the system are known. However, vehicles are subject to sensor measurement errors, and it can be difficult to obtain statistical information about the measurement noise in real operation, which is sometimes time-variable and uncertain. To overcome these shortcomings, an adaptive Kalman filter (AKF) is widely used. Hajiyev et al. [32] proposed a robust AKF algorithm with multiple adaptive factors. This algorithm combines the adaptive principle and adaptively adjusts for sensor/actuator failures, which can give accurate estimation in the case of sensor or actuator faults. Bahraini [33] proposed an algorithm to increase the estimation accuracy of UKF-SLAM by adjusting the scale parameter and designed an unscented KF with adaptive selection of the scaling parameter to improve the estimation accuracy of SLAM. If UKF is used on a model with a large state vector, numerical stability problems may occur. Mosconi et al. [34] considered the influence factors such as tire condition and road characteristics and composed an EKF state estimator with 14 states to provide the estimation of road slope and banking angle, which can accurately identify the physical characteristics of tire/road interaction in real time. This method relies on accurate vehicle models and detailed parameters. Wang et al. [35] proposed an improved Sage–Husa adaptive Kalman filter (SHAKF) scheme combined with fuzzy clustering method, which can change the Kalman filter structure through the adaptive mechanism, and the system is well estimated. The construction of AKF is of great significance for increasing the accuracy and stability of filtering.

Based on this, this paper combines the improved SHAEKF method and the two-layer MPC method to design the yaw stability control strategy, which can solve the safety and stability problem of the four-wheel-drive ESTV during operation. It has the following three advantages:

1. A two-layer control structure is adopted to decouple the dynamics between the yaw control system and the driving system. Compared with the integral control structure of yaw stability and torque distribution, the system dimension of the overall optimization strategy is reduced, the computational burden is reduced, and the computational efficiency of multi-objective optimization control with constraints is improved, which is easier to implement.
2. To improve the accuracy and stability of filtering, an improved SHAEKF method that can dynamically adjust the statistical characteristics of noise in a timely manner is used to optimize the state variables, which improves the accuracy of the state variables.
3. The upper-layer yaw stability controller uses MPC to solve the expected additional yaw moment and the front wheel steering angle. The lower-layer torque distribution controller uses the rolling time domain optimization method to track the additional yaw moment required for the upper-layer yaw stability.

The rest of this paper is organized as follows. In Section 2, the dynamics model and tire model of the ESTV are established. In Section 3, the improved SHAEKF algorithm is used to optimize the state of the ESTV. A two-layer MPC yaw stability controller is designed in Section 4. The effectiveness of the improved SHAEKF and the yaw stability control effect are simulated and analyzed in Section 5. Section 6 gives the conclusion.

2. Vehicle Dynamics Model

In this section, the two vehicle dynamics models are established to ensure the yaw stability of the ESTV. A two-degrees-of-freedom (2DOF) reference model is established to obtain the reference yaw rate γ^* and the reference sideslip angle β^* of the vehicle. The 7DOF dynamics model, which can comprehensively respond to the vehicle characteristics, is established considering the relationship between the actual dynamic parameters of the vehicle and its yaw stability, and then the tire model is introduced.

The prediction model needs to be simplified as much as possible on the basis of accurately reflecting the dynamic process of the ESTV to reduce the computing burden of the control algorithm. Therefore, the following idealized assumptions are made for the ESTV:

1. Assume that the vehicle is operating on a flat road, and ignore the vehicle's vertical movement.
2. Assume that the suspension system and the vehicle are rigid.
3. Assume that only pure lateral deflection tire characteristics are considered.
4. Assume that the lateral load transfer of the wheel is not considered.
5. Assume that the effect of lateral and longitudinal aerodynamics on the yaw characteristics of the vehicle is ignored.

2.1. Dynamics Model

The ideal reference model is required to accurately reflect the expected steering process of the vehicle, both to make the system have the same steering flexibility as a traditional vehicle and to ensure that the vehicle has a good posture. The 2DOF model, including the lateral and yaw motion of the ESTV, is shown in Figure 1.

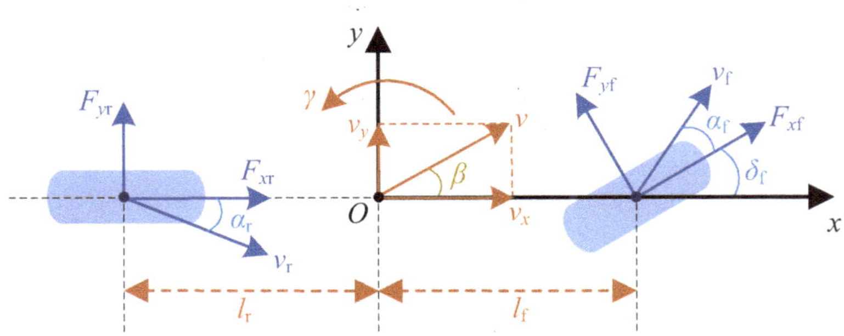

Figure 1. 2DOF reference model.

In the vehicle control system, β and γ represent the 2DOF of the simplified ESTV dynamics model, and δ_f is assumed to be small, that is $\cos\delta_f \approx 1$, $\sin\delta_f \approx \delta_f$. The 2DOF dynamics equation is:

$$\begin{cases} \dot{\beta}^* = \frac{F_{yf}+F_{yr}}{mv_x} - \gamma \\ \dot{\gamma}^* = \frac{l_f F_{yf} - l_r F_{yr} + M_x}{I_z} \end{cases}. \tag{1}$$

Figure 2 shows the 7DOF dynamics model used to verify the effectiveness of the controller design, including the longitudinal motion, lateral motion, yaw motion, and four-wheel rotation of the vehicle [36], which can truly express the dynamic characteristics of the ESTV.

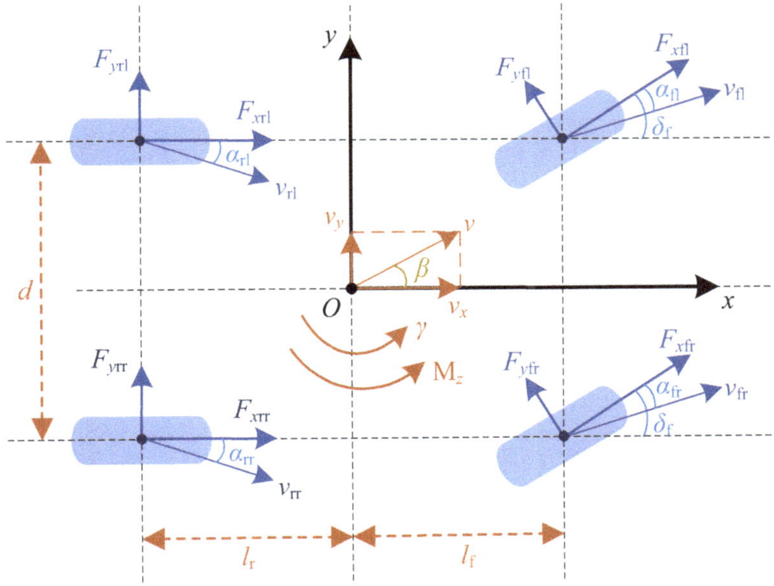

Figure 2. 7DOF reference model.

2.2. Tire Model

Given the structural characteristics of the tire, the dynamic performance is nonlinear. Selecting a tire model that is realistic and easy to use is the key to establishing a vehicle dynamics model. The tire longitudinal and lateral force model in this paper is a semi-empirical tire model based on the magic formula. The magic formula tire model is a relatively mature vehicle stability control model that considers the interactions between longitudinal and lateral force. In general, the magic formula is expressed as [37]:

$$F_{\mathrm{cij}} = D\sin\{C\arctan[B\alpha_{\mathrm{ij}} - E(B\alpha_{\mathrm{ij}} - \arctan(B\alpha_{\mathrm{ij}}))]\}, \tag{2}$$

where α_{ij} is the slip angle of the tire.

However, the magic model is too cumbersome to apply to controller design. Therefore, the model is considered to be simplified, and the cubic polynomial is obtained by Taylor expansion, such as in Equation (3):

$$\begin{cases} F_{y\mathrm{fj}} = -2C_{\mathrm{f}}\left(1 - K_a\alpha_{\mathrm{fj}}^2\right)\alpha_{\mathrm{fj}} \\ F_{y\mathrm{rj}} = -2C_{\mathrm{r}}\left(1 - K_b\alpha_{\mathrm{rj}}^2\right)\alpha_{\mathrm{rj}} \end{cases}, \tag{3}$$

where K_a and K_b are the fitting coefficients obtained from the tire magic formula. The dynamics relationship between β and α_{ij}, α_{ij} is calculated as follows:

$$\begin{cases} \alpha_{\mathrm{fj}} = \beta + \frac{l_{\mathrm{f}}\gamma}{v_x} - \delta_{\mathrm{f}} \\ \alpha_{\mathrm{rj}} = \beta - \frac{l_{\mathrm{r}}\gamma}{v_x} \end{cases}. \tag{4}$$

The longitudinal force of the tire is:

$$\begin{cases} F_{x\mathrm{fj}} = -2C_{l\mathrm{fj}}s_{\mathrm{fj}} \\ F_{x\mathrm{rj}} = -2C_{l\mathrm{rj}}s_{\mathrm{rj}} \end{cases}. \tag{5}$$

Tire slip rate is:

$$s_{\mathrm{ij}} = \frac{\omega_{\mathrm{ij}}R_\omega - v_x}{v_x}. \tag{6}$$

Under the driving conditions, the tire force analysis is shown in Figure 3.

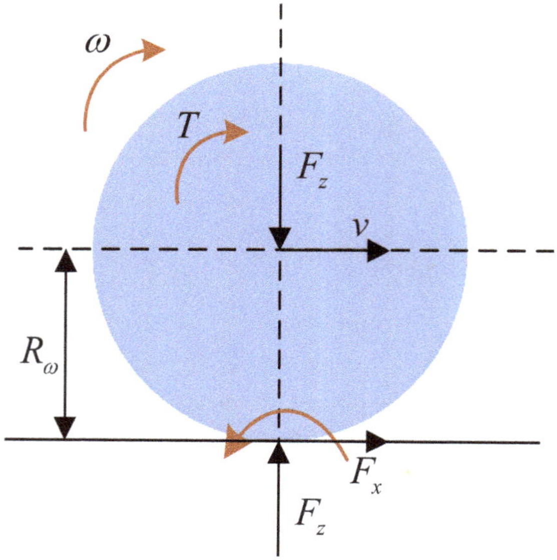

Figure 3. Tire force analysis.

The rotation dynamic equation of a single tire is:

$$I_\omega \dot{\omega}_{\mathrm{ij}} = T_{\mathrm{ij}} - F_{x\mathrm{ij}} R_\omega \quad (\mathrm{i} = \mathrm{f}, \mathrm{r}, \mathrm{j} = \mathrm{l}, \mathrm{r}). \tag{7}$$

The dynamics equation for the tire slip rate is [9]:

$$\dot{s}_{\mathrm{ij}} = (-\frac{R_\omega^2}{I_\omega v_x} - 4\frac{s_{\mathrm{ij}}+1}{m v_x}) C_{l\mathrm{ij}} s_{\mathrm{ij}} + \frac{R_\omega}{I_\omega v_x} T_{\mathrm{ij}}. \tag{8}$$

Ignoring the tire nonlinear characteristics, the 2DOF reference model is obtained as follows:

$$\begin{cases} \dot{\beta}^* = \frac{2C_\mathrm{f}(\delta_\mathrm{f} - \beta^* - \frac{l_\mathrm{f}\gamma^*}{v_x}) + 2C_\mathrm{r}(-\beta^* + \frac{l_\mathrm{r}\gamma^*}{v_x})}{m v_x} - \gamma^* \\ \dot{\gamma}^* = \frac{2l_\mathrm{f} C_\mathrm{f}(\delta_\mathrm{f}^* - \beta^* - \frac{l_\mathrm{f}\gamma^*}{v_x}) - 2l_\mathrm{r} C_\mathrm{r}(-\beta^* + \frac{l_\mathrm{r}\gamma^*}{v_x}) + M_x}{I_z} \end{cases}. \tag{9}$$

Since β and γ are closely related to the yaw stability of the vehicle, they are selected as the state variables $\boldsymbol{x} = \begin{bmatrix} \beta & \gamma \end{bmatrix}^\mathrm{T}$, while δ_f and M_x are selected as the control inputs $\boldsymbol{u} = \begin{bmatrix} \delta_\mathrm{f} & M_x \end{bmatrix}^\mathrm{T}$, and Equation (9) is written into the standard state space form as follows:

$$\begin{cases} \dot{\boldsymbol{x}} = \boldsymbol{A}\boldsymbol{x} + \boldsymbol{B}\boldsymbol{u} \\ y = \boldsymbol{C}\boldsymbol{x} \end{cases}, \tag{10}$$

where

$$\boldsymbol{A} = \begin{bmatrix} -\frac{2C_\mathrm{f}+2C_\mathrm{r}}{m v_x} & \frac{2l_\mathrm{r}C_\mathrm{r}-2l_\mathrm{f}C_\mathrm{f}}{m v_x^2} - 1 \\ \frac{2l_\mathrm{r}C_\mathrm{r}-2l_\mathrm{f}C_\mathrm{f}}{I_z} & -\frac{2l_f^2 C_\mathrm{f}+2l_r^2 C_\mathrm{r}}{I_z v_x} \end{bmatrix}, \boldsymbol{B} = \begin{bmatrix} \frac{2C_\mathrm{f}}{m v_x} & 0 \\ \frac{2l_\mathrm{f}C_\mathrm{f}}{I_z} & \frac{1}{I_z} \end{bmatrix}, \boldsymbol{C} = \begin{bmatrix} 1 & 0 \\ 0 & 1 \end{bmatrix}.$$

3. Improved SHAEKF

In the actual transportation of supplies, the operating conditions of the ESTV are complex and will be easily influenced by environmental factors. For example, factors including uphill and downhill, uneven road surface, and so on, lead to the continuous jitter of the vehicle during operation, and the detection accuracy of the sensor will be affected. In order to obtain more accurate and credible vehicle state information, it is necessary to filter and estimate the sensor data to reduce the influence of noise on state variables and improve the accuracy of measurement information.

In practice, the statistical characteristics of noise are unobtainable, so SHAEKF with simple principles and good real-time performance is used to solve this problem. The SHAKF is suitable for the state estimation problem with time-varying statistical characteristics of noise. It can predict and rectify the system process noise and measurement noise in a timely manner while using measurement data for recursive filtering when the statistical characteristics of the noise are unknown, so as to achieve the purpose of suppressing filter divergence and improving filtering accuracy. The SHAEKF algorithm can use the estimated parameters to estimate the state parameters of the EKF and obtain the state estimation value of the system [38], which improves the robustness of AKF to measurement noise.

Since the SHAKF algorithm cannot estimate both the system noise covariance matrix and the measurement noise covariance matrix [39], the measurement noise with large uncertainty is adaptively adjusted. The improved SHAEKF algorithm is used to correct the noise characteristics of the ESTV in real time to obtain better estimation accuracy. The state equation and the measurement equation of the system with noise are established by combining the ESTV dynamics model as:

$$\begin{cases} \boldsymbol{x}(k+1) = \boldsymbol{A}(k)\boldsymbol{x}(k) + \boldsymbol{B}(k)\boldsymbol{u}(k) + \boldsymbol{w}(k) \\ \boldsymbol{y}(k+1) = \boldsymbol{H}(k+1)\boldsymbol{x}(k+1) + \boldsymbol{v}(k+1) \end{cases}, \tag{11}$$

where $\boldsymbol{w}(k)$ is process noise, and $\boldsymbol{v}(k)$ is measurement noise.

The prediction state vector $\hat{\boldsymbol{x}}_{k+1,k}$ and the prediction error covariance matrix $\boldsymbol{P}_{k+1,k}$ are:

$$\hat{\boldsymbol{x}}_{k+1,k} = \boldsymbol{A}_k\hat{\boldsymbol{x}}_k + \boldsymbol{B}_k\boldsymbol{u}_k \tag{12}$$

$$\boldsymbol{P}_{k+1,k} = \boldsymbol{A}_k\boldsymbol{P}_{k,k}\boldsymbol{A}_k^{\mathrm{T}} + \hat{\boldsymbol{Q}}_k, \tag{13}$$

where $\hat{\boldsymbol{x}}_k$ is the state estimation vector, $\boldsymbol{A}_k$ is the Jacobian matrix after the partial derivative of the state equation for x, and $\boldsymbol{Q}_k$ is the process noise covariance matrix.

The Kalman measurement update includes the gain Equation (14), filter Equation (15), and error covariance update Equation (16):

$$\boldsymbol{K}_{k+1} = \boldsymbol{P}_{k+1,k}\boldsymbol{H}_{k+1}^{\mathrm{T}}(\boldsymbol{H}_{k+1}\boldsymbol{P}_{k+1,k}\boldsymbol{H}_{k+1}^{\mathrm{T}} + \hat{\boldsymbol{R}}_{k+1})^{-1} \tag{14}$$

$$\hat{\boldsymbol{x}}_{k+1} = \hat{\boldsymbol{x}}_{k+1,k} + \boldsymbol{K}_{k+1}\boldsymbol{\varepsilon}_{k+1} \tag{15}$$

$$\boldsymbol{P}_{k+1} = (\boldsymbol{I} - \boldsymbol{K}_{k+1}\boldsymbol{H}_{k+1})\boldsymbol{P}_{k+1,k}, \tag{16}$$

where $\boldsymbol{K}_{k+1}$ is the Kalman gain, $\boldsymbol{H}_{k+1}$ is the Jacobian matrix after the partial derivative of the measurement equation to x, $\boldsymbol{R}_{k+1}$ is the measurement noise covariance matrix, and $\boldsymbol{\varepsilon}_{k+1}$ is the residual. The residual $\boldsymbol{\varepsilon}_{k+1}$ is:

$$\boldsymbol{\varepsilon}_{k+1} = \boldsymbol{y}_{k+1} - \boldsymbol{H}_{k+1}\hat{\boldsymbol{x}}_{k+1,k} - \hat{\boldsymbol{r}}_{k+1}, \tag{17}$$

where $\boldsymbol{y}_{k+1}$ is the observation variable, and $\hat{\boldsymbol{r}}_{k+1}$ is the measurement noise variance.

Finally, the estimation of the noise and the equation for estimating the mean and covariance matrix of the noise are:

$$\hat{\boldsymbol{r}}_{k+1} = (1 - d_k)\hat{\boldsymbol{r}}_k + d_k(\boldsymbol{y}_{k+1} - \boldsymbol{H}_{k+1}\hat{\boldsymbol{x}}_{k+1,k}) \tag{18}$$

$$\hat{\boldsymbol{R}}_{k+1} = (1 - d_k)\hat{\boldsymbol{R}}_k + d_k(\boldsymbol{\varepsilon}_{k+1}\boldsymbol{\varepsilon}_{k+1}^{\mathrm{T}} - \boldsymbol{H}_{k+1}\boldsymbol{P}_{k+1,k}\boldsymbol{H}_{k+1}^{\mathrm{T}}), \tag{19}$$

where d_k is the weighting coefficient. The weighting coefficient d_k is:

$$d_k = \frac{1-b}{1-b^k}, \tag{20}$$

where b is the forgetting factor with a value range of [0.95, 0.995].

When there is a convergence trend in the filtering process, the error covariance $\boldsymbol{P}_{k+1}$ will decrease, and $\boldsymbol{H}_{k+1}$ is a fixed value; then, $\boldsymbol{H}_{k+1}\boldsymbol{P}_{k+1,k}\boldsymbol{H}_{k+1}^{\mathrm{T}}$ will gradually decrease until it tends to 0. Therefore, the effect of $\boldsymbol{H}_{k+1}\boldsymbol{P}_{k+1,k}\boldsymbol{H}_{k+1}^{\mathrm{T}}$ on the measurement noise covariance estimation can be ignored [40]. Therefore, $\hat{\boldsymbol{R}}_{k+1}$ of Equation (19) is replaced by:

$$\hat{\boldsymbol{R}}_{k+1} = (1 - d_k)\hat{\boldsymbol{R}}_k + d_k\boldsymbol{\varepsilon}_{k+1}\boldsymbol{\varepsilon}_{k+1}^{\mathrm{T}}. \tag{21}$$

The flowchart of the improved SHAEKF is shown in Figure 4.

Therefore, the two state variables of the system can be estimated to decrease the interference of noise and ensure the high accuracy of the calculation of δ_{f} and M_x of the vehicle.

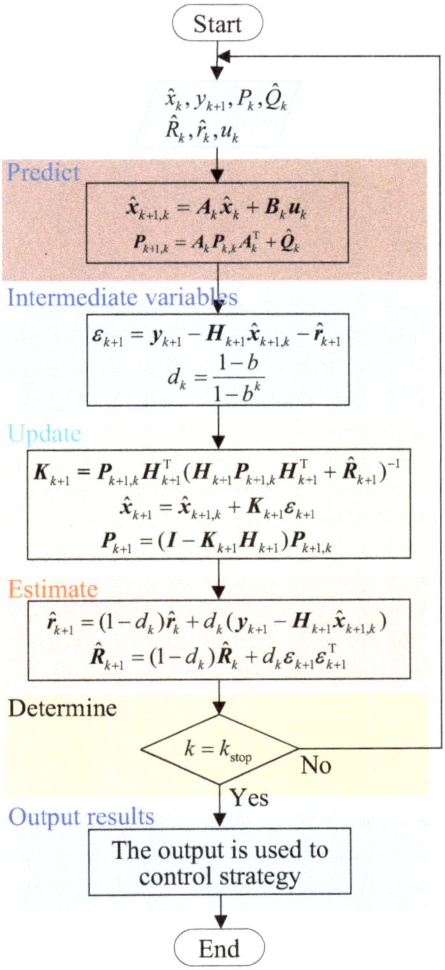

Figure 4. Improved SHAEKF algorithm flowchart.

4. Control Strategy

The β and γ of the vehicle are two important variables to describe the posture of the vehicle, which are closely related to vehicle stability. When the ESTV is steering, the operation stability of the vehicle is reflected by the β, and the γ reflects the steady-state steering characteristics and the vehicle manipulation stability. The control structure of the ESTV yaw stabilization system is shown in Figure 5.

As shown in the control structure diagram, the required β^* and γ^* are determined by the vehicle reference 2DOF model. The upper-layer yaw stability controller calculates δ_f and M_x and takes the yaw moment generated by the tire longitudinal force as the control target of the lower-layer torque distribution controller.

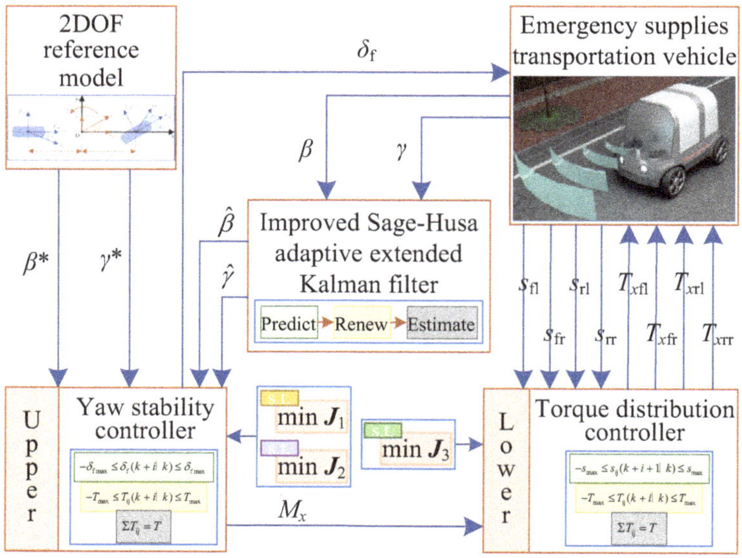

Figure 5. Two-layer controller structure diagram.

The upper-layer MPC controller calculates the optimal control inputs so that β and γ of the ESTV achieve the expected value as much as possible in the current operating state of the ESTV. The prediction model of yaw stability control is:

$$\begin{cases} \boldsymbol{x}(k+1|t) = \boldsymbol{A}_k\boldsymbol{x}(k|t) + \boldsymbol{B}_k\boldsymbol{u}(k|t) \\ \boldsymbol{y}(k|t) = \boldsymbol{C}_k\boldsymbol{x}(k|t) \end{cases}. \tag{22}$$

The controller is solved for the optimal input in each finite time domain using the quadratic programming method [41]. In MPC, N_p is the predictive time domain and $N_\mathrm{p} \geq 1$, while N_c is the control time domain and $1 \leq N_\mathrm{c} \leq N_\mathrm{p}$. The predicted control output sequence is [17]:

$$\boldsymbol{Y}_p(k+1|k) = \begin{bmatrix} \boldsymbol{y}(k+1|k) \\ \boldsymbol{y}(k+2|k) \\ \vdots \\ \boldsymbol{y}(k+N_\mathrm{p}|k) \end{bmatrix}_{N_\mathrm{p}\times 1}. \tag{23}$$

According to the 2DOF reference model of the ESTV, the control output reference sequence is:

$$\boldsymbol{R}(k+1|k) = \begin{bmatrix} \boldsymbol{r}(k+1|k) \\ \boldsymbol{r}(k+2|k) \\ \vdots \\ \boldsymbol{r}(k+N_\mathrm{p}|k) \end{bmatrix}_{N_\mathrm{p}\times 1}. \tag{24}$$

The control input increment sequence in N_p at time k is as follows:

$$\Delta\boldsymbol{U}(k) = \begin{bmatrix} \Delta\boldsymbol{u}(k) \\ \Delta\boldsymbol{u}(k+1) \\ \vdots \\ \Delta\boldsymbol{u}(k+N_\mathrm{c}-1) \end{bmatrix}_{N_\mathrm{c}\times 1}. \tag{25}$$

The upper-layer yaw stability controller obtains δ_f and M_x by tracking and calculating the reference state of the vehicle. The lower-layer control strategy adopts the torque optimization controller based on MPC to realize the stable operation of the vehicle by reasonably distributing the drive torque of four tires.

The yaw moment generated by the longitudinal force of the four tires is:

$$M_x = \frac{d}{2} \cdot (-F_{x\mathrm{fl}} + F_{x\mathrm{fr}} - F_{x\mathrm{rl}} + F_{x\mathrm{rr}}). \tag{26}$$

The vertical load of the vehicle's four tires is:

$$\begin{cases} F_{\mathrm{zfl}} = \frac{l_\mathrm{r}}{2(l_\mathrm{f}+l_\mathrm{r})}mg - \frac{ma_x h}{2(l_\mathrm{f}+l_\mathrm{r})} - \frac{ma_y h l_\mathrm{r}}{d(l_\mathrm{f}+l_\mathrm{r})} \\ F_{\mathrm{zfr}} = \frac{l_\mathrm{r}}{2(l_\mathrm{f}+l_\mathrm{r})}mg - \frac{ma_x h}{2(l_\mathrm{f}+l_\mathrm{r})} + \frac{ma_y h l_\mathrm{r}}{d(l_\mathrm{f}+l_\mathrm{r})} \\ F_{\mathrm{zrl}} = \frac{l_\mathrm{f}}{2(l_\mathrm{f}+l_\mathrm{r})}mg + \frac{ma_x h}{2(l_\mathrm{f}+l_\mathrm{r})} - \frac{ma_y h l_\mathrm{f}}{d(l_\mathrm{f}+l_\mathrm{r})} \\ F_{\mathrm{zrr}} = \frac{l_\mathrm{f}}{2(l_\mathrm{f}+l_\mathrm{r})}mg + \frac{ma_x h}{2(l_\mathrm{f}+l_\mathrm{r})} + \frac{ma_y h l_\mathrm{f}}{d(l_\mathrm{f}+l_\mathrm{r})} \end{cases}. \tag{27}$$

When the vehicle operates in conditions with a low adhesion coefficient and fluctuating roads, the vertical load transfer of each wheel plays an important role in the lateral stability of the vehicle. In the controller design, the tire vertical load transfer variable factor needs to be added to coordinate the drive torque distribution of each wheel. The vertical load transfer weight coefficients of the four wheels are [21]:

$$\begin{cases} \xi_{\mathrm{fl}} = \frac{F_{\mathrm{zfl}}}{F_{\mathrm{zfl}}+F_{\mathrm{zfr}}+F_{\mathrm{zrl}}+F_{\mathrm{zrr}}} \\ \xi_{\mathrm{fr}} = \frac{F_{\mathrm{zfr}}}{F_{\mathrm{zfl}}+F_{\mathrm{zfr}}+F_{\mathrm{zrl}}+F_{\mathrm{zrr}}} \\ \xi_{\mathrm{rl}} = \frac{F_{\mathrm{zrl}}}{F_{\mathrm{zfl}}+F_{\mathrm{zfr}}+F_{\mathrm{zrl}}+F_{\mathrm{zrr}}} \\ \xi_{\mathrm{rr}} = \frac{F_{\mathrm{zrr}}}{F_{\mathrm{zfl}}+F_{\mathrm{zfr}}+F_{\mathrm{zrl}}+F_{\mathrm{zrr}}} \end{cases}. \tag{28}$$

The lower-layer controller state variable is the tire slip ratio $\boldsymbol{x}_\mathrm{L} = [s_\mathrm{fl} \quad s_\mathrm{fr} \quad s_\mathrm{rl} \quad s_\mathrm{rr}]^\mathrm{T}$, the control input is the drive torque of four tires $\boldsymbol{u}_\mathrm{L} = [T_\mathrm{fl} \quad T_\mathrm{fr} \quad T_\mathrm{rl} \quad T_\mathrm{rr}]^\mathrm{T}$, and the output is $\boldsymbol{y}_\mathrm{L} = M_x$. Convert Equation (8) into the state space equation as:

$$\begin{cases} \dot{\boldsymbol{x}}_\mathrm{L} = (-\frac{R_\omega^2}{I_\omega v_x} - \frac{(\xi_\mathrm{fl}+\xi_\mathrm{fr}+\xi_\mathrm{rl}+\xi_\mathrm{rr})(x+1)}{\xi_\mathrm{ij} m v_x})C_{l\mathrm{ij}}\boldsymbol{x}_\mathrm{L} + \frac{R_\omega}{I_\omega v_x}\boldsymbol{u}_\mathrm{L} \\ \boldsymbol{y}_\mathrm{L} = \frac{d}{2}[-\xi_\mathrm{fl}C_{l\mathrm{fl}} \quad \xi_\mathrm{fr}C_{l\mathrm{fr}} \quad -\xi_\mathrm{rl}C_{l\mathrm{rl}} \quad \xi_\mathrm{rr}C_{l\mathrm{rr}}]\boldsymbol{x}_\mathrm{L} \end{cases}. \tag{29}$$

The state Equation (29) is discretized:

$$\begin{cases} \boldsymbol{x}_\mathrm{L}(k+1) = \boldsymbol{\phi}\boldsymbol{x}_\mathrm{L}(k) + \boldsymbol{D}\boldsymbol{u}_\mathrm{L}(k) \\ \boldsymbol{y}_\mathrm{L}(k) = \boldsymbol{C}_l\boldsymbol{x}_\mathrm{L}(k) \end{cases}. \tag{30}$$

The upper-layer controller ensures that the ESTV has good stability during operation so that $\hat{\beta}$ and $\hat{\gamma}$ estimated by the SHAEKF can quickly track β^* and γ^*. To ensure that the ESTV accurately tracks the reference trajectory, $\boldsymbol{Y}_p$ should closely track $\boldsymbol{R}$ to maintain the tracking accuracy of the system. The fluctuation of the system control variables affects the smoothness of the vehicle's operation. Therefore, we need to consider the stability of control input variables. The lower-layer controller is required not only to track M_x but also to minimize energy consumption while ensuring the stability of the ESTV. The objective function is:

$$\begin{aligned}J_1 = & \sum_{i=1}^{N_p}\left[\Gamma_{y1}\left(\hat{\beta}(k+i\mid k)-\beta^*(k)\right)^2+\Gamma_{y2}\left(\hat{\gamma}(k+i\mid k)-\gamma^*(k)\right)^2\right] \\ & + \sum_{i=0}^{N_c-1}\left[\Gamma_{\Delta 1}(\Delta\delta_f^2(k+i\mid k))+(\Gamma_{\Delta 2}(\Delta T_{fl}^2(k+i\mid k)+\Delta T_{fr}^2(k+i\mid k)\right. \\ & + \Delta T_{rl}^2(k+i\mid k)+\Delta T_{rr}^2(k+i\mid k))] \\ & + \Gamma_y\|Y(k+1\mid k)-M_x(k+1)\|^2+\Gamma_u\|U_L(k)\|^2,\end{aligned} \quad (31)$$

where $U_L(k)$ is the control variables sequence, and $Y(k+1\mid k)$ is the N_p-step prediction output at moment k, and Γ_y, Γ_Δ, and Γ_u are the weight coefficient matrices of the system output, the controlled input variables, and the inputs, respectively. To achieve the expected control performance during the actual operation of the vehicle, the capacity range constraint of the actuator should be considered in the design of the MPC [20]. Due to the saturation characteristics of the actuator, δ_f and the drive torque of the system are constrained by physical conditions. To ensure the yaw stability while avoiding the operational stability of the vehicle caused by tire slip, the following boundary constraints are imposed:

$$\begin{aligned}& -\delta_{fmax} \le \delta_f(k+i\mid k) \le \delta_{fmax}, \\ & -T_{max} \le T_{ij}(k+i\mid k) \le T_{max} \\ & -s_{max} \le s_{ij}(k+i\mid k) \le s_{max}(i=f,r,j=l,r).\end{aligned} \quad (32)$$

At the same time, the sum of each drive torque should be equal to the total torque required:

$$\Sigma T_{ij} = T_t, \quad (33)$$

where $i = 1,2,\cdots,N_p$. Therefore, the objective function expression of the controller's k-th control period is:

$$\begin{aligned}\min \quad & J_k = \sum_{i=1}^{N_p} Y_e^T(k)\Gamma_y Y_e(k) + \sum_{i=0}^{N_c-1}\left[\Delta U^T(k)\Gamma_\Delta \Delta U(k) + U_L^T(k)\Gamma_u U_L(k)\right] \\ s.t. \quad & \begin{array}{l} -\delta_{fmax} \le \delta_f(k+i|k) \le \delta_{fmax} \\ -s_{max} \le s_{ij}(k+i|k) \le s_{max} \\ -T_{max} \le T_{ij}(k+i|k) \le T_{max} \\ \Sigma T_{ij} = T_t \end{array}\end{aligned} \quad (34)$$

5. Analysis and Validation

In this paper, a MATLAB/Simulink 2018a simulation model is built to simulate and verify the estimation effect of the improved SHAEKF and the two-layer MPC controller. To verify the performance of the designed yaw stability controller, the stability is simulated and tested under multiple operating conditions. The ESTV model used in this experiment is a 7DOF dynamics model, and the controller prediction model adopts a 2DOF model. The parameters of the ESTV defined by simulation are shown in Table 1, and the controller parameters are shown in Table 2, where the δ_f constraint is 0.1 rad, the maximum driving torque of a single wheel is 100 Nm, and the boundary value of the safe range of tire slip rate is 0.08.

Table 1. ESTV parameters.

Definition	Value
Vehicle mass m	930 kg
Moment of inertia I_z	$1372\ \text{kg} \cdot \text{m}^2$
Distance from the center of gravity to front axle l_f	0.986 m
Distance from the center of gravity to rear axle l_r	1.253 m
Wheel tread d	1.14 m
Tire rolling radius R_ω	0.26 m

Table 2. Controller parameters.

Definition	Controller Value
Predictive time domain N_p	15
Control time domain N_c	5
Sampling time T	0.05 s
Weight coefficient matrix $\mathbf{\Gamma}_y$	$\text{diag}(100, 200)$
Weight coefficient matrix $\mathbf{\Gamma}_\Delta$	60
Weight coefficient matrix $\mathbf{\Gamma}_u$	100

5.1. Simulation of Improved SHAEKF

To verify the effectiveness of SHAEKF, Gaussian white noise is added. The trajectories of the vehicle before and after the improved SHAEKF are shown in Figure 6. Figure 7 shows the state error before and after the KF, AKF, and the improved SHAEKF. The comparison between the measured values and the estimated values of the state variables are shown in Table 3. The γ_e and β_e of the original data represent the difference between the measured value and the reference value, and the γ_e and β_e after filtering represent the difference between the estimated value and the reference value.

Table 3. Measured values and estimated values.

State Variables	Original Data	KF	AKF	SHAEKF
γ_e maximum value	0.1659 m	0.1249 m	0.0968 m	0.0847 m
β_e maximum value	0.2666 m	0.1618 m	0.1016 m	0.0882 m
$\lvert\gamma_e\rvert$ average value	0.0038 m	0.0027 m	0.0017 m	0.0013 m
$\lvert\beta_e\rvert$ average value	0.0049 m	0.0034 m	0.0022 m	0.0011 m
γ_e variance	$1.7210 \times 10^{-4}\ \text{m}^2$	$9.0315 \times 10^{-5}\ \text{m}^2$	$4.6964 \times 10^{-5}\ \text{m}^2$	$2.6667 \times 10^{-5}\ \text{m}^2$
β_e variance	$2.2990 \times 10^{-4}\ \text{m}^2$	$8.7293 \times 10^{-5}\ \text{m}^2$	$3.7698 \times 10^{-5}\ \text{m}^2$	$2.6103 \times 10^{-5}\ \text{m}^2$

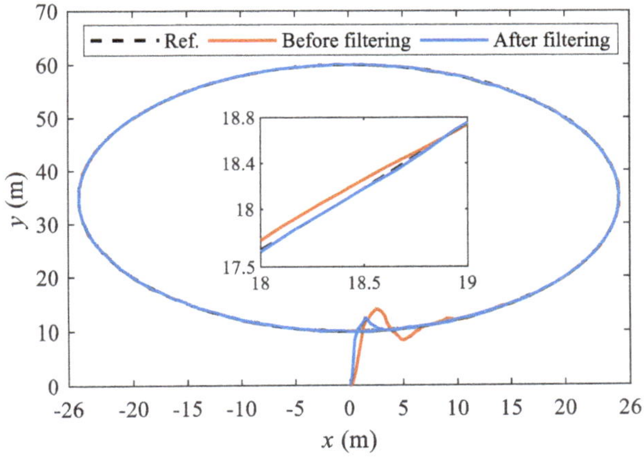

Figure 6. Trajectory.

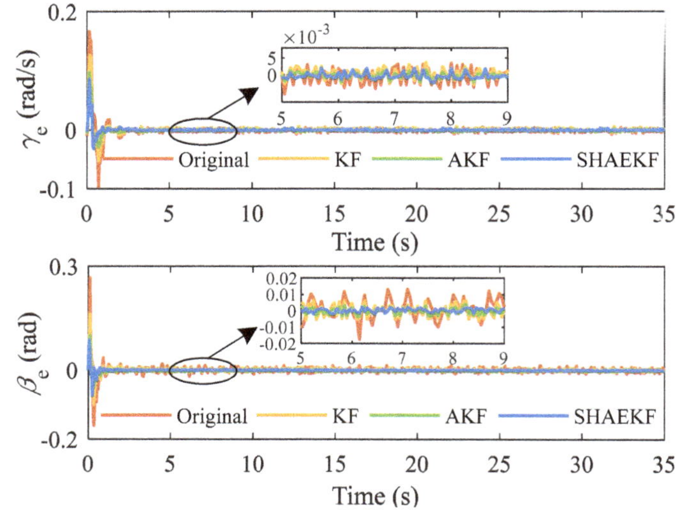

Figure 7. State variables error.

According to Figures 6 and 7, it can be seen that the improved SHAEKF is closer to the reference trajectory without the prior knowledge of the measurement noise, and the average value of the absolute value of the state variables estimation error is reduced by more than 78% compared with the measured value. It can be seen that the improved SHAEKF reduces the influence of noise on the ESTV.

5.2. Double Shift Trajectory Simulation

5.2.1. Comparison of Different Control Methods

To verify the effectiveness of the two-layer MPC control strategy in improving the operation stability of ESTV, a comparative simulation is conducted to verify the two-layer MPC control strategy, no control, and traditional MPC control. The velocity is set to 15 m/s, and the road adhesion coefficient is set to 0.4. The simulation results are shown in Figures 8–10.

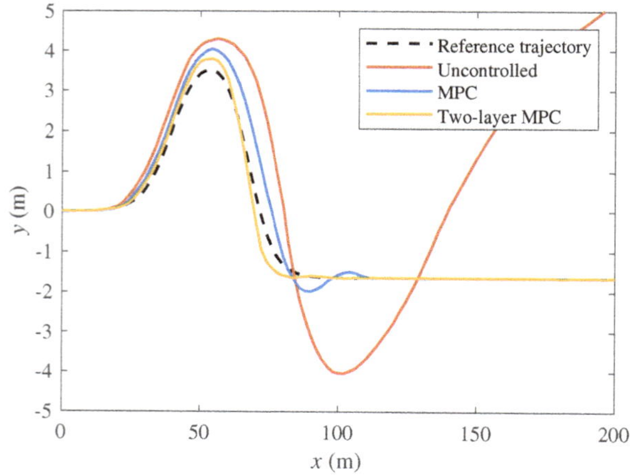

Figure 8. Trajectory.

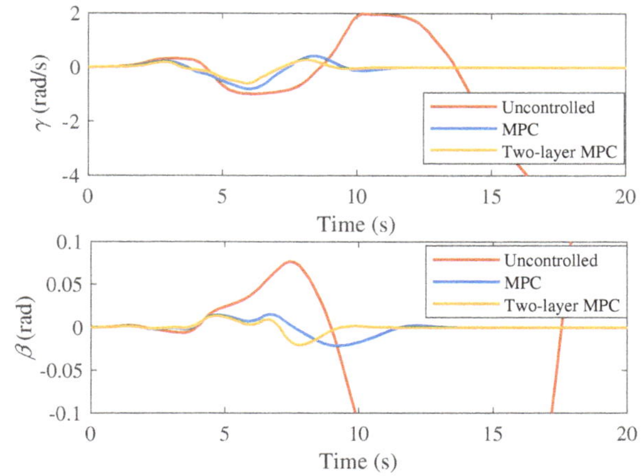

Figure 9. Yaw rate and sideslip angle.

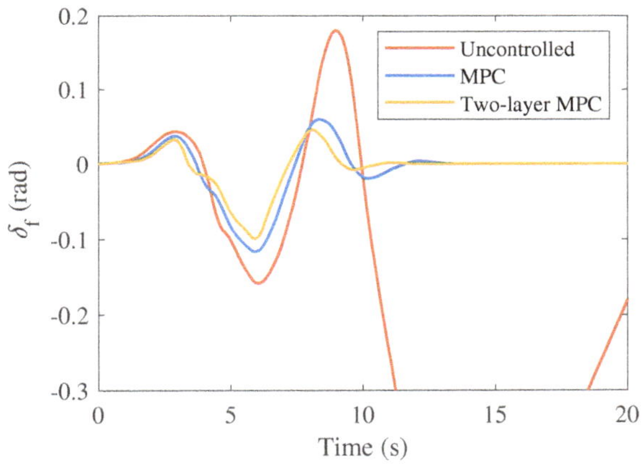

Figure 10. Front wheel steering angle.

Figures 8–10 show the trajectory tracking ability of the ESTV under three control methods. From Figure 8, we can see that the uncontrolled vehicle has a large sideslip, the error between the operation trajectory and the reference trajectory is too large, and the vehicle is out of control. The vehicle based on traditional MPC can basically ensure safe operation, but there is also a large overshoot at the corner. Based on the two-layer MPC control strategy, the accuracy of the ESTV tracking is significantly improved, it has a strong ability to correct the turning attitude, and there is basically no excessive sideslip and tail flick. Figure 9 is the comparison of the ESTV's γ and β under three control methods. Compared with the traditional MPC, the γ and β of the ESTV based on the two-layer MPC control strategy are smoother, which can ensure the stable operation of the ESTV and significantly improve the vehicle stability. The uncontrolled vehicle has completely exceeded the stable range under this experimental condition, and there is a risk of instability. From Figure 10, we can see that, compared with the traditional MPC, δ_f based on the two-layer MPC control strategy is smaller and smoother, which avoids excessive steering and ensures the stability of transporting supplies.

5.2.2. Robustness of the Control System to Road Adhesion Conditions

When the ESTV operates on roads with different adhesion conditions (such as dry roads or wet and slippery roads), the dynamic parameters of the vehicle itself, such as tire cornering stiffness, will change, and the lateral force provided by the ground will also be insufficient. The adhesion coefficient of the vehicle is $\mu = 0.7\sim1$ on the dry driving road and about $\mu = 0.4$ on the slippery road. Therefore, two road conditions with μ of 0.4 and 0.8 are adopted for experiments. The vehicle operating velocity is set to 15 m/s. The simulation results are shown in Figures 11–14.

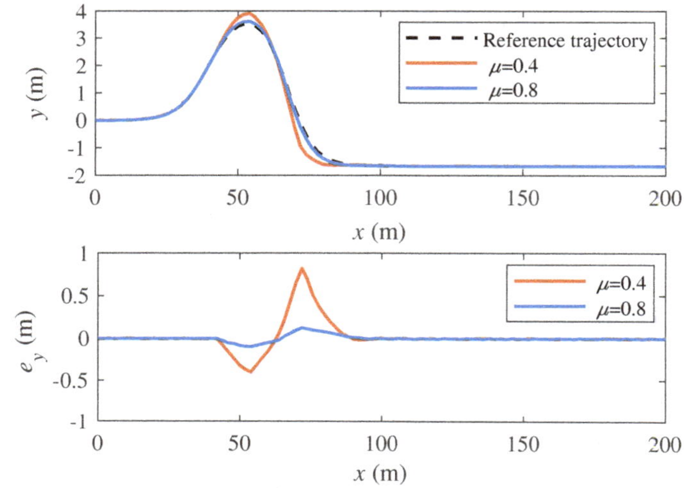

Figure 11. Trajectory.

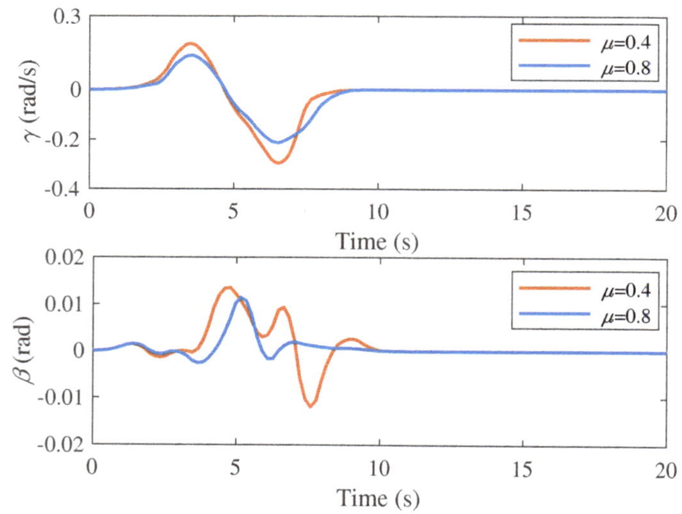

Figure 12. Yaw rate and sideslip angle.

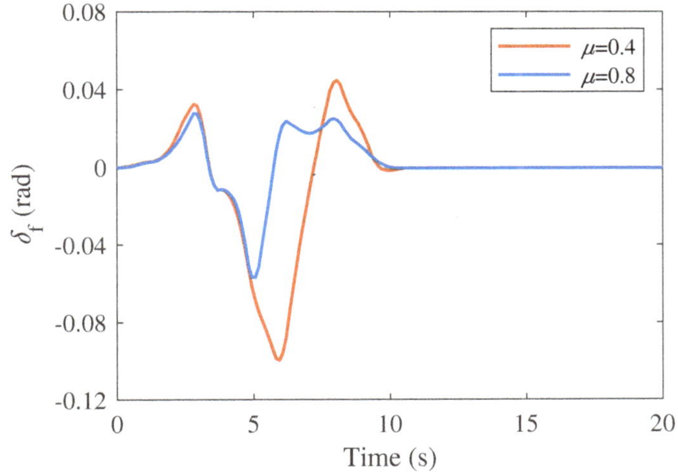

Figure 13. Front wheel steering angle.

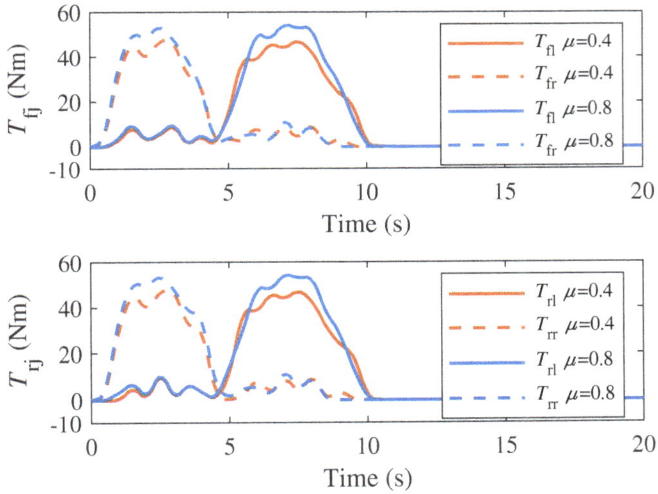

Figure 14. Drive torque.

Figure 11 shows the trajectory of the ESTV, and it reflects that the controller can operate along the reference trajectory on the road with different adhesion conditions, but the tracking error of ESTV is smaller when the adhesion condition is good. The ground cannot provide sufficient lateral force, and the yaw angle will have a large deviation when the vehicle is turning under the condition of poor road adhesion; however, the controller can correct the deviation in a timely manner, and finally the error tends to 0. Figure 12 is the yaw rate and sideslip angle of the vehicle. It reflects that the γ and β will increase when the ESTV operates on a road with adhesion conditions that are poor but still within a certain range, reflecting that the controller can maintain the stability of the ESTV under low adhesion conditions. Figure 13 shows the front wheel steering angle of ESTV, and it can be seen that, when the road adhesion condition is poor, δ_f of the ESTV is still within the constraint range, which ensures the normal operation of the actuator. Figure 14 is the drive torque of the vehicle, and it can be seen that there is no drive torque when there is no steering signal input. The MPC torque distribution control strategy can effectively maintain the yaw moment of the upper-layer yaw stability, and each motor torque is within the constraint, ensuring that the ESTV can operate smoothly and safely while reducing energy loss.

5.2.3. Robustness of the Control System to Operating Velocity

The control algorithm needs to determine different control parameters for different operating velocities, while the MPC has strong robustness to the change in operating velocity. In this section, different operating velocities are used to achieve stable control of the ESTV under the same conditions, and the robustness of the controller to different operating velocities is analyzed.

The simulation experiments are conducted at the velocities of 10 m/s, 15 m/s, and 20 m/s, respectively, and the road adhesion coefficient is 0.8. The simulation results are shown in Figures 15–19.

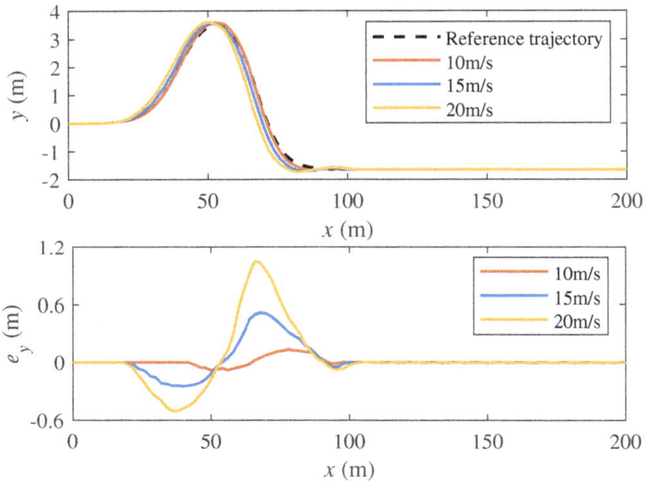

Figure 15. Trajectory.

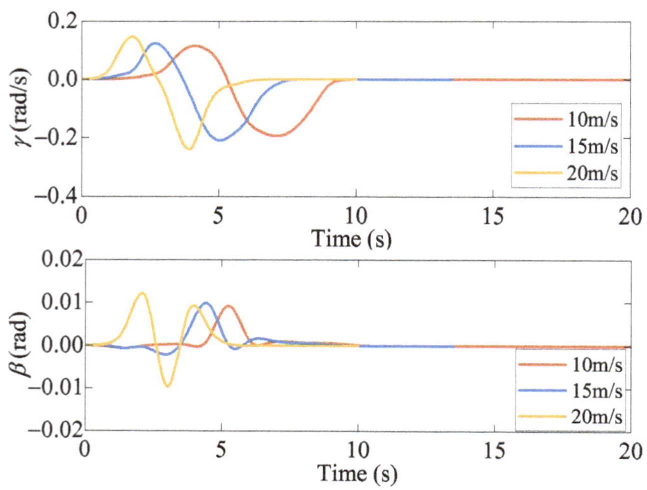

Figure 16. Yaw rate and sideslip angle.

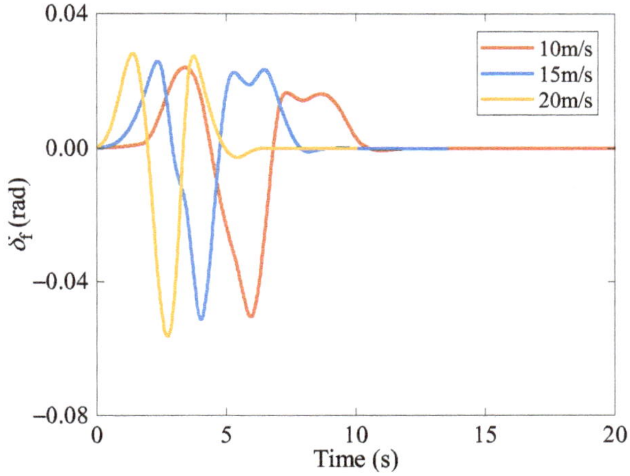

Figure 17. Front wheel steering angle.

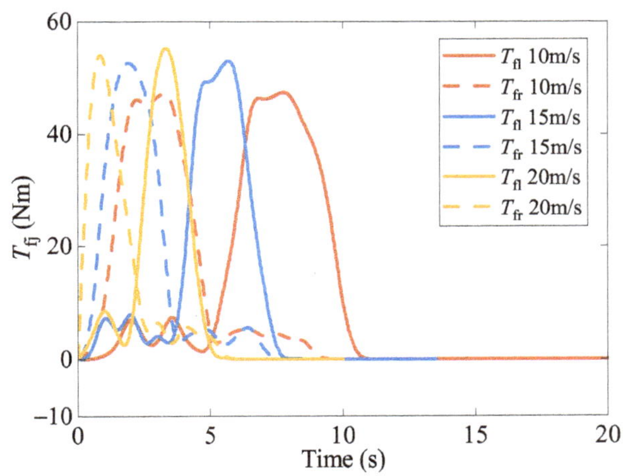

Figure 18. Front wheel drive torque.

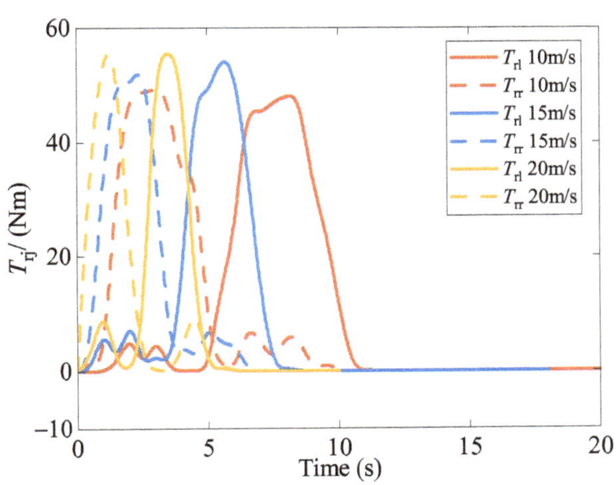

Figure 19. Rear wheel drive torque.

Figure 15 is the trajectory of the vehicle at three different operating velocities. The simulation results verify that the vehicle, operating at different velocities, has good trajectory tracking performance, reflecting the robustness to velocity. Figure 16 shows the ESTV's yaw rate and sideslip angle. It reflects that the ESTV, operating at different velocities, has good yaw stability under the same control parameters, reflecting that the two-layer MPC controller has strong robustness to velocity. Figure 17 is the front-wheel steering angle, and it can be seen that the increment of the control variable is different at different velocities, but always within the constraint range. Figure 18 is the drive torque of the two front wheels of the ESTV, and Figure 19 is the drive torque of the two rear wheels of the ESTV. It reflects that the higher the ESTV velocity, the greater the drive force obtained by the wheels. The vehicle can flexibly obtain the optimal value of the drive torque during operation, which improves the mobility and yaw stability of the vehicle. The designed controller can maintain good stability at different operating velocities.

6. Conclusions

Aiming at the problem of yaw stability control of an ESTV, the improved SHAEKF algorithm and two-layer MPC method are adopted to realize the yaw stability control of a four-wheel-drive ESTV. The improved SHAEKF is used to obtain the estimated state of the system under external disturbances and measurement noise, which improves the accuracy of the state variables. The two-layer MPC is used to realize the yaw stability control of the ESTV. The upper-layer yaw stability controller solves the additional yaw moment and the front wheel steering angle of ESTV, and the lower-layer torque distribution controller optimizes the torque distribution of the four tires of the vehicle. It improves the control accuracy, ensures the stability of the ESTV, and reduces the computational burden of the controller. The simulation experiments verify that the improved SHAEKF can effectively estimate the state variables, and the yaw stability controller based on the two-layer MPC can prevent instability and maintain the stability of the vehicle.

The control method proposed in this paper has a positive effect on improving the stability of the emergency supplies transportation vehicle, but it still needs to be improved. This paper does not consider its inherent complex nonlinear characteristics and only considers the different road adhesion conditions and different velocities for the operation of the ESTV. Further, it does not consider the influence of road inclination and curvature on the yaw stability control. Therefore, in the future work, we will continue to improve the controller design to further expand its application range.

Author Contributions: Conceptualization, M.T.; methodology, M.T. and Y.Z.; software, Y.Z., W.W. and Y.Y.; validation, W.W. and B.A.; formal analysis, Y.Z. and Y.Y.; investigation, Y.Z., W.W. and B.A.; resources, M.T., Y.Z. and Y.Y.; data curation, Y.Z.; writing—original draft preparation, Y.Z.; writing—review and editing, M.T., B.A. and Y.Y.; visualization, Y.Z. and W.W.; supervision, M.T. and B.A.; project administration, M.T.; funding acquisition, M.T. All authors have read and agreed to the published version of the manuscript.

Funding: This work was financially supported by the National Natural Science Foundation of China (grant numbers: 62363022, 61663021, 71763025, and 61861025); the Natural Science Foundation of Gansu Province (grant number: 23JRRA886); and the Gansu Provincial Department of Education: Industrial Support Plan Project (grant number: 2023CYZC-35).

Data Availability Statement: Data are contained within the paper.

Conflicts of Interest: The authors declare no conflicts of interest.

Abbreviations

The following abbreviations are used in this paper:

m	Vehicle mass
h	Height of center of gravity
d	Wheel tread
l_f	Distance from the center of gravity to front axle
l_r	Distance from the center of gravity to rear axle
δ_f	Front wheel steering angle
γ	Yaw rate
β	Sideslip angle
v_x/v_y	Longitudinal and lateral velocity
$\dot{v}_x/\dot{v}_y$	Longitudinal and lateral acceleration
M_x	Total yaw moment
I_z	Moment of inertia of vehicle
I_ω	Moment of inertia of wheel
$\dot{\omega}_{ij}$	Wheel angular acceleration
R_ω	Tire rolling radius
C_f/C_r	Front and rear wheel cornering stiffness
C_{lf}/C_{lr}	Front and rear wheel longitudinal stiffness
$F_{xfl}/F_{xfr}/F_{xrl}/F_{xrr}$	Longitudinal force of the four tires
$F_{yfl}/F_{yfr}/F_{yrl}/F_{yrr}$	Lateral force of the four tires
T_{ij}	Drive torque

References

1. Hu, Z. A container multimodal transportation scheduling approach based on immune affinity model for emergency relief. *Expert Syst. Appl.* **2011**, *38*, 2632–2639. [CrossRef]
2. Hu, X.; Yan, S.; Lin, J.; Liu, Q.; Wang, P.; Chen, H. Lateral motion control and analysis for autonomous vehicles considering road curvature. In Proceedings of the 2021 China Automation Congress, CAC 2021, Beijing, China, 22–24 October 2021.
3. Li, R.; Guo, C.; Li, Y.; Zhao, Y. Trajectory tracking of intelligent vehicle based on steering and braking coordination control. *Syst. Eng. Electron.* **2023**, *45*, 1185–1192.
4. Wu, X.; Wei, C.; Zhai, J.; Yuan, S. Study on the optimization of autonomous vehicle on path-following considering yaw stability. *J. Mech. Eng.* **2022**, *58*, 130–142.
5. Park, M.; Yim, S. Comparative study on coordinated control of path tracking and vehicle stability for autonomous vehicles on low-friction roads. *Actuators* **2023**, *12*, 398. [CrossRef]
6. Liu, L.; Zhang, L.; Pan, G.; Zhang, S. Robust yaw control of autonomous underwater vehicle based on fractional-order PID controller. *Ocean Eng.* **2022**, *257*, 111493. [CrossRef]
7. Li, H.; Wang, X.; Song, S.; Li, H. Vehicle control strategies analysis based on PID and fuzzy logic control. *Procedia Eng.* **2016**, *137*, 234–243. [CrossRef]
8. Patil, H.; Devika, K.B.; Vivekanandan, G.; Sivaram, S.; Subramanian, S.C. Direct yaw-moment control integrated with wheel slip regulation for heavy commercial road vehicles. *IEEE Access* **2022**, *10*, 69883–69895. [CrossRef]
9. Yu, S.; Li, W.; Liu, Y.; Chen, H. Steering stability control of four-wheel-drive electric vehicle. *Control Theory Appl.* **2021**, *38*, 719–730.
10. Zhou, X.; Shen, H.; Wang, Z.; Ahn, H.; Kung, Y.; Wang, J. Ground Vehicle generalized forces and moment governor design via noncertainty-equivalent adaptive prescribed performance control. In Proceedings of the 2023 IEEE International Automated Vehicle Validation Conference, IAVVC 2023, Austin, TX, USA, 16–18 October 2023.
11. Hussain, N.A.A.; Ali, S.S.A.; Ovinis, M.; Arshad, M.R.; Al-Saggaf, U.M. Underactuated coupled nonlinear adaptive control synthesis using U-Model for multivariable unmanned marine robotics. *IEEE Access* **2020**, *8*, 1851–1865. [CrossRef]
12. Park, B.S.; Yoo, S.J. Quantized-communication-based neural network control for formation tracking of networked multiple unmanned surface vehicles without velocity information. *Eng. Appl. Artif. Intell.* **2022**, *114*, 105160. [CrossRef]
13. Taghavifar, H.; Hu, C.; Taghavifar, L.; Qin, Y.; Na, J.; Wei, C. Optimal robust control of vehicle lateral stability using damped least-square backpropagation training of neural networks. *Neurocomputing* **2020**, *384*, 256–267. [CrossRef]
14. Gros, S.; Zanon, M.; Quirynen, R.; Bemporad, A.; Diehl, M. From linear to nonlinear MPC: Bridging the gap via the real-time iteration. *Int. J. Control* **2020**, *93*, 62–80. [CrossRef]
15. Zhang, Z.; Xie, L.; Lu, S.; Wu, X.; Su, H. Vehicle yaw stability control with a two-layered learning MPC. *Veh. Syst. Dyn.* **2023**, *61*, 423–444. [CrossRef]
16. Bossi, L.; Rottenbacher, C.; Mimmi, G.; Magni, L. Multivariable predictive control for vibrating structures: An application. *Control Eng. Pract.* **2011**, *19*, 1087–1098. [CrossRef]
17. Xu, F.; Guo, Z.; Yu, S.; Chen, H.; Liu, Q. Fast realization of stability prediction controller for four-wheel drive electric vehicles. *Control Theory Appl.* **2022**, *39*, 777–787.

18. Qu, Y.; Xu, F.; Yu, S.; Chen, H.; Li, Z. Model predictive control based on extended state observer for vehicle yaw stability. *Control Theory Appl.* **2020**, *37*, 941–949.
19. Wang, L.; Pang, H.; Wang, P.; Liu, M.; Hu, C. A yaw stability-guaranteed hierarchical coordination control strategy for four-wheel drive electric vehicles using an unscented Kalman filter. *J. Frankl. Inst.* **2023**, *360*, 9663–9688. [CrossRef]
20. Chowdhri, N.; Ferranti, L.; Iribarren, F.S.; Shyrokau, B. Integrated nonlinear model predictive control for automated driving. *Control Eng. Pract.* **2021**, *106*, 104654. [CrossRef]
21. Liu, C.; Liu, H.; Han, L.; Chen, K. High-speed obstacle avoidance and stability control of distributed electric drive vehicle under extreme off-road conditions. *Acta Armamentarii* **2021**, *42*, 2102–2113.
22. Zhang, C.; Feng, Y.; Wang, J.; Gao, P.; Qin, P. Vehicle sideslip angle estimation based on radial basis neural network and unscented Kalman filter algorithm. *Actuators* **2023**, *12*, 371. [CrossRef]
23. Slimi, H.; Sentouh, C.; Mammar, S.; Nouveliere, L. Sensor position identification and vehicle state estimation using the extended Kalman filter. *Intell. Syst. Autom.* **2008**, *1019*, 41–46.
24. You, J.; Zhang, Z.; Zhang, F.; Wu, W. Cooperative filtering and parameter identification for advection-diffusion processes using a mobile sensor network. *IEEE Trans. Control Syst. Technol.* **2023**, *31*, 527–542. [CrossRef]
25. Dong, Z.; Yang, X.; Zheng, M.; Song, L.; Mao, Y. Parameter identification of unmanned marine vehicle manoeuvring model based on extended Kalman filter and support vector machine. *Int. J. Adv. Robot. Syst.* **2019**, *16*, 1–10. [CrossRef]
26. Khalkhali, M.B.; Vahedian, A.; Yazdi, H.S. Situation assessment-augmented interactive Kalman filter for multi-vehicle tracking. *IEEE Trans. Intell. Transp. Syst.* **2022**, *23*, 3766–3776. [CrossRef]
27. Khalkhali, M.B.; Vahedian, A.; Yazdi, H.S. Multi-target state estimation using interactive Kalman filter for multi-vehicle tracking. *IEEE Trans. Intell. Transp. Syst.* **2020**, *21*, 1131–1144. [CrossRef]
28. Smieszek, M.; Dobrzanska, M. Application of Kalman filter in navigation process of automated guided vehicles. *Metrol. Meas. Syst.* **2015**, *22*, 443–454. [CrossRef]
29. Ge, L.; Zhao, Y.; Ma, F.; Guo, K. Towards longitudinal and lateral coupling control of autonomous vehicles using offset free MPC. *Control Eng. Pract.* **2022**, *121*, 105074. [CrossRef]
30. Ahangarnejad, A.H.; Baslamisli, S.C. Adap-tyre: DEKF filtering for vehicle state estimation based on tyre parameter adaptation. *Int. J. Veh. Des.* **2016**, *71*, 52–74. [CrossRef]
31. Cheng, C.; Cebon, D. Parameter and state estimation for articulated heavy vehicles. *Veh. Syst. Dyn.* **2011**, *49*, 399–418. [CrossRef]
32. Hajiyev, C.; Soken, H.E. Robust adaptive Kalman filter for estimation of UAV dynamics in the presence of sensor/actuator faults. *Aerosp. Sci. Technol.* **2013**, *28*, 376–383. [CrossRef]
33. Bahraini, M.S. On the efficiency of SLAM using adaptive unscented Kalman filter. *Iran. J. Sci. Technol. Trans. Mech. Eng.* **2020**, *44*, 727–735. [CrossRef]
34. Mosconi, L.; Farroni, F.; Sakhnevych, A.; Timpone, F.; Gerbino, F. Adaptive vehicle dynamics state estimator for onboard automotive applications and performance analysis. *Veh. Syst. Dyn.* **2023**, *61*, 3244–3268. [CrossRef]
35. Wang, X.; Wang, A.; Wang, D.; Xiong, Y.; Liang, B.; Qi, Y. A modified Sage-Husa adaptive Kalman filter for state estimation of electric vehicle servo control system. *Energy Rep.* **2022**, *8*, 20–27. [CrossRef]
36. Ding, S.; Sun, J. Direct yaw-moment control for 4WID electric vehicle via finite-time control technique. *Nonlinear Dyn.* **2017**, *88*, 239–254. [CrossRef]
37. Ren, B.; Chen, H.; Zhao, H.; Yuan, L. MPC-based yaw stability control in in-wheel-motored EV via active front steering and motor torque distribution. *Mechatronics* **2016**, *38*, 103–114. [CrossRef]
38. Li, G.; Zhao, D.; Xie, R.; Han, H.; Zong, C. Vehicle state estimation based on improved Sage-Husa adaptive extended Kalman filtering. *Automot. Eng.* **2015**, *37*, 1426–1432.
39. Zhang, C. Approach to adaptive filtering algorithm. *Acta Aeronaut. Astronaut. Sin.* **1998**, *S1*, 97–100.
40. Xu, L.; Li, S.; Qu, X. Robust adaptive algorithm for transfer alignment. *J. Syst. Simul.* **2012**, *24*, 2324–2328.
41. Fnadi, M.; Du, W.; Plumet, F.; Benamar, F. Constrained model predictive control for dynamic path tracking of a bi-steerable rover on slippery grounds. *Control Eng. Pract.* **2021**, *107*, 104693. [CrossRef]

actuators

MDPI

Article

Coordinated Control for the Trajectory Tracking of Four-Wheel Independent Drive–Four-Wheel Independent Steering Electric Vehicles Based on the Extension Dynamic Stability Domain

Yiran Qiao [1], Xinbo Chen [1,*] and Dongxiao Yin [2]

1 School of Automotive Studies, Tongji University, Shanghai 201804, China; qiaoyiran@tongji.edu.cn
2 Tongji Automobile Design and Research Institute Co., Ltd., Shanghai 201804, China; 2011666@tongji.edu.cn
* Correspondence: chenxinbo@tongji.edu.cn; Tel.: +86-021-6958-9120

Abstract: In order to achieve multi-objective chassis coordination control for 4WID-4WIS (four-wheel independent drive–four-wheel independent steering) electric vehicles, this paper proposes a coordinated control strategy based on the extension dynamic stability domain. The strategy aims to improve trajectory tracking performance, handling stability, and economy. Firstly, expert PID and model predictive control (MPC) are used to achieve longitudinal speed tracking and lateral path tracking, respectively. Then, a sliding mode controller is designed to calculate the expected yaw moment based on the desired vehicle states. The extension theory is applied to construct the extension dynamic stability domain, taking into account the linear response characteristics of the vehicle. Different coordinated allocation strategies are devised within various extension domains, providing control targets for direct yaw moment control (DYC) and active rear steering (ARS). Additionally, a compound torque distribution strategy is formulated to optimize driving efficiency and tire adhesion rate, considering the vehicle's economy and stability requirements. The optimal wheel torque is calculated based on this strategy. Simulation tests using the CarSim/Simulink co-simulation platform are conducted under slalom test and double-lane change to validate the control strategy. The test results demonstrate that the proposed control strategy not only achieves good trajectory tracking performance but also enhances handling stability and economy during driving.

Keywords: 4WID-4WIS EVs; direct yaw moment control (DYC); active rear steering (ARS); multi-objective coordinated control; extension dynamic stability domain

1. Introduction

4WID-4WIS electric vehicles, which incorporate four-wheel steering technology, are built on the foundation of four-wheel distributed drive [1]. This design simplifies the spatial structure by reducing mechanical components like differentials and half-shafts, providing the vehicle with multiple controllable degrees of freedom. It offers unique advantages in vehicle dynamic control [2–4], which can enable the realization of multi-objective coordinated control of the chassis [5,6]. In [7], Haonan Peng et al. adopted a torque coordinated control strategy that comprehensively considered stability and economy. They employed MPC to solve generalized forces, aiming to minimize control distribution errors, tire utilization rates, and drive system energy consumption. Feixiang Xu et al. [8] investigated steering mode switching strategies for a four-wheel independent steering rescue vehicle, using a multi-objective genetic algorithm to optimize two objectives: the maximum change rate of lateral acceleration and tire dissipation energy. Bohan Zhang et al. [9] proposed an actuator fault-tolerant control strategy based on cooperative game theory. While meeting the stability requirements, they utilized a two-dimensional game controller to minimize tire energy dissipation for economic efficiency. Junnian Wang et al. [10] introduced an energy management strategy based on multi-objective online optimization of torque distribution. Considering drive system efficiency, tire slip energy consumption, wheel torque fluctuation, and yaw rate tracking error, they dynamically adjusted weights using fuzzy control. To sum up, stability and economy are crucial performance indicators for the multi-objective control of steer-by-wire chassis.

The development of electrification and intelligence has led to the maturing of autonomous driving technology, which has become a key focus of the automobile industry [11]. Combined with autonomous driving technology [12,13], the coordination of chassis control in 4WID-4WIS EVs with a focus on trajectory tracking performance has garnered significant interest and research among academics [14]. There are two main categories for chassis coordinated control methods: centralized and hierarchical structures. Centralized control methods involve the design of a central controller that directly addresses control objectives for individual subsystems based on desired trajectory information, thus achieving multi-objective coordination of the chassis. For instance, in [15], a tube-based model predictive control (MPC) algorithm is proposed to design an integrated controller, considering control vector constraints, lateral stability constraints, anti-rollover constraints, and path tracking error constraints. Similarly, in [16], an integrated controller is developed using LTV-MPC to achieve speed and path tracking while ensuring vehicle handling stability. On the other hand, hierarchical control methods utilize multiple controllers to establish a multi-layered control structure. The top layer calculates the vehicle's target inputs based on trajectory tracking objectives, while the lower layer computes control objectives for subsystems. In [17], the top layer employs MPC to track the desired trajectory, while the lower layer utilizes a sliding mode controller to compute the target yaw moment, thereby achieving multi-objective coordination control. Additionally, in [18], the top layer establishes an HFDB path tracking controller considering lateral offset constraints, while the lower layer employs an adaptive robust LQR controller to achieve coordinated control of the drive and steering subsystems, ensuring handling stability. Centralized control methods are straightforward but suffer from high computational complexity, intricate problem solving, limited real-time performance, and challenges in accurately describing the coupling issues of the chassis system. Conversely, hierarchical control methods are comparatively complex but offer simpler problem solving, the effective decoupling of multi-objective coordination issues, and utilize modular theoretical approaches, thereby facilitating the resolution of coupling problems among subsystems. However, the current focus of chassis coordinated control is primarily on trajectory tracking and stability control, lacking attention to multi-objective coordinated control, including economy. This research aims to bridge this gap and lay the groundwork for further investigation.

In addition to considering the trade-off issues among multiple control objectives, it is also crucial to consider the distribution of control weights among multiple subsystems of the chassis during execution. The active rear steering system (ARS) directly controls the vehicle's lateral dynamics by adjusting the rear wheel angle, thereby changing the rear lateral forces. On the other hand, the direct yaw moment control system (DYC) adjusts the distribution of longitudinal forces among the four wheels, indirectly achieving lateral dynamics control [19]. Both systems have some redundancy in controlling the vehicle's lateral dynamics, especially under extreme conditions where ARS and DYC may be highly coupled. The function domains and control weights of subsystems have a close relationship with the vehicle stability boundaries [20]. One of the most classic and effective methods for nonlinear stability analysis is based on the phase plane [21]. In [22], Liang et al. designed an integrated controller for a 4WID-4WIS EV that combines the active front steering system (AFS) and DYC. A penalty function was used to allocate interventions of AFS and DYC and a transient layer was introduced in the phase plane, further improving the controller's performance. In [23], Tian et al. designed a control strategy for AFS and DYC based on the phase plane, which controlled subsystems separately in different regions to achieve good handling stability.

However, the traditional logic threshold method for establishing stability domain boundaries categorizes lateral states as either stable or unstable without taking into account the critical states in between. This approach leads to overly conservative stability boundaries. Moreover, frequent switching between subsystems during the coordinated control process can cause oscillation interference in the controller. To address this issue, researchers have conducted studies on extension theory. Extension theory has the ability to overcome the limitations of traditional control methods and is not restricted to any specific control approach. It has the potential to expand the original control area and partition the global control area. Different control strategies can be applied in different ranges based on specific control functions, enabling the achievement of control effects that are not possible with any particular conventional control method. In [24], Chen et al. proposed an extension controller which divided the classical domain, extension domain, and non-domain based on the phase plane, and designed different control strategies in different areas to improve the vehicle's handling stability. In [25], Zheng et al. proposed an extension coordination control for distributed drive electric vehicles based on evolutionary game theory. They played a game on the control weights of AFS and DYC and verified the effectiveness of the strategy via vehicle tests. One of the crucial steps in the application of extension theory involves dividing the boundaries between the classical domain, extension domain, and non-domain. However, current research on boundary values has certain limitations. Firstly, the boundaries of the extension domain are primarily determined based on simulation data and engineering experience, lacking a relevant theoretical basis. As a result, accurately obtaining the boundary values becomes challenging, and it cannot guarantee the optimal solution. Secondly, the division of boundary values cannot be adaptively adjusted according to the status of vehicles and roads, which hampers the adaptability of the extension domain boundary and subsequently affects the effectiveness of control in extension theory.

In summary, there are two major challenges in the chassis coordinated control that need to be addressed. The first challenge is the coupling problem of multiple control objectives, which include trajectory tracking performance, handling stability, and economy. The second challenge is the coupling problem between the steering and drive subsystems, which involves the division of stability domains, dynamic adjustment of stability domain boundaries, and adaptive tuning of control weights. To tackle these challenges, this paper proposes a hierarchical control architecture. The top layer focuses on trajectory tracking, the middle layer handles stability control, and the bottom layer considers both economy and stability for torque allocation. Additionally, the extension theory is employed to construct dynamic stability domains and dynamically adjust boundary values, which effectively resolves the coupling problem between the steering and drive subsystems. This strategy is simple, efficient, and demonstrates good real-time performance. It showcases strong adaptability to various driving conditions, thereby providing a solid theoretical foundation for multi-objective coordination control of the chassis.

The main contributions of this paper are as follows:

- A hierarchical chassis coordinated control architecture is proposed, consisting of a trajectory tracking layer, coordinated control layer, and optimal distribution layer, taking into account trajectory tracking performance, handling stability, and economy.
- The extension theory is employed to extend the traditional phase plane, constructing an extension dynamic stability domain based on the vehicle's linear response characteristics. The boundary values of the extension domain are adaptively adjusted according to vehicle speed and road adhesion coefficient, determining the control weights for ARS and DYC. This method is simple and efficient, overcoming the limitations of traditional stability domain boundaries that cannot be adjusted and are difficult to accurately obtain.
- A compound torque distribution strategy is developed that combines economic distribution with stability distribution, taking into account driving efficiency and tire adhesion rate as indicators. The real-time optimal distribution of wheel torque is achieved using the mutant particle swarm algorithm (MPSO) and the quadratic programming algorithm, respectively. It demonstrates good real-time performance and enables multi-objective optimization of stability and economy.

The paper is organized as follows: Section 2 proposes a hierarchical chassis coordinated control strategy. Section 3 performs simulation validation. Section 4 compares and discusses the simulation results. Section 5 reaches the conclusions.

2. Chassis Coordinated Control

2.1. Vehicle Model

To investigate the chassis coordinated control strategy of 4WID-4WIS EVs, the dynamic modeling of the vehicle, tires, and motor is performed based on the nonlinearity of the tire and the efficiency characteristics of the motor.

2.1.1. Vehicle Dynamic Model

Various scholars have proposed different linear and nonlinear vehicle models of varying complexity to investigate vehicle dynamics control problems [26]. This research primarily focuses on the control of lateral and longitudinal coupling for trajectory tracking. A 7-DOF vehicle dynamic model is utilized, which includes longitudinal motion, lateral motion, yaw motion, and the rotation of the four wheels based on the distributed drive system.

The vehicle under study has a small mass, compact dimensions, and a low center of gravity. Therefore, when establishing the vehicle dynamic model, it is assumed that the vehicle is symmetric about the center plane and is traveling on a flat horizontal road. The vertical motion of the body is neglected. Additionally, the suspension system is regarded as a rigid structure, ignoring the pitch and roll movements. After these simplifications, the dynamic model is shown in Figure 1.

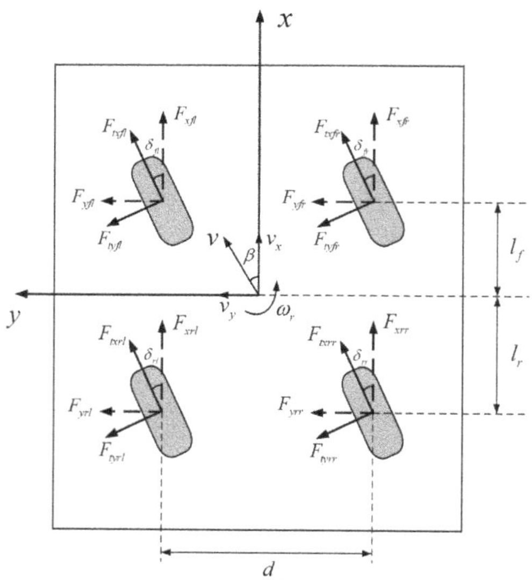

Figure 1. The 4WID-4WIS vehicle dynamic model.

The longitudinal motion equation is

$$m(\dot{v}_x - v_y\omega_r) = F_{xfl} + F_{xfr} + F_{xrl} + F_{xrr} \quad (1)$$

The lateral motion equation is

$$m(\dot{v}_y + v_x\omega_r) = F_{yfl} + F_{yfr} + F_{yrl} + F_{yrr} \quad (2)$$

The equation of yaw motion is

$$I_z\dot{\omega}_r = \left(F_{xfr} + F_{xrr} - F_{xfl} - F_{xrl}\right)\frac{d}{2} + \left(F_{yfl} + F_{yfr}\right)l_f - \left(F_{yrl} + F_{yrr}\right)l_r \tag{3}$$

where

$$F_{xij} = F_{txij}\cos\delta_{ij} - F_{tyij}\sin\delta_{ij} \tag{4}$$

$$F_{yij} = F_{txij}\sin\delta_{ij} + F_{tyij}\cos\delta_{ij} \tag{5}$$

$$i \in \{f, r\}, j \in \{l, r\} \tag{6}$$

The equation of wheel motion is

$$I_w\dot{\omega}_{ij} = T_{dij} - F_{txij}R_{eff} - T_{bij} \tag{7}$$

where v_x and v_y are the longitudinal velocity and lateral velocity; ω_r is the yaw rate; F_{xij} and F_{yij} are the longitudinal force and lateral force of each wheel, respectively; F_{txij} and F_{tyij} are the longitudinal and lateral forces of each wheel in the tire coordinate system; δ_{ij} is the wheel angle; I_w is the wheel moment of inertia; ω_{ij} is the wheel speed; T_{dij} is the wheel driving torque; and T_{bij} is the wheel braking torque.

To validate the effectiveness of the constructed 7-DOF model, a comparison is made with the vehicle model in the CarSim 2019. The simulation parameters are set as follows: the vehicle speed is set to 60 km/h, the road adhesion coefficient is set to 0.8, and the steering wheel input is set to a sinusoidal steering angle signal.

The model validation results are shown in Figure 2. The yaw rate and sideslip angle results obtained from the 7-DOF model are basically the same as those from the CarSim vehicle model, meeting the model accuracy requirements. This confirms the effectiveness of the 7-DOF vehicle model.

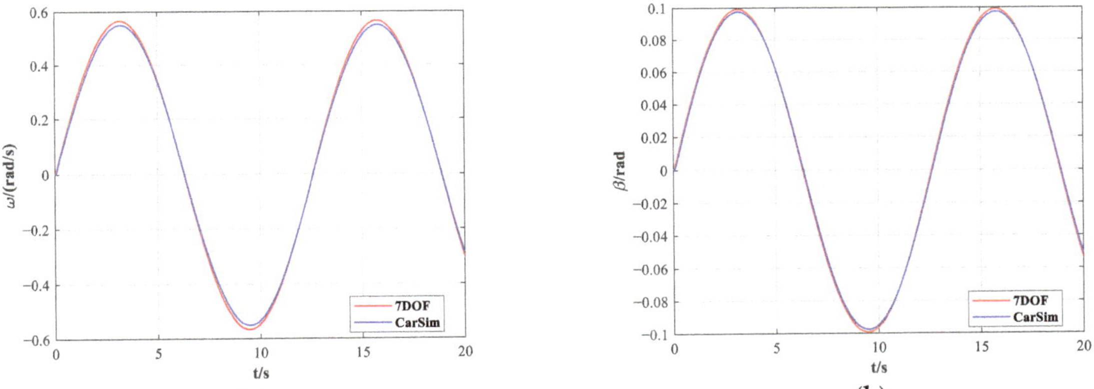

(a) (b)

Figure 2. Model validation results: (**a**) results of yaw rate; (**b**) results of sideslip angle.

2.1.2. Tire Model

The tire model describes the relationship between the tire's motion parameters and tire forces. It can be broadly classified into three types: theoretical models, empirical models, and semi-empirical models [27]. Considering the tire's side slip characteristics, this research selects the widely used UniTire semi-empirical tire model. The UniTire tire model takes into account various factors such as longitudinal slip, lateral slip, camber, turning angle, and hysteresis effects on tire force generation. It is characterized by high simulation accuracy, strong predictive capability, and well-defined theoretical boundaries. The longitudinal slip ratio and lateral slip ratio are defined as follows:

$$\begin{cases} S_x = \frac{v_{sx}}{\omega r} \\ S_y = \frac{v_{sy}}{\omega r} = (1 + S_x)\tan\alpha \end{cases} \tag{8}$$

where α is the tire slip angle.

The dimensionless relative longitudinal slip ratio, lateral slip ratio, and total slip ratio are defined as follows:

$$\begin{cases} \phi_x = \frac{K_x S_x}{\mu_x F_z} \\ \phi_y = \frac{K_y \tan\alpha}{\mu_y F_z} \\ \phi_n = \sqrt{(k_u\phi_x)^2 + \phi_y^2} \end{cases} \tag{9}$$

where K_x and K_y are the longitudinal and lateral stiffness of the tire, respectively; μ_x and μ_y are the longitudinal and lateral friction coefficients between the tire and the ground, respectively; and k_u is the correction factor that accounts for the difference in longitudinal and lateral stiffness of the tire.

In combined longitudinal and lateral slip conditions, the UniTire tire model accurately describes the dimensionless total shear force using the exponential form

$$\overline{F}_n = 1 - \exp(-\phi_n - E\cdot\phi_n^2 - (E^2 + \frac{1}{12})\cdot\phi_n^3) \tag{10}$$

The dimensionless total shear force is distributed into dimensionless longitudinal force and dimensionless lateral force based on the proportion of dimensionless relative longitudinal and lateral slip ratios in the dimensionless total slip ratio.

$$\begin{cases} \overline{F}_x = \overline{F}_n \frac{k_u\phi_x}{\phi_n} \\ \overline{F}_y = \overline{F}_n \frac{\phi_y}{\phi_n} \end{cases} \tag{11}$$

The longitudinal and lateral forces on the tire are

$$\begin{cases} F_x = \overline{F}_x \mu_x F_z \\ F_y = \overline{F}_y \mu_y F_z \end{cases} \tag{12}$$

2.1.3. Motor Model

The 4WID-4WIS EV chassis system consists of in-wheel motors and steering motors. Based on the vehicle's economy and handling stability performance requirements, and combined with the efficiency and dynamic characteristics of the motor, separate models are established for in-wheel motors and steering motors.

1. In-wheel motor model:

In order to investigate the energy efficiency of the chassis, this study develops a motor energy efficiency model that takes into account the efficiency characteristics of the in-wheel motor. To simplify the computation, this research focuses solely on the steady-state characteristics of the motor, disregarding the transient characteristics and simplifying the charging and discharging efficiency. Therefore, the efficiency model can be represented as Equation (13).

$$\eta_{em} = \eta(n_{em}, T_{em}) \tag{13}$$

where n_{em} is the motor speed, and T_{em} is the motor torque.

The efficiency characteristics of the in-wheel motor's driving efficiency are depicted using external characteristic curves and efficiency maps, as illustrated in Figure 3. The external characteristic curve sets the upper limit of the motor's maximum output power.

Additionally, the motor's discharge efficiency can be assessed in real time by referring to a two-dimensional lookup table of the motor efficiency map. This allows for a comprehensive study of the economy of the drive system.

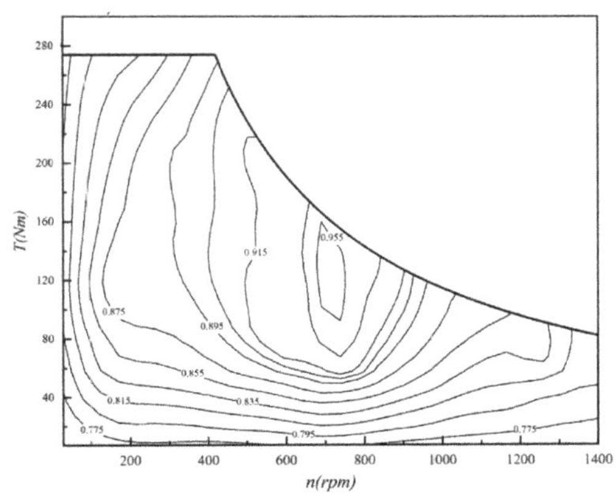

Figure 3. In-wheel motor efficiency map.

2. Steering motor model

In order to meet the dynamic response requirements during vehicle driving, the steering motor model is simplified into a corner tracking model, the delay caused by the steering mechanism is considered, and a model shown in Equation (14) is established:

$$\delta_{ij\_out} = \delta_{ij\_req} \cdot \frac{1}{1+\tau s} \tag{14}$$

where δ_{ij_out} is the actual output angle, δ_{ij_req} is the required output angle, and τ is the response time constant.

2.2. Chassis Control Architecture

Intelligent driving vehicles receive trajectory information from the planning layer and use rational control strategies to achieve trajectory tracking control, fulfilling the functional goals of autonomous driving. Vehicles with four-wheel independent drive and four-wheel independent steering (4WID-4WIS EVs) have multiple degrees of freedom, which significantly expands the control margin of chassis dynamics and increases the upper limit of the chassis stability domain. This potential allows for the improvement in vehicle performance and serves as a basis for the multi-objective coordinated control of the chassis. Therefore, this research aims to leverage the control advantages of the fully wired chassis to enhance handling stability and economy during the trajectory tracking process.

The designed chassis coordinated control architecture Is shown in Figure 4, employing a layered strategy structure that includes a trajectory tracking layer, coordinated control layer, and optimal distribution layer.

Based on a decentralized control structure, the trajectory tracking layer separates the longitudinal and lateral control targets, converting the desired trajectory information into vehicle input information. The expert PID control method is employed to track the desired vehicle speed and generate the expected longitudinal acceleration. The MPC controller is responsible for tracking the desired path and producing the expected front wheel angle.

The coordinated control layer calculates the expected actions of the subsystems based on the anticipated vehicle state. Sliding mode control is used to compute the generalized yaw moment and incorporates an adaptive adjustment strategy for the control weights of ARS and DYC. This strategy is designed considering the extension dynamic stability domain and allocates the desired yaw moment, determining the control objectives for the ARS and DYC.

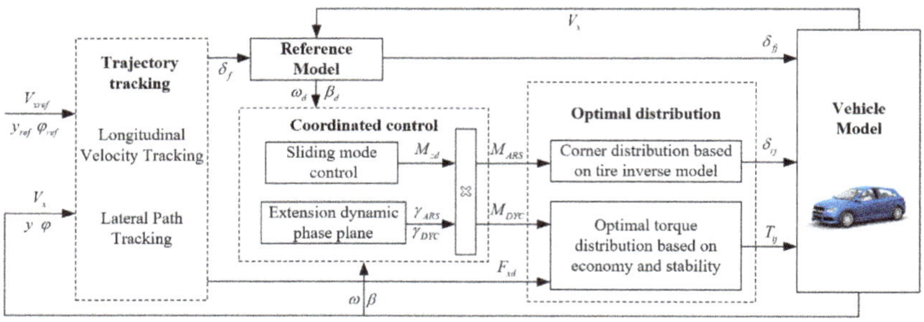

Figure 4. The chassis coordinated control architecture.

The optimal distribution layer designs torque distribution and corner distribution strategies, converting the expected subsystem actions into control commands for actuators. The optimal torque distribution control strategy based on economy and stability is adopted, using driving efficiency and tire adhesion rate as control indicators for economy and stability, respectively. It computes the optimal driving torque for each wheel via a combination of MPSO and DP. The corner distribution strategy employs the tire inverse model to calculate the tire slip angle. Combined with the Ackermann steering theory, the steering angle for each wheel is output to achieve the vehicle's closed-loop control.

2.3. Trajectory Tracking Layer

The trajectory tracking layer's main objective is to replace the driver in following the desired trajectory by providing the desired longitudinal acceleration and front wheel angle. This layer takes into account the desired path and vehicle speed, which are the combined desired longitudinal and lateral motion states of the vehicle. As a result, the trajectory tracking problem is divided into two separate problems: longitudinal velocity tracking and lateral path tracking. By calculating the expected longitudinal acceleration and front wheel angle independently, this layer is able to provide the necessary inputs for the coordinated control of the chassis.

2.3.1. Longitudinal Velocity Tracking

Proportional integral derivative (PID) control is widely applied in longitudinal vehicle speed tracking due to its simplicity, practicality, and independence from precise system models. Given the uncertainty of system parameters, external disturbances, and nonlinear couplings between subsystems, many researchers have explored PID control based on intelligent algorithms [28,29]. In this study, based on expert system theory, the expert PID is employed for longitudinal velocity tracking. The specific algorithm flow can be found in [30]. The calculation formula for longitudinal acceleration is as follows:

$$a_{xd} = u(k) = \begin{cases} a_{x\max} & (|e(k)| > M_{max}) \\ u(k-1) + k_1\{k_i e(k) + k_p \Delta e(k) + k_d \Delta e(k)\} & (e(k) \times \Delta e(k) > 0 \text{ or } \Delta e(k) = 0) \\ u(k-1) & (e(k)\Delta e(k) < 0,\ \Delta e(k)\Delta e(k-1) > 0,\ e(k) = 0) \\ u(k-1) + k_2 k_i e(k) & (e(k)\Delta e(k) < 0,\ \Delta e(k)\Delta e(k-1) < 0) \\ u(k-1) + k_i e(k) + k_p \Delta e(k) & (|e(k)| < M_{min}) \end{cases} \tag{15}$$

where $e(k) = v(k) - v_d(k)$ is the velocity error, and $e(k) = v(k) - v_d(k)$ is the velocity error increment. The maximum deviation value is set as M_{max}, the middle deviation value is set as M_{mid}, and the minimum deviation value is set as M_{min}.

Ignoring the road gradient resistance and considering rolling resistance, aerodynamic resistance, and acceleration resistance, a feedforward controller is designed according to the longitudinal vehicle dynamics equation as follows:

$$F_{xd} = F_f + F_{air} + F_j = mgf + \frac{C_D A v_x^2}{21.15} + m a_{xd} \tag{16}$$

where F_{xd} is the generalized longitudinal force, F_f is the rolling resistance, F_{air} is the air resistance, F_j is the acceleration resistance, f is the rolling resistance coefficient, C_D is the air resistance coefficient, and A is the windward area.

2.3.2. Lateral Path Tracking

In order to navigate complex and ever-changing driving environments, autonomous vehicles need to take into account road constraints, safety constraints, and actuator constraints while tracking their trajectory. Model predictive control (MPC) is a widely used feedback control algorithm that predicts future system outputs based on the current model and state and solves a finite-time open-loop optimization problem on a rolling basis to achieve control objectives. MPC has proven to be highly effective in handling multi-constraint optimization problems. As a result, scholars have conducted extensive research on the application of MPC-based lateral path tracking control in recent years [29]. This study utilizes MPC to achieve lateral path tracking.

The accuracy of the prediction model directly impacts the control effectiveness of MPC. To balance the requirements of accuracy and complexity, a nonlinear 3-DOF vehicle model is chosen as the prediction model to forecast the system's output. The state variable is $X = \left[v_x, v_y, \varphi, \dot{\varphi}, x, y\right]^T$, the control variable is $u = \left[\delta_f\right]^T$, and the output variable is $Y = [\varphi, y]^T$. Assuming constant longitudinal velocity during path tracking and to meet the real-time requirements of high-speed control, this paper linearizes the nonlinear system using Taylor's formula. Simultaneously, a discretization process is applied to obtain the control system model:

$$\begin{aligned} X(k+1) &= A_{dyn}(k)X(k) + B_{dyn}(k)u(k) \\ Y(k) &= CX(k) \end{aligned} \tag{17}$$

where

$$A_{dyn}(k) = \begin{bmatrix} \frac{C_{\alpha f}+C_{\alpha r}}{m\dot{x}} & -\dot{\varphi} - \frac{(C_{\alpha f}+C_{\alpha r})\dot{y} - (l_r C_{\alpha r} - l_f C_{\alpha f})\dot{\varphi}}{m\dot{x}^2} & 0 & -\dot{x} + \frac{l_f C_{\alpha f} - l_r C_{\alpha r}}{m\dot{x}} & 0 & 0 \\ \dot{\varphi} + \frac{C_{\alpha f}\delta_f}{m\dot{x}} & -\frac{C_{\alpha f}\delta_f(\dot{y}+l_f\dot{\varphi})}{m\dot{x}^2} & 0 & \dot{y} + \frac{l_f C_{\alpha f}\delta_f}{m\dot{x}} & 0 & 0 \\ 0 & 0 & 0 & 0 & 0 & 0 \\ \frac{l_f C_{\alpha f} - l_r C_{\alpha r}}{I_z\dot{x}} & -\frac{(l_f C_{\alpha f} - l_r C_{\alpha r})\dot{y} + (l_f^2 C_{\alpha f} + l_r^2 C_{\alpha r})\dot{\varphi}}{I_z\dot{x}^2} & 0 & \frac{l_f^2 C_{\alpha f} + l_r^2 C_{\alpha r}}{I_z\dot{x}} & 0 & 0 \\ \cos\varphi & \sin\varphi & \dot{x}\cos\varphi - \dot{y}\sin\varphi & 0 & 0 & 0 \\ -\sin\varphi & \cos\varphi & -\dot{x}\sin\varphi - \dot{y}\cos\varphi & 0 & 0 & 0 \end{bmatrix} \tag{18}$$

$$B_{dyn}(k) = \left[-\frac{C_{\alpha f}}{m} \; \frac{C_{\alpha f}\left[\left(\dot{y} + l_f\dot{\varphi}\right) - 2\delta_f\dot{x}\right]}{m\dot{x}} \; 0 \; -\frac{l_f C_{\alpha f}}{I_z} \; 0 \; 0 \right]^T \tag{19}$$

$$C = \begin{bmatrix} 0 & 0 & 1 & 0 & 0 & 0 \\ 0 & 0 & 0 & 0 & 1 & 0 \end{bmatrix} \tag{20}$$

where φ is the heading angle, $C_{\alpha f}$ and $C_{\alpha r}$ are the equivalent cornering stiffnesses of the front axle and the rear axle, respectively, and x and y are the longitudinal and lateral coordinates of the centroid in the geodetic coordinate system, respectively.

The control objectives are defined based on the tracking accuracy of the reference path and the smooth variation in control inputs. The tracking accuracy is assessed using lateral offset and heading angle error as evaluation metrics. Therefore, the design of the performance evaluation function is as follows:

$$J = \left[Y_p(k) - Y_{ref}\right]^T Q_Q \left[Y_p(k) - Y_{ref}\right] + \Delta U(k)^T R_R \Delta U(k) \tag{21}$$

where $Q_Q = I_p \otimes Q$, $R_R = I_c \otimes R$, Q is the weight coefficient matrix of the output of the control system, R is the weight coefficient matrix of the control increment, and $\otimes$ represents the Kronecker product.

During the solving process, three aspects need to be considered: control increment constraints, control quantity constraints, and stability constraints.

To ensure a smoother control process and enhance stability and comfort while tracking the desired path, constraints are applied to the control increment, as shown in Equation (22):

$$\Delta U_{\min}(k) \le \Delta U(k) \le \Delta U_{\max}(k) \tag{22}$$

Due to limitations in the actuators, there are constraints on the control quantity for the front wheel angle, represented as

$$u_{\min}(k) \le u(k) \le u_{\max}(k) \tag{23}$$

To ensure that the vehicle has good stability, constraints are imposed on the lateral yaw rate and tire slip angle. The constraint on the lateral yaw rate due to road conditions is as follows:

$$-\frac{\mu g}{v_x} \le \omega_r \le \frac{\mu g}{v_x} \tag{24}$$

When the vehicle becomes unstable, the consequences of rear axle sideslip are more severe. So, the rear wheel slip angle needs to be restrained; that is

$$\alpha_{r,\min} \le \alpha_r \le \alpha_{r,\max} \tag{25}$$

By solving the QP problem, the optimal control increment sequence $\Delta U^*(k)$ is obtained. The first element $\Delta u^*(k)$ is applied to the controlled system and converted into the control quantity at the present moment, which provides the desired front wheel angle.

$$\delta_f = u(k) = u(k-1) + \Delta u^*(k) \tag{26}$$

2.4. Coordinated Control Layer

The coordinated control layer calculates the expected vehicle state by using the output of the trajectory tracking layer, which provides the expected vehicle action. To track the desired state, a sliding mode controller is designed to generate the expected vehicle effect. Furthermore, this research explores a coordinated control strategy based on the theory of extension dynamic phase plane. This strategy aims to improve the vehicle's handling stability and coordinate the coupling between the steering and drive subsystems. It achieves this by allocating the expected yaw moment and calculating the desired control objectives for the steering and drive systems.

2.4.1. Reference States

The handling stability of a vehicle's lateral motion is determined by two important parameters: yaw rate and sideslip angle. The yaw rate affects the quality of the vehicle's dynamic response. If the yaw rate is too large, it may cause oversteer, while if it is too small, it may result in understeer. On the other hand, the sideslip angle reflects how much the vehicle deviates from its intended path during driving. A larger sideslip angle indicates a greater deviation in path tracking. Therefore, to ensure handling stability, it is essential to set appropriate values for the yaw rate and sideslip angle.

Based on the 2-DOF model of a front wheel steering vehicle, the expected yaw rate can be obtained by calculating the steady-state response of the yaw rate:

$$\omega_d = \frac{\delta_f V_x}{L + \frac{mV_x^2(l_f C_{\alpha f} - l_r C_{\alpha r})}{L C_{\alpha f} C_{\alpha r}}} \tag{27}$$

Considering the road adhesion constraints, the expected yaw rate can be represented as

$$\omega_d = \min\left\{ |\omega_d|, \left|0.85\frac{\mu g}{v_x}\right| \right\} \mathrm{sgn}(\delta_f) \tag{28}$$

The expected side slip angle can also be calculated via the steady-state response of the 2-DOF vehicle model. However, to simplify the calculation, the expected side slip angle is set to be 0 rad.

2.4.2. Additional Yaw Moment

For both the 4WID system and the 4WIS system, the control objectives for handling stability include tracking the expected yaw rate and the expected side slip angle. Therefore, this paper adopts sliding mode control to track these two reference states. It calculates the additional yaw moment required to meet the tracking task, transforming the desired vehicle state into the desired vehicle action, thus achieving the objectives of improving chassis maneuverability and stability.

The sliding surface of the tracking error is set as follows:

$$s = \omega - \omega_d + \rho(\beta - \beta_d) \tag{29}$$

where ρ represents the weight coefficient of the sideslip angle.

The yaw rate and side slip angle are control objectives that characterize vehicle stability. However, in many cases, it is challenging to simultaneously achieve both objectives. The yaw rate indicates whether the vehicle is understeering or oversteering, which directly impacts its handling performance. On the other hand, the side slip angle measures the deviation of the vehicle's trajectory during steering, thereby affecting its lateral stability. Consequently, the coupling between the yaw rate and side slip angle becomes a crucial aspect when calculating the desired vehicle action. Currently, the phase plane method is considered one of the effective approaches to address this coupling problem. The phase plane index is defined as follows:

$$X_{\text{region}} = \left| B_1 \dot{\beta} + B_2 \beta \right| \tag{30}$$

where B_1 and B_2 are the boundary model coefficients.

Within the stable region of the phase plane, the phase trajectory of the side slip angle can automatically converge to zero, which imparts trajectory-following characteristics and lateral stability to the car. As a result, the primary control objective is to achieve the desired yaw rate. However, outside the stable region, the phase trajectory of the side slip angle diverges, leading to vehicle instability. In such cases, both the desired yaw rate and side slip angle become the main control objectives. Furthermore, as the phase plane index increases, the weight of the side slip angle gradually increases. So, the weight coefficient is defined as

$$\rho = \begin{cases} 0 & X_{\text{region}} \leqslant 0.8 \\ (X_{\text{region}} - 0.8) \times 5k & 0.8 \leqslant X_{\text{region}} \leqslant 1 \\ k & X_{\text{region}} \geqslant 1 \end{cases} \tag{31}$$

where k is a positive number.

The vehicle 2-DOF model is used as the control model:

$$\begin{cases} \dot{\omega} = \frac{l_f C_{\alpha f}}{I_z}\left(\beta + \frac{l_f \omega}{V_x} - \delta_f\right) - \frac{l_r C_{\alpha r}}{I_z}\left(\beta - \frac{l_r \omega}{V_x}\right) + \frac{M}{I_z} \\ \dot{\beta} = -\omega + \frac{C_{\alpha f}}{m V_x}(\beta + \frac{l_f \omega}{V_x} - \delta_f) + \frac{C_{\alpha r}}{m V_x}(\beta - \frac{l_r \omega}{V_x}) \end{cases} \tag{32}$$

During the control process, the presence of inertia and hysteresis can lead to chattering, which has a significant impact on the system's performance. To attenuate chattering and enhance system stability, this research employs an exponential reaching law with a saturation function, as shown in the following equation:

$$\dot{s} = -\varepsilon \cdot sat(s/\phi) - ks \tag{33}$$

where, $\varepsilon > 0, k > 0, \phi > 0$, and

$$sat\left(\frac{s}{\phi}\right) = \begin{cases} 1, & s > \phi \\ \frac{s}{\phi}, & |s| \leq \phi \\ -1, & s < -\phi \end{cases} \tag{34}$$

According to Equations (30), (32) and (33), the additional yaw moment can be calculated as follows:

$$\begin{aligned} M_{zd} = {} & I_z\{(-\varepsilon \cdot sat(s/\phi) - ks) + l_f C_{\alpha f}\left(\delta_f - \beta - \frac{l_f \omega}{V_x}\right) - l_r C_{\alpha r}\left(-\beta + \frac{l_r \omega}{V_x}\right) + I_z \dot{\omega}_{rd} \\ & - I_z \rho\{(-\omega + \frac{C_{\alpha f}}{m V_x}(\beta + \frac{l_f \omega}{V_x} - \delta_f) + \frac{C_{\alpha r}}{m V_x}(\beta - \frac{l_r \omega}{V_x}) - \dot{\beta}_d\} \end{aligned} \tag{35}$$

2.4.3. Coordinated Strategy

To address the limitations of the traditional phase plane, this research introduces the extension theory to expand the stability domain of the original phase portrait. Additionally, the dynamic boundaries for the extension domain are designed by considering the linear response characteristics of the vehicle. These boundary values can be dynamically adjusted based on factors such as vehicle speed and road adhesion coefficient, which helps overcome the issues associated with fixed domain boundaries and poor adaptability observed in existing research. By utilizing the extension dynamic phase portraits, the lateral states of the vehicle are categorized into three regions: classical domain, extension domain, and non-domain. Different allocation strategies are applied to each region to distribute the expected additional yaw moments. Furthermore, the control weights of DYC and ARS are adaptively adjusted, and the target yaw moments for DYC and ARS are calculated and outputted.

The control strategy based on the extension dynamic stability domain is mainly divided into several parts, including characteristic state extraction, domain division, correction calculation, and measurement pattern recognition.

1. Characteristic State Extraction:

According to the definition expansion of the extension primitive theory, the characteristic quantity can represent the state characteristic variables that describe the system control objectives. The phase plane of the side slip angle and its angular velocity can reflect the vehicle's lateral stability, and it is widely used. Therefore, $S(\beta, \mathrm{d}\beta)$ is selected as the characteristic state. A two-dimensional extension set is established to describe the stable driving state.

2. Domain Division

In this research, a dual-line method is used to divide the original phase portrait into three regions: classical domain, extension domain, and non-domain, as shown in Figure 5. The classical domain represents the stable state, the extension domain represents the critical stable state, and the non-domain represents the unstable state. Therefore, accurately determining the boundary values for the extension domain and non-domain is crucial for constructing the extension phase portrait.

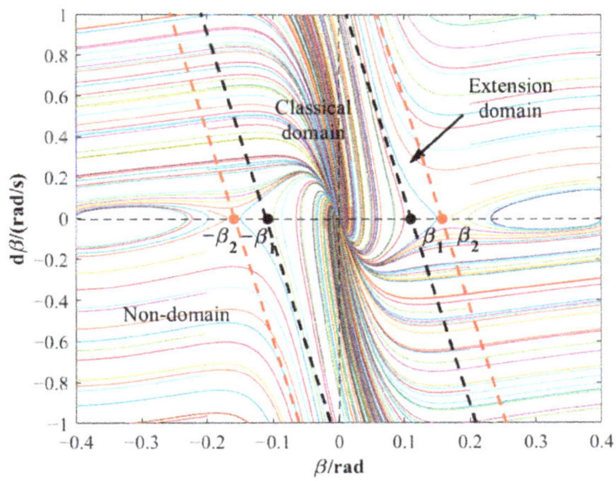

Figure 5. The extension phase portrait.

1. Non-domain Boundary

The non-domain describes the unstable state during driving, which corresponds to the instability domain of the traditional phase plane. Therefore, the unstable boundary of the phase plane is chosen as the non-domain boundary:

$$\left|B_1\dot{\beta} + B_2\beta\right| \geq 1 \tag{36}$$

The stable domain of the phase plane is determined by the vehicle speed, road adhesion coefficient, and front wheel angle. When the front wheel angle increases, the stable boundary of the phase plane is shifted to one side. However, within a range of 10 degrees of front wheel angle, the stable domain boundary is relatively close to the boundary when the front wheel angle is 0 degrees. Therefore, the impact of the front wheel angle on the phase portrait boundary values can be disregarded. On the other hand, as the vehicle speed increases or the road adhesion coefficient decreases, the stable domain gradually decreases, and the unstable boundary expands outward. Therefore, the values of B_1 and B_2 are adaptively adjusted according to the vehicle speed and road adhesion coefficient. Simulation data establish the mapping relationship under different driving speeds and road adhesion coefficients as follows:

$$\begin{cases} B_1 = B_1(V_x, \mu) \\ B_2 = B_2(V_x, \mu) \end{cases} \tag{37}$$

B_1 and B_2 are dynamically adjusted according to the vehicle speed and road adhesion coefficient to achieve dynamic adjustment of the non-domain boundary.

$$\beta_2 = \beta_2(B_1, B_2) = \beta_2(V_x, \mu) \tag{38}$$

2. Extension Domain Boundary

Currently, several scholars have examined the extension domain boundary of a vehicle's stable state. In a previous study [24], the extension domain boundary was determined via the fixed-proportion scaling of the traditional phase portrait boundary using extensive simulation calculations. However, this method lacks real-time adjustability according to external conditions and vehicle states. Additionally, the scaling ratio has limited adaptability to operating conditions and lacks a theoretical basis. Another study [25] determined the critical front wheel angle via the simulation analysis of tire lateral characteristics under specific conditions and calculated the extension domain boundary values using a dynamics model. While this method has some theoretical basis, it cannot dynamically adjust the limit value of the front wheel angle, which depends on the predetermined simulation conditions, making it challenging to meet the vehicle's requirements under different conditions. Moreover, accurately obtaining the limit value of the front wheel angle is also a challenge. To address these issues, this paper proposes a dynamic boundary method based on the vehicle's linear response characteristics. This method adaptively adjusts the critical front wheel angle based on the vehicle's speed and road adhesion coefficient.

The goal of the vehicle dynamics stability control is to maintain the linear response characteristics of the vehicle, specifically the linear relationship between the yaw rate and the front wheel angle. When the vehicle operates within its linear region, where the tire slip characteristics are also linear, it exhibits a large stability margin and remains stable. However, when the vehicle enters the nonlinear region, where the tire slip characteristics become nonlinear, the lateral force is easily constrained by the adhesion ellipse, leading to vehicle instability. To address this, this paper proposes considering the boundary between the linear and nonlinear regions as the extension domain boundary. This ensures that the tire slip characteristics in the classical domain always remain in the linear region, thereby maintaining vehicle stability.

To determine the dynamic boundary of the vehicle's linear region, the calibration method used in this research is as follows: First, a sinusoidal steering wheel angle signal is applied to the vehicle. Then, the relationship between yaw rate and front wheel angle is calculated, as shown in Figure 6. If the change in yaw rate corresponding to a unit front wheel angle is lower or higher than a certain threshold relative to near zero angle, it indicates that the vehicle has entered the nonlinear region. The front wheel angle at this critical point is recorded, which is referred to as the critical angle δ_{f0}. This calibration method is suitable not only for simulation analysis but also for vehicle testing, as it allows for the easier determination of the critical angle of the linear region. Simulation tests were conducted in this research using the built vehicle model to determine the critical angle.

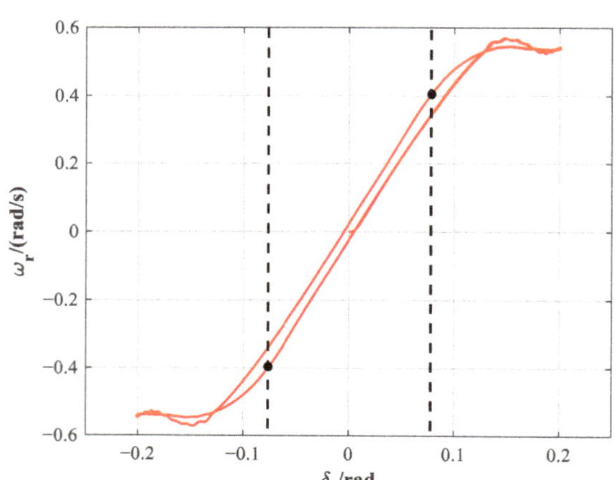

Figure 6. The relationship between yaw rate and front wheel angle.

The tire's slip characteristics, as illustrated in Figure 7, demonstrate that when the slip angle of the tire surpasses the linear region, the lateral force becomes nonlinear and less responsive to the slip angle. As the tire slip angle increases, the lateral force experiences a slight increase, gradually transitioning from a nonlinear state to a saturation state. Consequently, given a specific road adhesion coefficient, the relationship between the critical angle, vehicle speed, and road adhesion can be described as follows [31]:

$$V_x^2 \cdot \delta_{f0} = C_\mu \tag{39}$$

where C_μ is a constant, indicating that the product of the square of the vehicle speed and the critical angle is a constant under the corresponding adhesion coefficient. Based on this conclusion, further calibration of C_μ under different adhesion coefficients is performed. Furthermore, the mapping relationship of the critical angle under different vehicle speeds and adhesion coefficients is established, as shown in Figure 8.

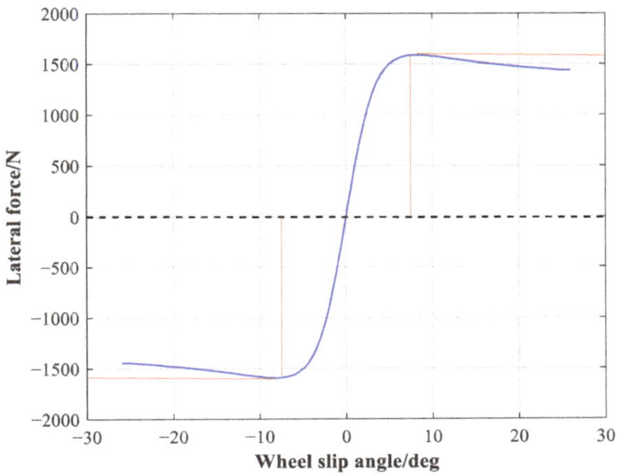

Figure 7. Lateral tire force curve.

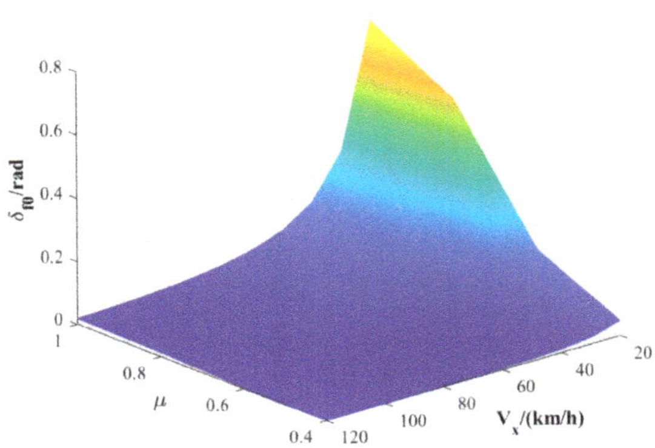

Figure 8. The critical angle under different speeds and adhesion coefficients.

The critical angle δ_{f0} is dynamically adjusted according to vehicle speed and adhesion coefficient. The steady-state sideslip angle β_1 under this steering angle is solved via the 2-DOF model. The straight line passing through $(\pm\beta_1, 0)$ and parallel to the non-domain boundary is regarded as the boundary of the extension domain.

$$\beta_1 = \frac{l_r - \frac{l_f m V_x^2}{2C_{\alpha r} L}}{L + \frac{m V_x^2 (l_r C_{\alpha r} - l_f C_{\alpha f})}{2C_{\alpha f} C_{\alpha r} L}} \delta_{f0} \tag{40}$$

3. Correction Calculation

To simplify the calculation process, the characteristic quantity $\psi(S)$ is introduced to reduce the dimensionality of the two-dimensional domain and transform it into a one-dimensional domain as follows:

$$\psi(S) = \beta + E\dot{\beta} \tag{41}$$

where E is mainly determined based on B_1 and B_2 in the equation. The characteristic quantity is determined by the characteristic state and represents the driving stability of the vehicle.

In extension theory, the extension distance refers to the distance between a point and an interval, which can be used to describe the relationship between points and regions in the phase plane. Therefore, the extension distance from any characteristic quantity to the classical domain $X = [-\beta_1, \beta_1]$ can be expressed as follows:

$$\rho(\psi, X) = |\psi| - \beta_1 \tag{42}$$

The extension distance from any characteristic quantity to the non-domain $X_0 = [-\beta_2, \beta_2]$ is expressed as follows:

$$\rho(\psi, X_0) = |\psi| - \beta_2 \tag{43}$$

The correlation function can be determined as

$$K(\psi) = \frac{\rho(\psi, X_0)}{\rho(\psi, X_0) - \rho(\psi, X)} \tag{44}$$

The correlation function $K(\psi)$ describes the correlation degree of the characteristic quantity with respect to the extension set of the system state. it determines the lateral stable state of the current vehicle state, thereby determining the control weight of the drive and steering subsystems.

4. Measurement Pattern Recognition

The yaw moment generated by tire lateral forces is generally greater than that produced by longitudinal forces of the same size due to the wheelbase of a car being typically greater than its track width. By actively adjusting the rear wheel slip angle, the vehicle's linear handling range can be expanded when the rear wheels actively participate in steering, effectively preventing the vehicle from entering the unstable area prematurely. Furthermore, it can be observed from the tire adhesion ellipse that the limit of tire longitudinal force is greater than that of tire lateral force. Even when the tire lateral force reaches its extreme boundary, there is still some space for tire longitudinal force. Therefore, a system that controls the vehicle's handling stability by adjusting tire longitudinal force can play a more significant role when the tire lateral force is saturated.

Based on the correlation function calculated above, the recognition of measurement patterns is performed, and the control weight allocation strategy is devised in the corresponding patterns:

M1: When $K(\psi) \geq 1$, the measurement pattern is M1, and the corresponding characteristic state belongs to the classic domain. The current vehicle state is in an easily controllable state, indicating that the tires have linear slip characteristics and are within the linear response range. In order to maximize the benefits of rapid response in tire lateral forces for the vehicle's yaw motion, the primary control method used is ARS. ARS effectively manages the steering motion by controlling the rear wheel angle. It can be expressed as follows:

$$\begin{cases} \gamma_{ARS} = 1 \\ \gamma_{DYC} = 0 \end{cases} \tag{45}$$

M2: When $0 < K(\psi) < 1$, the measurement pattern is M2, and the corresponding characteristic state belongs to the extension domain. In this case, the characteristic state is in a controllable state, indicating that the current vehicle state is within the nonlinear response range. The tires do not exhibit linear slip characteristics and have limited control capability over lateral force. As a result, the DYC system starts to intervene to compensate for the limitations of the ARS system. Furthermore, as $K(\psi)$ decreases, the corresponding characteristic state gradually moves away from the classical domain and tends toward uncontrollability. The vehicle state is gradually approaching an unstable state. Consequently, the control weight of the ARS system gradually decreases while the control weight of the DYC system gradually increases. It can be expressed as

$$\begin{cases} \gamma_{ARS} = K(\psi) \\ \gamma_{DYC} = 1 - K(\psi) \end{cases} \tag{46}$$

M3: When $K(\psi) \leq 0$, the measurement pattern is M3, and the corresponding characteristic state belongs to the non-domain. The current vehicle state is in an unstable condition, as indicated by the challenging-to-control state. The tire lateral forces are approaching their limits, resulting in the withdrawal of ARS and prioritizing DYC. This enables the tires to fully utilize their longitudinal adhesion. It can be expressed as

$$\begin{cases} \gamma_{ARS} = 0 \\ \gamma_{DYC} = 1 \end{cases} \tag{47}$$

By calculating the control weights of the drive and steering subsystems, the control objectives for ARS and DYC can be determined as follows:

$$\begin{cases} M_{ARS} = \gamma_{ARS} M_{zd} \\ M_{DYC} = \gamma_{DYC} M_{zd} \end{cases} \tag{48}$$

2.5. Optimal Distribution Layer

The optimal distribution layer receives the longitudinal force and the target yaw moment from the upper layer. It then distributes these inputs and calculates the output torque and corner for each wheel. These values are subsequently sent to the in-wheel motors and steering motors. The optimal distribution layer is composed of two main parts: torque distribution and corner distribution. Depending on the execution subsystem, these two parts are responsible for distributing torque and corner values, respectively.

2.5.1. Torque Distribution

The torque distribution control system receives the longitudinal force and the target yaw moment for DYC from the upper layer. It then calculates the driving torque for each wheel. To ensure both stability and economy [32], this study evaluates the driving efficiency and tire adhesion ratio as criteria. The optimal torque control law is determined using a combination of Mutation Particle Swarm Optimization (MPSO) and Quadratic Programming (QP), enabling real-time torque optimization. Figure 9 illustrates the process of torque distribution control, where different torque distribution strategies are designed based on the extension dynamic phase portrait.

1. Classical Domain

In the classical domain, the vehicle remains in a stable state, with the Dynamic Yaw Control (DYC) system inactive. As a result, the torque distribution strategy focuses on achieving economy by optimally distributing the total longitudinal force. Considering the driving efficiency characteristics of the in-wheel motors and aiming to improve the computational efficiency of the algorithm for real-time torque distribution, this paper utilizes the MPSO algorithm to offline optimize the distribution coefficients for the front and rear axle torques. This establishes a mapping relationship for the optimal distribution law, allowing the motors to operate in the most efficient region as much as possible.

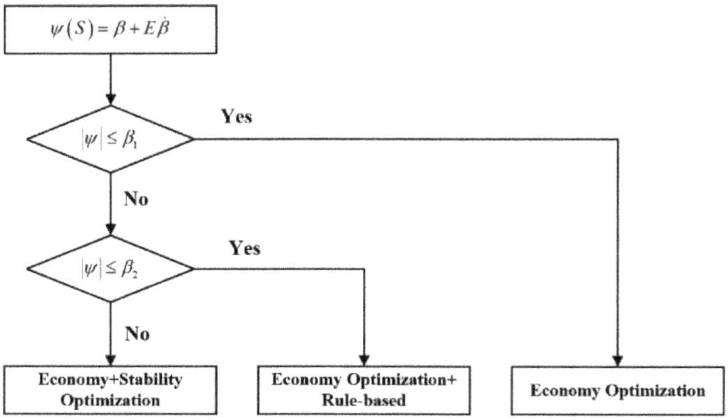

Figure 9. The process of torque distribution control.

Building upon the literature [30], the MPSO algorithm combines the traditional PSO algorithm [33] with the mutation concept from genetic algorithms. This incorporation of mutation during the update of population positions enables the algorithm to escape locally optimal solutions, increasing the chances of finding a global optimum and reducing the likelihood of premature convergence.

To simplify the optimization problem, this paper introduces the front axle torque distribution coefficient, which is the ratio of the front axle torque to the total required torque, as follows:

$$\lambda = \frac{T_f}{T_{xd}} = \frac{T_f}{F_{xd}R_{eff}} \tag{49}$$

where T_f is the front axle output torque, and T_{xd} is the total longitudinal demand torque of the vehicle.

In order to optimize the efficiency of the drive system, the overall efficiency of the motor is used as the economic indicator:

$$J = \frac{T_f n_f + T_r n_r}{\frac{T_f n_f}{\eta_f} + \frac{T_r n_r}{\eta_r}} \tag{50}$$

where T_r is the output torque of the rear axle; n_f and n_r are the output speeds of the front and rear axle motors, respectively; and η_f and η_r are the output efficiencies of the front and rear axle motors, respectively.

The driving efficiency can be expressed as

$$J = \frac{T_f n_f + T_r n_r}{\frac{T_f n_f}{\eta_f} + \frac{T_r n_r}{\eta_r}} = \frac{\lambda n_f + (1-\lambda)n_r}{\frac{\lambda n_f}{\eta_f} + \frac{(1-\lambda)n_r}{\eta_r}} = \frac{1}{\frac{\lambda}{\eta_f} + \frac{1-\lambda}{\eta_r}} \tag{51}$$

Therefore, the optimization problem can be expressed as

$$\min J = \frac{\lambda}{\eta_f} + \frac{1-\lambda}{\eta_r} \tag{52}$$

Considering that the vehicle under study has a center of gravity closer to the front axle, with the front axle bearing a greater vertical load than the rear axle, the limit of the longitudinal force on the front axle is larger. Therefore, the constraint for the front axle distribution coefficient is as follows:

$$0.5 \le \lambda \le 1 \tag{53}$$

The efficiency of the in-wheel motor is mainly determined by the speed and torque. The front axle torque distribution coefficient is influenced by factors such as vehicle speed and demand torque. In this study, vehicle speed is represented by wheel speed. The front axle torque distribution coefficient is optimized using the MPSO algorithm in MATLAB. Figure 10 illustrates the optimal relationship between the front axle distribution coefficient, wheel speed, and demand torque.

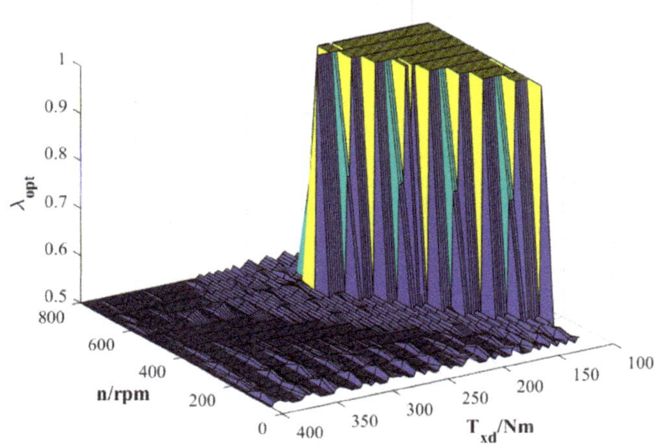

Figure 10. Optimal front axle distribution coefficient.

Based on the optimal distribution coefficient of the front axle, the torque of the left and right wheels is evenly distributed to obtain the driving torque of each wheel:

$$\begin{cases} T_{fl} = T_{fr} = \lambda_{opt} T_{xd}/2 \\ T_{rl} = T_{rr} = (1-\lambda_{opt}) T_{xd}/2 \end{cases} \tag{54}$$

2. Extension Domain

In the extension domain, the vehicle is in a critical stable state, and DYC is active. However, the vehicle has not become unstable. Therefore, the torque distribution strategy continues to prioritize the economy. It optimally allocates the total longitudinal force and additional yaw moment. Different from the classical domain method, the target additional yaw moment distribution of DYC needs to be considered. A distribution strategy is proposed that combines the rule-based method with economy-optimal distribution.

Based on the rule of average distribution, the target demanded torques on the left and right sides are calculated as follows:

$$\begin{aligned} T_{dl} &= \tfrac{F_{xd}}{2} R_{eff} - \tfrac{M_{zd}}{d} R_{eff} \\ T_{dr} &= \tfrac{F_{xd}}{2} R_{eff} + \tfrac{M_{zd}}{d} R_{eff} \end{aligned} \tag{55}$$

where T_{dl} and T_{dr} are the left and right demand torques, respectively, which are obtained by considering the generalized longitudinal force and generalized yaw moment.

Since the four in-wheel motors are the same, the torque offline optimization problems on the left and right sides are essentially the same. Therefore, the same economy distribution strategy used in the classical domain can be applied. It involves separately solving for the optimal front axle distribution coefficients for the left and right sides, resulting in the driving torques for each wheel:

$$\begin{cases} T_{fl} = \lambda_{opt,l} T_{dl} \\ T_{rl} = \left(1 - \lambda_{opt,l}\right) T_{dl} \\ T_{fr} = \lambda_{opt,r} T_{dr} \\ T_{rr} = \left(1 - \lambda_{opt,r}\right) T_{dr} \end{cases} \quad (56)$$

where $\lambda_{opt,l}$ and $\lambda_{opt,r}$ are the optimal front axle distribution coefficients of the left and right sides, respectively.

3. Non-domain

In the non-domain, the vehicle is in an unstable state. To meet safety requirements, the torque distribution strategy prioritizes stability as the primary objective, with the economy being secondary. Therefore, a compound distribution control strategy is employed, which combines the optimal distribution of stability and economy. The utilization of tire adhesion coefficients is considered as the optimal control target for stability. Simultaneously, the optimal front-rear axle distribution results of economy serve as constraints for the longitudinal driving force, ensuring a vehicle stability margin and maximizing the improvement in the vehicle's handling stability.

The optimization performance index in this paper is the optimal tire adhesion ratio of four wheels. The objective function for the torque optimization distribution problem is defined as follows:

$$J = \sum_{i=1}^{4} \frac{F_{yij}^2 + F_{xij}^2}{\mu^2 F_{zij}^2} \quad (57)$$

where μ is the road adhesion coefficient, and F_{xij}, F_{yij}, and F_{zij} are the longitudinal force, lateral force, and vertical force of each wheel.

The control variable of the optimization problem is the longitudinal force of each wheel $u = \left[F_{xfl}, F_{xfr}, F_{xrl}, F_{xrr}\right]$. The equality constraints for the optimization problem include front and rear axle driving force constraints and additional yaw moment constraints. The inequality constraints consist of tire friction ellipse constraints and motor actuator constraints.

The total demand driving force of the front and rear axles needs to be the upper-layer longitudinal force requirement. Based on the calculated optimal front axle driving force distribution coefficient, it is expressed as follows:

$$\begin{cases} T_{fl} + T_{fr} = \lambda_{opt} T_{xd} \\ T_{rl} + T_{rr} = \left(1 - \lambda_{opt}\right) T_{xd} \end{cases} \quad (58)$$

At the same time, the driving torque distribution of four wheels needs to meet the calculated demand yaw moment as follows:

$$\begin{cases} T_{fl} + T_{fr} = \lambda_{opt} T_{xd} \\ T_{rl} + T_{rr} = \left(1 - \lambda_{opt}\right) T_{xd} \end{cases} \quad (59)$$

To prevent tire forces from reaching saturation and causing slippage, each wheel must satisfy the tire friction ellipse constraint. Therefore, the driving force of each wheel needs to satisfy the following inequality constraint:

$$\begin{cases} F_{xfl} \le \sqrt{\mu^2 \cdot F_{zfl}^2 - F_{yfl}^2} \\ F_{xfr} \le \sqrt{\mu^2 \cdot F_{zfr}^2 - F_{y,fr}^2} \\ F_{xrl} \le \sqrt{\mu^2 \cdot F_{zrl}^2 - F_{y,rl}^2} \\ F_{xrr} \le \sqrt{\mu^2 \cdot F_{zrr}^2 - F_{y,rr}^2} \end{cases} \quad (60)$$

The torque distribution problem is transformed into a constrained quadratic programming problem. In situations where the optimization problem cannot be solved, it becomes necessary to relax the constraints. Since the yaw moment has a significant effect on vehicle stability, when there is no solution, the torque constraints are converted into performance functions, giving priority to meeting yaw moment requirements. By solving the optimization problem, the optimal driving torques for each wheel can be calculated.

2.5.2. Corner Distribution

The coordinated control layer calculates the target yaw moment for ARS, so the lateral force of the rear axle can be obtained:

$$F_{yr} = \frac{M_{ARS}}{l_r} \tag{61}$$

Based on the built tire model, the inverse tire model of the mapping relationship between tire lateral force and tire slip angle is obtained via transformation. Then, the slip angle at the rear axle center is calculated based on the rear lateral force:

$$\alpha_r = F_y^{-1}(F_{yr}) \tag{62}$$

The equivalent rear wheel angle can be calculated as

$$\delta_r = \frac{V_y - l_r\omega}{V_x} - \alpha_r = \beta - \frac{l_r\omega}{V_x} - \alpha_r \tag{63}$$

Combined with the obtained front wheel angles from lateral path tracking, according to the Ackermann steering principle, the steering angles for each wheel can be determined:

$$\begin{cases} \tan\delta_{fl} = \frac{\tan\delta_f}{1-\frac{d}{2l}(\tan\delta_f-\tan\delta_r)} \\ \tan\delta_{fr} = \frac{\tan\delta_f}{1+\frac{d}{2l}(\tan\delta_f-\tan\delta_r)} \\ \tan\delta_{rl} = \frac{\tan\delta_r}{1-\frac{d}{2l}(\tan\delta_f-\tan\delta_r)} \\ \tan\delta_{rr} = \frac{\tan\delta_r}{1+\frac{d}{2l}(\tan\delta_f-\tan\delta_r)} \end{cases} \tag{64}$$

where δ_{ij} is the angle of each wheel; α_{ij} is the side slip angle of each wheel; and δ_f and δ_r are the equivalent angles of the front and rear wheels, respectively.

3. Simulation and Results

3.1. Environment and Configuration

In this section, we develop a simulation model using the CarSim 2019 software produced by MSC (USA) and MATLAB R2022a software produced by MathWorks (Natick, MA, USA), as shown in Figure 11. The vehicle model employed in this study is based on the CarSim vehicle model. To evaluate the effectiveness of the control strategy on different road surfaces, two different driving conditions, namely the slalom test and double-lane change, are simulated. The slalom test is conducted on a high-adhesion road surface, while the double-lane change is performed on a low-adhesion road surface. By comparing the experimental results of AFS, AFS + DYC, 4WS, and coordinated control, we analyze and verify the efficacy of the proposed control strategy. The AFS strategy involves only MPC trajectory tracking control. The AFS + DYC strategy is a strategy that adds DYC control on the basis of AFS and sets the control weight of DYC to 1. The 4WS strategy is a strategy that adds ARS control on the basis of AFS and sets the ARS control weight to 1. Coordinated control is the strategy based on the extension dynamic stability region proposed in this article.

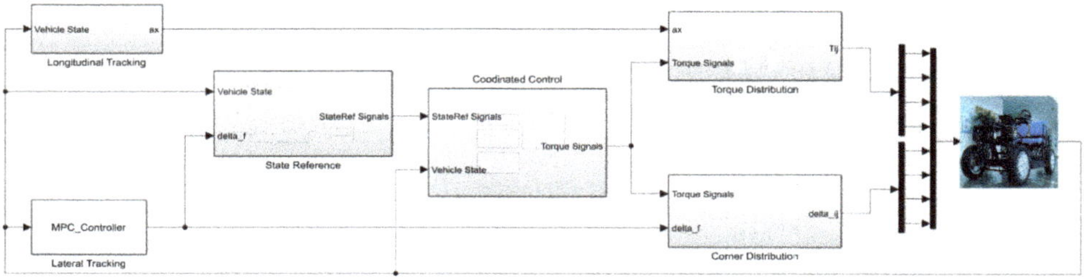

Figure 11. The 4WID-4WIS simulation model.

To further validate the effects of the coordinated control strategy on economy optimization, this study compared it with the equal torque distribution strategy. The equal torque distribution strategy is commonly used in practical engineering due to its simplicity and efficiency. In this strategy, the generalized longitudinal force and generalized yaw moment output from the upper layer are evenly distributed to each wheel. The torque distribution for each wheel can be expressed as follows:

$$\begin{cases} T_{fl} = T_{rl} = \frac{F_{xd}}{4} R_{eff} - \frac{M_{zd}}{2d} R_{eff} \\ T_{fr} = T_{rr} = \frac{F_{xd}}{4} R_{eff} + \frac{M_{zd}}{2d} R_{eff} \end{cases} \tag{65}$$

3.2. *Results and Analysis*

3.2.1. Slalom Test

The vehicle speed is set to 80 km/h, the road adhesion coefficient is set to 0.8, and the simulation time is set to 20 s. The proposed coordinated control strategy is verified from three aspects: trajectory tracking results, handling stability results, and economy results.

The longitudinal tracking results are presented in Figure 12. During the slalom test, the longitudinal velocity tracking based on the expert PID performs well in tracking the desired velocity, with a tracking error of less than 0.5 km/h. This meets the accuracy requirements for longitudinal tracking. Moreover, when compared with the control results of AFS, the longitudinal tracking results of AFS + DYC and coordinated control exhibit significant fluctuations and relatively reduced tracking accuracy.

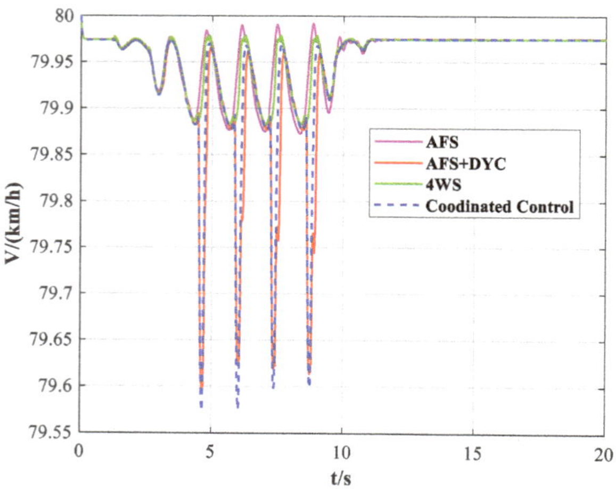

Figure 12. Velocity tracking results under slalom test.

The results of the lateral tracking are presented in Figure 13 and Table 1. The AFS strategy, based on lateral path tracking using MPC, demonstrates good tracking performance with minimal errors, meeting the accuracy requirements. Comparing the lateral offsets and heading angle errors under different strategies, it is observed that AFS + DYC, 4WS, and coordinated control outperform AFS. Among these, 4WS exhibits the best lateral displacement tracking performance, with a maximum error of 0.1053 m and an RMS value of 0.0435 m. Coordinated control closely follows, with a maximum error of 0.1062 m and an RMS value of 0.0445 m. In terms of heading angle tracking, coordinated control performs the best, with a maximum error of 0.0286 rad and an RMS value of 0.0067 rad, followed by 4WS, with a maximum error of 0.0302 rad and an RMS value of 0.0070 rad. These results indicate that 4WS and the proposed coordinated control strategy have the best tracking performance, while AFS + DYC shows relatively poorer tracking performance but still outperforms the pure tracking control strategy of AFS. In summary, the lateral tracking results can be summarized as follows: 4WS ≈ coordinated control > AFS + DYC > AFS.

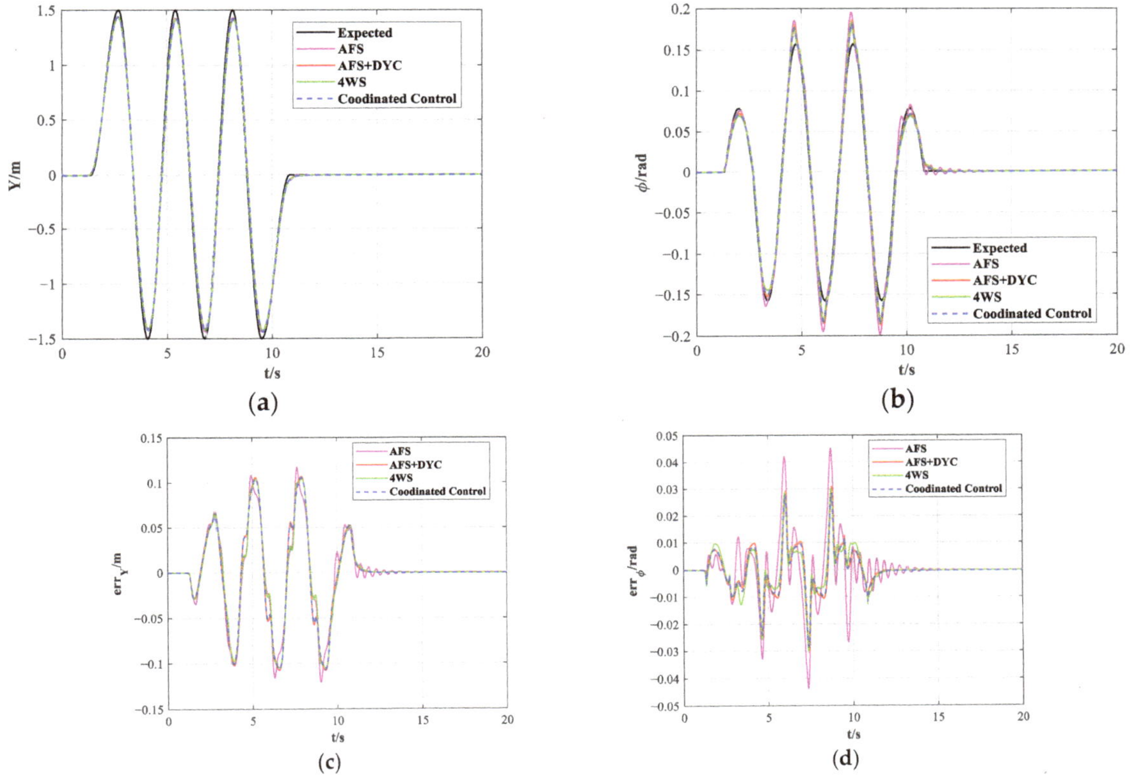

Figure 13. Path tracking results under slalom test: (**a**) results of lateral displacement; (**b**) results of heading angle; (**c**) results of lateral offset; (**d**) results of heading angle error.

Table 1. Comparison of path tracking errors under slalom test.

Control Strategy	Lateral Offset (m)		Heading Angle Error (Rad)	
	Max	RMS	Max	RMS
AFS	0.1206	0.0439	0.0453	0.0107
AFS + DYC	0.1076	0.0453	0.0308	0.0071
4WS	0.1053	0.0435	0.0302	0.0070
Coordinated Control	0.1062	0.0445	0.0286	0.0067

The handling stability results are presented in Figures 14 and 15 and Table 2. During the slalom test, AFS + DYC, 4WS, and coordinated control all exhibit better handling stability compared to the simulation results of AFS. Notably, coordinated control demonstrates the highest level of handling stability, with a maximum yaw rate error of 0.2037 rad/s and an RMS value of 0.0258 rad/s. The maximum sideslip angle error is 0.0326 rad, with an RMS value of 0.0108 rad. In summary, the handling stability results can be ranked as follows: coordinated control > AFS + DYC > 4WS > AFS.

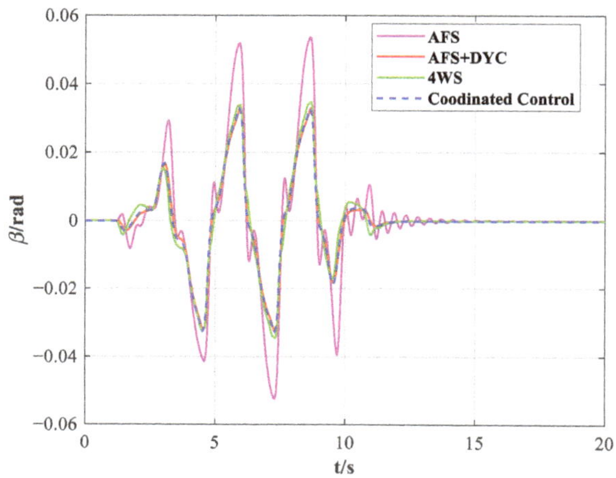

Figure 14. Side slip angle under slalom test.

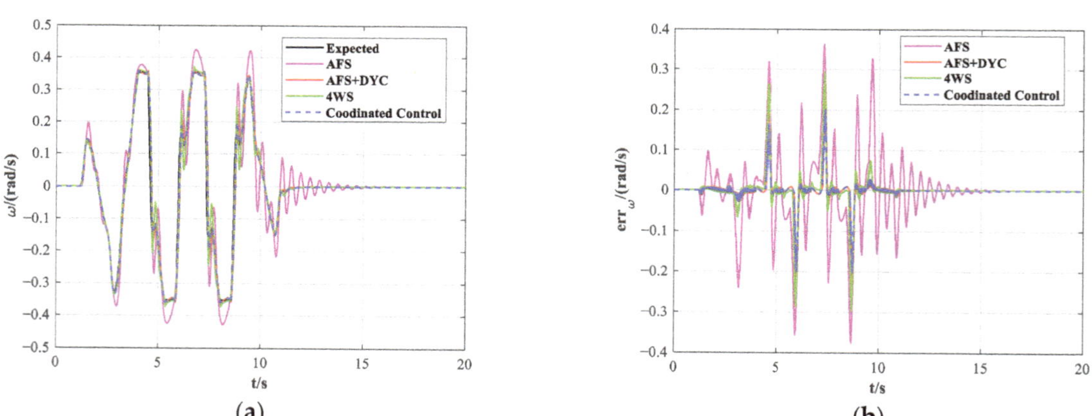

Figure 15. Yaw rate results under slalom test: (**a**) results of yaw rate; (**b**) results of yaw rate error.

Table 2. Comparison of handling stability errors under slalom test.

Control Strategy	Yaw Rate Error (Rad/s)		Sideslip Angle Error (Rad)	
	Max	RMS	Max	RMS
AFS	0.3752	0.0878	0.0536	0.0171
AFS + DYC	0.2393	0.0306	0.0328	0.0112
4WS	0.2957	0.0433	0.0347	0.0118
Coordinated Control	0.2037	0.0258	0.0326	0.0108

Under the slalom test, comparing the path tracking results and handling stability results of AFS + DYC and 4WS, it can be seen that the path tracking results of 4WS are superior to those of AFS + DYC, while the handling stability results are worse. Therefore, the 4WS strategy can achieve better trajectory tracking performance than the AFS + DYC strategy but sacrifices a certain degree of handling stability accuracy. Conversely, the AFS + DYC strategy exhibits the opposite trend.

In order to take advantage of both, the coordinated controller adjusts the control weights of the drive and steering subsystems, as shown in Figure 16. When the vehicle is still in the linear region, the controller maximizes the effect of tire lateral forces, with 4WS being the primary contributor. As the vehicle gradually deviates from the linear region and tire lateral forces approach saturation, the weight of DYC increases. This explains why the coordinated controller can achieve improved trajectory tracking control while enhancing handling stability.

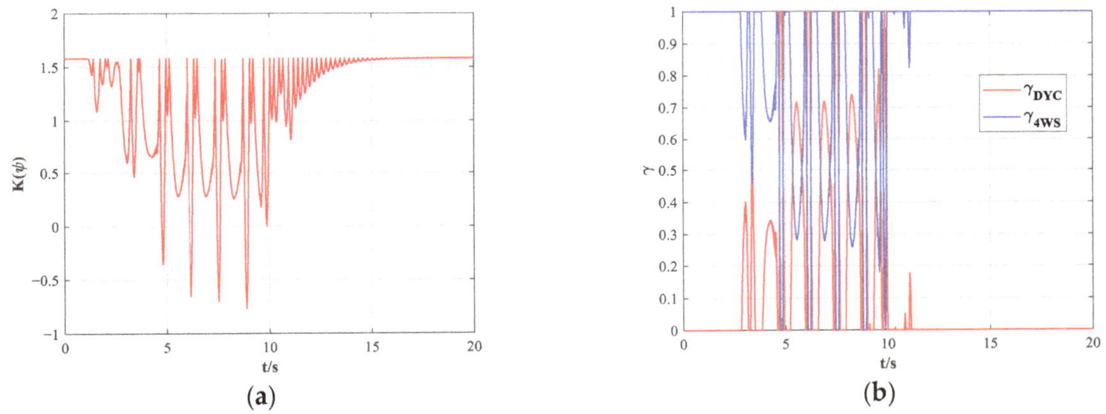

Figure 16. Coordinated control results under slalom test: (**a**) results of correlation function; (**b**) results of control weights.

In this study, the simulation results are compared between the equal distribution strategy and the proposed compound distribution strategy using the system's comprehensive efficiency and battery energy consumption as economic evaluation indicators. Figure 17 shows that under the slalom test, the system comprehensive efficiency of the proposed compound distribution strategy is superior to that of the equal distribution strategy. Additionally, the energy consumption of the battery is reduced from 240.8691 kJ to 234.4290 kJ, representing a 2.67% decrease when using the compound distribution strategy. This confirms the effectiveness of the proposed compound distribution strategy in terms of energy consumption economy.

3.2.2. Double-Lane Change

The vehicle speed is set to 60 km/h, the road adhesion coefficient is set to 0.4, and the simulation time is set to 10 s. The proposed coordinated control strategy is verified from three aspects: trajectory tracking results, handling stability results, and economy results.

The longitudinal tracking results are shown in Figure 18. Under the double-lane change scenarios, longitudinal velocity tracking based on the performs well in tracking the desired velocity, with a tracking error of less than 0.2 km/h. It meets the accuracy requirements for longitudinal tracking. In addition, comparing the tracking results under different strategies, the longitudinal tracking results are basically the same.

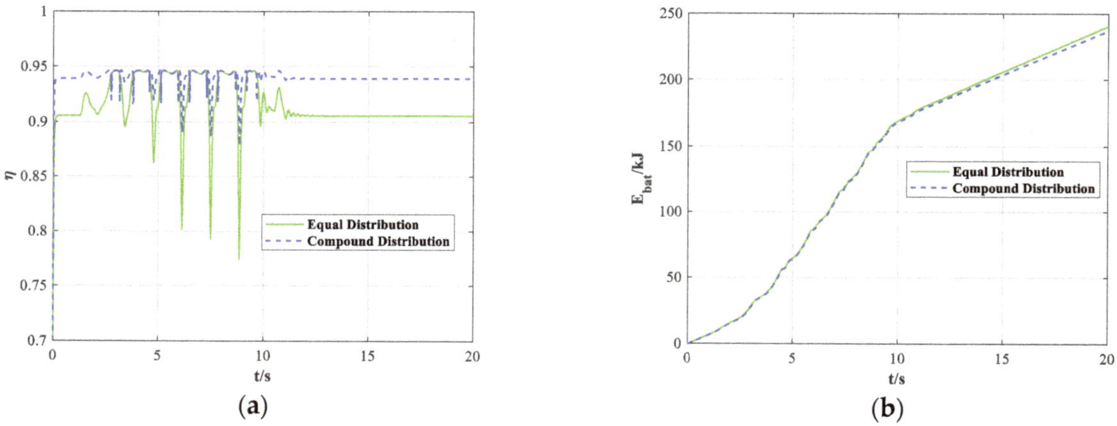

Figure 17. Economic simulation results under slalom test: (**a**) results of system comprehensive efficiency; (**b**) results of battery energy consumption.

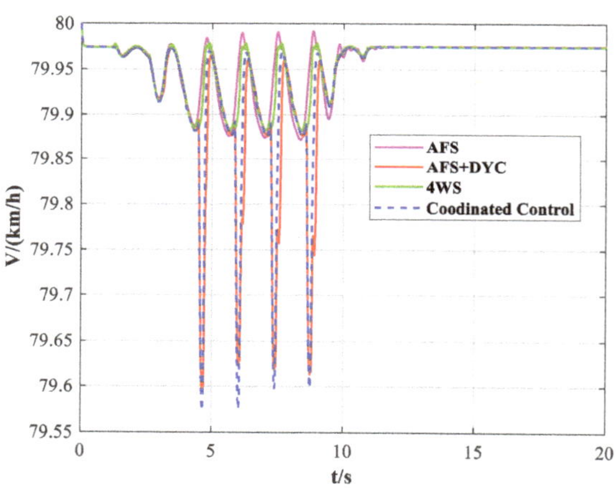

Figure 18. Velocity tracking results under double-lane change.

The lateral tracking results are presented in Figure 19 and Table 3. In the double-lane change scenarios, the AFS strategy demonstrates effective lateral path tracking based on MPC, even on low-adhesion roads. When comparing the lateral offsets and heading angle errors among different strategies, it is evident that AFS + DYC, 4WS, and coordinated control outperform AFS in terms of lateral tracking results. Among these strategies, 4WS exhibits the best lateral displacement tracking and heading angle tracking effects, with a maximum lateral offset of 0.0811 m and an RMS value of 0.0320 m. The maximum heading angle error is 0.0273 rad, with an RMS value of 0.0127 rad. Coordinated control closely follows, with a maximum lateral offset of 0.0814 m and an RMS value of 0.0321 m. The maximum heading angle error is 0.0276 rad, with an RMS value of 0.0127 rad. These results indicate that 4WS demonstrates the highest tracking performance, followed by coordinated control. AFS + DYC shows relatively poorer tracking performance but still outperforms the pure tracking control strategy of AFS. In summary, the ranking of the lateral tracking results is as follows: 4WS > coordinated control > AFS + DYC > AFS.

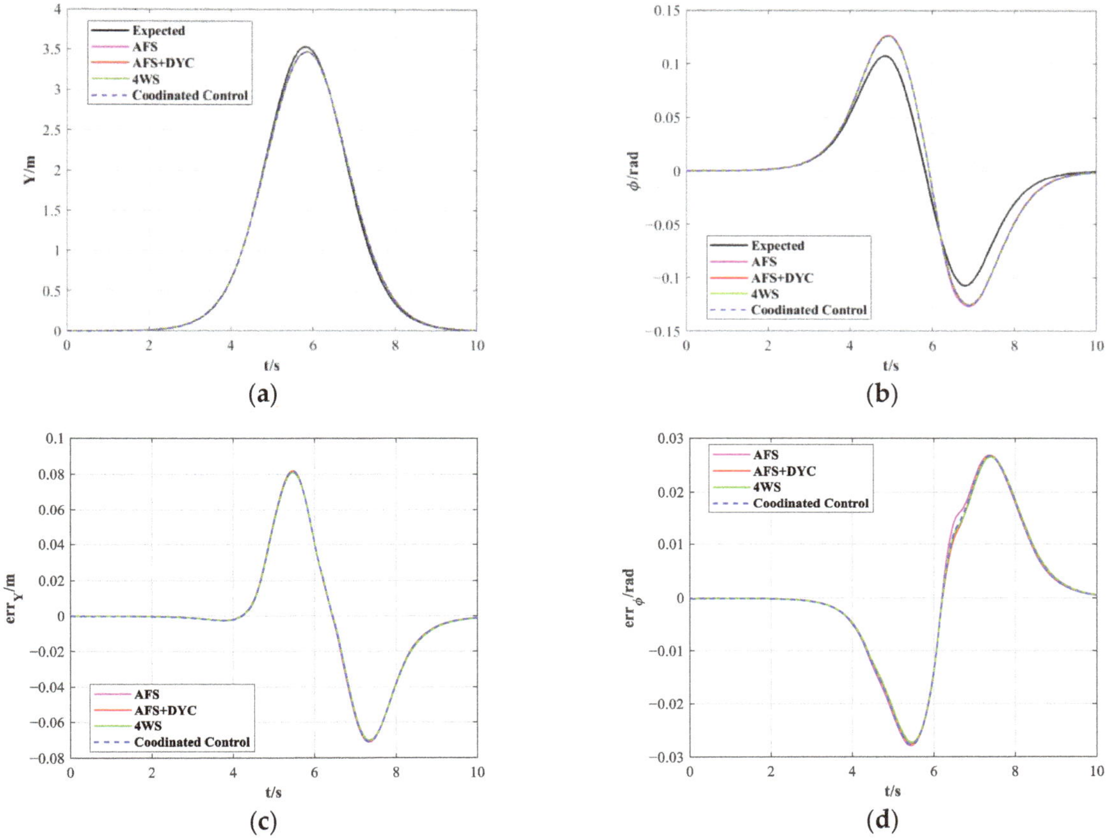

Figure 19. Path tracking results under double-lane change: (**a**) results of lateral displacement; (**b**) results of heading angle; (**c**) results of lateral offset; (**d**) results of heading angle error.

Table 3. Comparison of path tracking errors under double-lane change.

Control Strategy	Lateral Offset (m)		Heading Angle Error (Rad)	
	Max	RMS	Max	RMS
AFS	0.0821	0.0324	0.0278	0.0130
AFS + DYC	0.0819	0.0322	0.0276	0.0128
4WS	0.0811	0.0320	0.0273	0.0127
Coordinated Control	0.0814	0.0321	0.0276	0.0127

The handling stability results are presented in Figures 20 and 21, and Table 4. In the double-lane change scenarios, AFS + DYC, 4WS, and coordinated control all exhibit superior handling stability compared to the simulation results of AFS. Among these, coordinated control demonstrates the highest level of handling stability, with a maximum yaw rate error of 0.0597 rad/s and an RMS value of 0.0115 rad/s. The maximum sideslip angle error is 0.032 rad, with an RMS value of 0.014 rad. In summary, the handling stability results can be ranked as follows: coordinated control > AFS + DYC > 4WS > AFS.

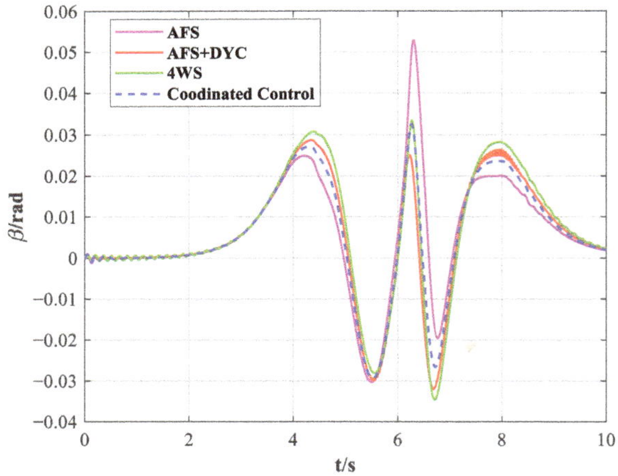

Figure 20. Side slip angle under double-lane change.

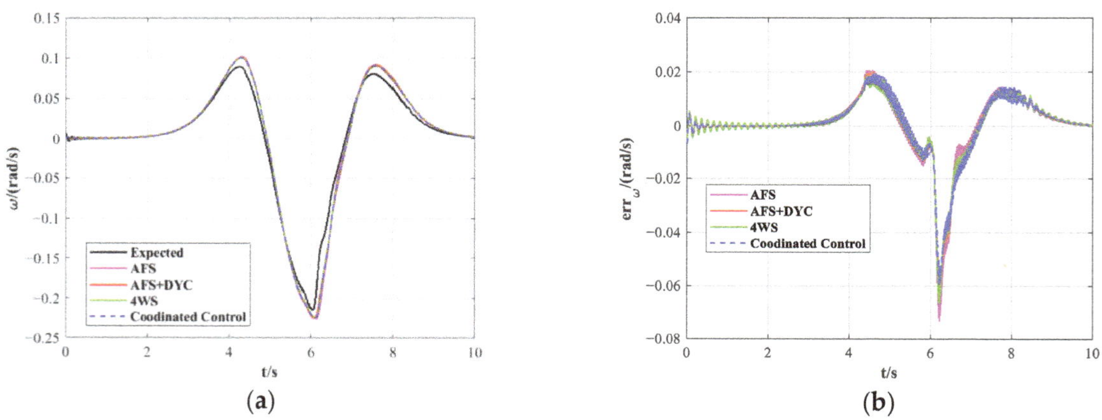

Figure 21. Yaw rate results under double-lane change: (**a**) results of yaw rate; (**b**) results of yaw rate error.

Table 4. Comparison of handling stability errors under double-lane change.

Control Strategy	Yaw Rate Error (Rad/s)		Sideslip Angle Error (Rad)	
	Max	RMS	Max	RMS
AFS	0.3752	0.0878	0.0536	0.0171
AFS + DYC	0.2393	0.0306	0.0328	0.0112
4WS	0.2957	0.0433	0.0347	0.0118
Coordinated Control	0.2037	0.0258	0.0326	0.0108

In the context of double-lane change scenarios, a comparison was made between the path tracking results and handling stability results of the AFS + DYC and 4WS strategies. It was observed that the 4WS strategy outperforms in terms of path tracking, while the AFS + DYC strategy exhibits better handling stability. These findings align with the results obtained from the slalom test.

To achieve both trajectory tracking control and good handling stability, the coordinated controller adjusts the control weights, as depicted in Figure 22. A trade-off is made between the two objectives. Due to a lower road adhesion coefficient, the friction ellipse of the tires becomes smaller, leading to a higher likelihood of lateral forces reaching saturation. To address this, the coordinated controller relies more on DYC control and adjusts the longitudinal forces of the tires to enhance stability. Consequently, the lateral path tracking performance under the coordinated control strategy is slightly inferior to that of 4WS, but the handling stability surpasses other strategies.

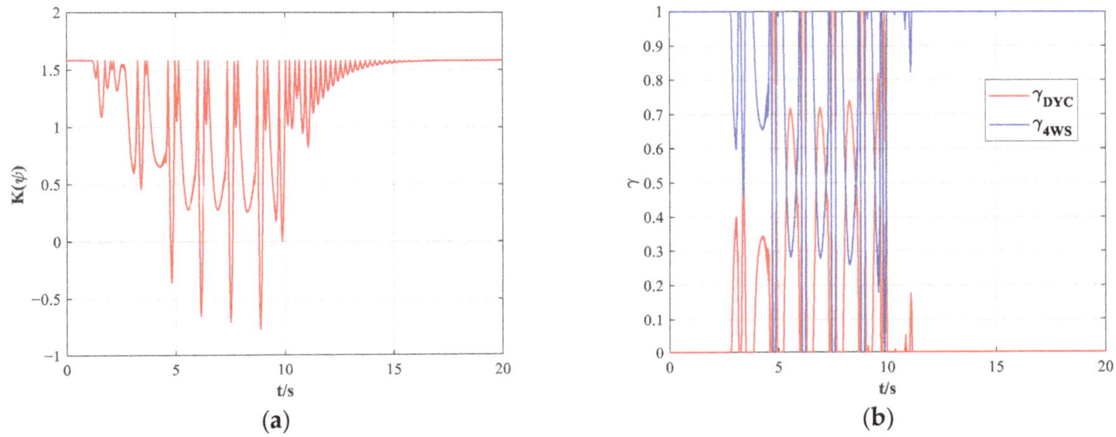

Figure 22. Coordinated control results under double-lane change: (**a**) results of correlation function; (**b**) results of control weights.

The economic simulation results for the double-lane change scenario are presented in Figure 23. The comprehensive efficiency of the proposed compound distribution strategy is higher than that of the equal distribution strategy. Additionally, the energy consumption of the battery is 39.901 kJ under the equal distribution strategy, whereas it is reduced to 38.2123 kJ under the compound distribution strategy, resulting in a 4.23% decrease in energy consumption. These findings confirm the effectiveness of the proposed compound distribution strategy in terms of energy consumption economy.

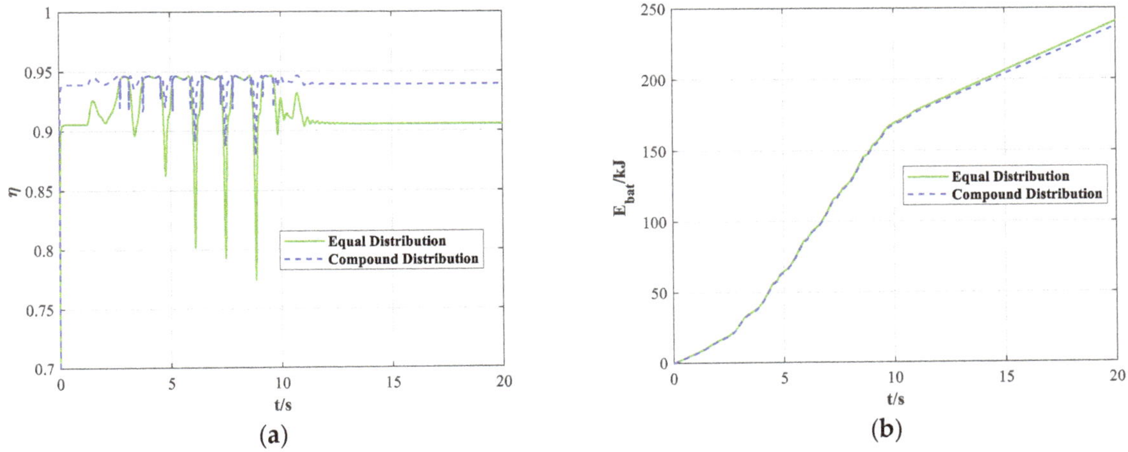

Figure 23. Economic simulation results under double-lane change: (**a**) results of system comprehensive efficiency; (**b**) results of battery energy consumption.

4. Discussion

Comparing the slalom test results on a high-adhesion road and the double-lane change results on a low-adhesion road, the following conclusions can be summarized:

- The designed PID-based longitudinal velocity tracking controller and MPC-based lateral path tracking controller in this study both achieve good tracking performance, meeting the accuracy requirements of trajectory tracking control.
- When comparing the 4WS and AFS + DYC strategies, the 4WS strategy provides better path tracking performance, while the AFS + DYC strategy offers better handling stability.
- Stability and path tracking accuracy interact with each other. Improved stability leads to a decrease in path tracking accuracy, and the increase in path tracking accuracy reduces vehicle stability. The proposed coordinated control strategy maximizes the advantages of the steering and drive subsystems. By combining 4WS with DYC, it comprehensively controls stability and path tracking accuracy. This strategy can achieve better trajectory tracking control while improving the vehicle's handling stability, realizing coordinated control with dual objectives.
- The proposed compound torque distribution strategy can enhance the vehicle's economy while maintaining trajectory tracking performance and stability.

To further clarify the characteristics of the proposed control strategy in this study, a comparison is made with chassis coordinated control strategies proposed in References [24,25]. Both references utilized chassis coordinated control strategies that were based on extension theory and dynamically adjusted the control weights for the driving and steering subsystems. Hence, by combining the findings from these two studies, an analysis is conducted from two perspectives, as shown in Table 5.

Table 5. Comparison of different control strategies.

Control Strategy	Stability Domain Division		Control Weights	
	Extension Domain	Non-Domain	System	Strategy
Ref [24]	Phase diagram boundaries	Scaling factor	AFS + DYC	Correlation function
Ref [25]	Tire's linear zone	Phase diagram boundaries	AFS + DYC	Game theory
Proposed strategy	Vehicle's linear zone	Phase diagram boundaries	4WS + DYC	Correlation function

- Division of Stability Domain:

Reference [24] selected traditional phase diagram boundaries as the extension domain boundaries and expanded them to obtain non-domain boundaries using a fixed scaling factor. However, this strategy lacks a theoretical basis and cannot be adjusted according to external conditions, resulting in overly conservative extension domain boundaries and potential instability. In contrast, Reference [25] utilized the tire's linear zone limit as the extension domain boundary and the traditional phase diagram boundary as the non-domain boundary. However, this approach cannot accurately determine the critical angle of the tire's linear zone and cannot adaptively adjust to external environmental conditions. In this paper, we propose a strategy that defines the vehicle's linear zone as the extension domain boundary and the traditional phase diagram boundary as the non-domain boundary. Both extension domain and non-domain boundaries can be adaptively adjusted based on the vehicle's speed and road adhesion conditions, enabling a more accurate assessment of the vehicle's lateral motion state.

- Adaptive Adjustment of Control Weights:

Reference [24] developed allocation coefficients for AFS and DYC control weights based on the correlation function of extension theory. In a similar vein, Reference [25] optimized the control allocation coefficients for AFS and DYC control weights using a combination of correlation functions and game theory. In contrast, this paper's control strategy directly determines control weights for 4WS and DYC based on the correlation function. While References [24,25] focused on the control weights of the AFS system, our study takes into account the impact of ARS on vehicle lateral motion, fully utilizing the steering system's role in handling stability. Furthermore, both Reference [24] and our strategy design control weights based on the correlation function, which is a simple, efficient, and practical approach. However, it heavily relies on the precise boundary definition of the stability domain. On the other hand, Reference [25] incorporates game theory to dynamically optimize the weights of the driving and steering systems, potentially leading to superior control outcomes. Nonetheless, this approach involves more adjustment parameters, increases computational complexity, and reduces real-time performance, posing challenges for practical engineering applications.

By considering both aspects, the control strategy proposed in this paper can dynamically adjust the boundaries of the stability domain, resulting in a more accurate assessment of the vehicle's lateral motion state. Additionally, the allocation strategy for control weights is relatively simple and, when combined with more precise stability domain boundaries, can yield excellent control results.

5. Conclusions

This paper presents a study on the coordinated control problem of 4WID-4WIS electric vehicles, focusing on trajectory tracking performance, handling stability, and economy. Additionally, it addresses the coupling allocation issue between the driving and steering subsystems. To tackle these challenges, the paper proposes a hierarchical chassis control architecture. The expert PID algorithm is used for longitudinal velocity tracking, while the MPC algorithm is employed for lateral path tracking. To achieve optimal control of trajectory tracking performance and handling stability, a sliding mode controller is utilized to calculate the additional yaw moment. Additionally, a coordinated control strategy based on the extension dynamic stability domain is designed, where the control weights of DYC and ARS are adaptively adjusted to output the optimal control objectives. Taking into consideration the demand for energy efficiency, the optimal wheel drive torques are calculated using MPSO and QP with driving efficiency and tire adhesion ratio as control objectives. Simulation test results demonstrate that the coordinated control strategy improves trajectory tracking performance, handling stability, and economy, achieving multi-objective coordinated control of the chassis. This method has high real-time performance, simple implementation, and strong adaptability to working conditions, laying the foundation for further research on chassis-integrated control.

In future work, vehicle tests will be conducted to validate the effectiveness of the control strategy. Furthermore, the impact of vertical dynamics will be considered, and a deeper investigation into the coupling characteristics between subsystems will be carried out to achieve multi-dimensional dynamic optimization control of the chassis subsystems.

Author Contributions: Conceptualization, Y.Q. and D.Y.; Validation, Y.Q., and D.Y.; Investigation, Y.Q.; Data curation, D.Y.; Supervision, X.C.; Writing—original draft preparation, Y.Q. All authors have read and agreed to the published version of the manuscript.

Funding: This research was funded by the Jiangsu Provincial Science and Technology Project, grant number BE2021006-3, and the National Natural Science Foundation of China, grant number 52275123.

Data Availability Statement: Data are contained within the article.

Conflicts of Interest: Author Dongxiao Yin was employed by the company Tongji Automobile Design and Research Institute Co. The remaining authors declare that the research was conducted in the absence of any commercial or financial relationships that could be construed as a potential conflict of interest.

References

1. He, S.W.; Fan, X.B.; Wang, Q.W.; Chen, X.B.; Zhu, S.W. Review on Torque Distribution Scheme of Four-Wheel In-Wheel Motor Electric Vehicle. *Machines* **2022**, *10*, 619. [CrossRef]
2. Peng, H.; Chen, X.B. Active Safety Control of X-by-Wire Electric Vehicles: A Survey. *SAE Int. J. Veh. Dyn. Stab. NVH* **2022**, *6*, 115–133. [CrossRef]
3. Zhang, L.; Zhang, Z.Q.; Wang, Z.P.; Deng, J.J.; Dorrell, D.G. Chassis Coordinated Control for Full X-by-Wire Vehicles—A Review. *Chin. J. Mech. Eng.* **2021**, *34*, 42. [CrossRef]
4. Hang, P.; Chen, X.B. Towards Autonomous Driving: Review and Perspectives on Configuration and Control of Four-Wheel Independent Drive/Steering Electric Vehicles. *Actuators* **2021**, *10*, 184. [CrossRef]
5. Song, P.; Tomizuka, M.; Zong, C.F. A novel integrated chassis controller for full drive-by-wire vehicles. *Veh. Syst. Dyn.* **2015**, *53*, 215–236. [CrossRef]
6. Lai, F.; Huang, C.Q.; Jiang, C.Y. Comparative Study on Bifurcation and Stability Control of Vehicle Lateral Dynamics. *SAE Int. J. Veh. Dyn. Stab. NVH* **2022**, *6*, 35–52. [CrossRef]
7. Peng, H.N.; Wang, W.D.; Xiang, C.L.; Li, L.; Wang, X.Y. Torque Coordinated Control of Four In-Wheel Motor Independent-Drive Vehicles With Consideration of the Safety and Economy. *IEEE Trans. Veh. Technol.* **2019**, *68*, 9604–9618. [CrossRef]
8. Xu, F.X.; Liu, X.H.; Chen, W.; Zhou, C. Dynamic Switch Control of Steering Modes for Four Wheel Independent Steering Rescue Vehicle. *IEEE Access* **2019**, *7*, 135595–135605. [CrossRef]
9. Zhang, B.H.; Lu, S.B.; Zhao, L.; Xiao, K.X. Fault-tolerant control based on 2D game for independent driving electric vehicle suffering actuator failures. *Proc. Inst. Mech. Eng. Part D J. Automob. Eng.* **2020**, *234*, 3011–3025. [CrossRef]
10. Wang, J.N.; Gao, S.L.; Wang, K.; Wang, Y.; Wang, Q.S. Wheel torque distribution optimization of four-wheel independent-drive electric vehicle for energy efficient driving. *Control Eng. Pract.* **2021**, *110*, 104779. [CrossRef]
11. Li, D.F.; Liu, A.; Pan, H.; Chen, W.T. Safe, Efficient and Socially-Compatible Decision of Automated Vehicles: A Case Study of Unsignalized Intersection Driving. *Automot. Innov.* **2023**, *6*, 281–296. [CrossRef]
12. Zhang, X.; Li, J.; Ma, Z.; Chen, D.; Zhou, X. Lateral Trajectory Tracking of Self-Driving Vehicles Based on Sliding Mode and Fractional-Order Proportional-Integral-Derivative Control. *Actuators* **2023**, *13*, 7. [CrossRef]
13. Liu, K.; Wen, G.; Fu, Y.; Wang, H. A Hierarchical Lane-Changing Trajectory Planning Method Based on the Least Action Principle. *Actuators* **2023**, *13*, 10. [CrossRef]
14. Tong, Y.W.; Li, C.; Wang, G.; Jing, H. Integrated Path-Following and Fault-Tolerant Control for Four-Wheel Independent-Driving Electric Vehicles. *Automot. Innov.* **2022**, *5*, 311–323. [CrossRef]
15. Hang, P.; Xia, X.; Chen, G.; Chen, X. Active Safety Control of Automated Electric Vehicles at Driving Limits: A Tube-Based MPC Approach. *IEEE Trans. Transp. Electrif.* **2022**, *8*, 1338–1349. [CrossRef]
16. Guo, N.; Zhang, X.; Zou, Y. Real-Time Predictive Control of Path Following to Stabilize Autonomous Electric Vehicles Under Extreme Drive Conditions. *Automot. Innov.* **2022**, *5*, 453–470. [CrossRef]
17. Ren, Y.; Zheng, L.; Khajepour, A. Integrated model predictive and torque vectoring control for path tracking of 4-wheel-driven autonomous vehicles. *IET Intell. Transp. Syst.* **2019**, *13*, 98–107. [CrossRef]
18. Hu, C.; Wang, R.; Yan, F.; Chen, N. Output Constraint Control on Path Following of Four-Wheel Independently Actuated Autonomous Ground Vehicles. *IEEE Trans. Veh. Technol.* **2016**, *65*, 4033–4043. [CrossRef]
19. Mousavinejad, E.; Han, Q.L.; Yang, F.W.; Zhu, Y.; Vlacic, L. Integrated control of ground vehicles dynamics via advanced terminal sliding mode control. *Veh. Syst. Dyn.* **2017**, *55*, 268–294. [CrossRef]
20. Mirzaeinejad, H.; Mirzaei, M.; Rafatnia, S. A novel technique for optimal integration of active steering and differential braking with estimation to improve vehicle directional stability. *ISA Trans.* **2018**, *80*, 513–527. [CrossRef]
21. Zhang, H.; Li, X.S.; Shi, S.M.; Liu, H.F.; Guan, R.; Liu, L. Phase Plane Analysis for Vehicle Handling and Stability. *Int. J. Comput. Intell. Syst.* **2011**, *4*, 1179–1186. [CrossRef]
22. Liang, Y.; Li, Y.; Yu, Y.; Zheng, L. Integrated lateral control for 4WID/4WIS vehicle in high-speed condition considering the magnitude of steering. *Veh. Syst. Dyn.* **2019**, *58*, 1711–1735. [CrossRef]
23. Tian, J.; Wang, Q.; Ding, J.; Wang, Y.Q.; Ma, Z.S. Integrated Control With DYC and DSS for 4WID Electric Vehicles. *IEEE Access* **2019**, *7*, 124077–124086. [CrossRef]
24. Chen, W.W.; Liang, X.T.; Wang, Q.D.; Zhao, L.F.; Wang, X. Extension coordinated control of four wheel independent drive electric vehicles by AFS and DYC. *Control Eng. Pract.* **2020**, *101*, 104504. [CrossRef]
25. Zheng, Z.C.; Zhao, X.; Wang, S.; Yu, Q.; Zhang, H.C.; Li, Z.K.; Chai, H.; Han, Q. Extension coordinated control of distributed-driven electric vehicles based on evolutionary game theory. *Control Eng. Pract.* **2023**, *137*, 105583. [CrossRef]
26. Liu, R.Q.; Wei, M.X.; Zhao, W.Z. Trajectory tracking control of four wheel steering under high speed emergency obstacle avoidance. *Int. J. Veh. Des.* **2018**, *77*, 1–21. [CrossRef]
27. Xu, N.; Zhou, J.F.; Li, X.Y.; Li, F. Analysis of the Effect of Inflation Pressure on Vehicle Handling and Stability under Combined Slip Conditions Based on the UniTire Model. *SAE Int. J. Veh. Dyn. Stab. NVH* **2021**, *5*, 259–277. [CrossRef]
28. Yu, Z.; Wang, J. Automatic Vehicle Trajectory Tracking Control with Self-calibration of Nonlinear Tire Force Function. In Proceedings of the 2017 American Control Conference (ACC), Seattle, WA, USA, 24–26 May 2017; pp. 985–990.
29. Attia, R.; Orjuela, R.; Basset, M. Combined longitudinal and lateral control for automated vehicle guidance. *Veh. Syst. Dyn.* **2014**, *52*, 261–279. [CrossRef]

30. Qiao, Y.R.; Chen, X.B.; Liu, Z. Trajectory Tracking Coordinated Control of 4WID-4WIS Electric Vehicle Considering Energy Consumption Economy Based on Pose Sensors. *Sensors* **2023**, *23*, 5496. [CrossRef] [PubMed]
31. Wu, X.J.; Zhou, B.; Wen, G.L.; Long, L.F.; Cui, Q.J. Intervention criterion and control research for active front steering with consideration of road adhesion. *Veh. Syst. Dyn.* **2018**, *56*, 553–578. [CrossRef]
32. Wong, A.; Kasinathan, D.; Khajepour, A.; Chen, S.K.; Litkouhi, B. Integrated torque vectoring and power management framework for electric vehicles. *Control Eng. Pract.* **2016**, *48*, 22–36. [CrossRef]
33. Marini, F.; Walczak, B. Particle swarm optimization (PSO). A tutorial. *Chemom. Intell. Lab. Syst.* **2015**, *149*, 153–165. [CrossRef]

actuators

MDPI

Article

Adaptive Fuzzy Power Management Strategy for Extended-Range Electric Logistics Vehicles Based on Driving Pattern Recognition

Changyin Wei [1,*], Xiaodong Wang [2], Yunxing Chen [1], Huawei Wu [1] and Yong Chen [3,4]

[1] Hubei Key Laboratory of Power System Design and Test for Electrical Vehicle, School of Automotive and Traffic Engineering, Hubei University of Arts and Science, Xiangyang 441053, China; chenyunxing@chd.edu.cn (Y.C.); whw_xy@hbuas.edu.cn (H.W.)
[2] School of Mechanical Automation, Wuhan University of Science and Technology, Wuhan 430081, China; 202203601016@wust.edu.cn
[3] School of Mechanical Engineering, Guangxi University, Nanning 530004, China; chenyong1585811@163.com
[4] Tianjin Key Laboratory of New Energy Automobile Power Transmission and Safety Technology, School of Mechanical Engineering, Hebei University of Technology, Tianjin 300130, China
* Correspondence: weiag2@163.com

Abstract: The primary objective of an energy management strategy is to achieve optimal fuel economy through proper energy distribution. The adoption of a fuzzy energy management strategy is hindered due to different reasons, such as uncertainties surrounding its adaptability and sustainability compared to conventional energy control methods. To address this issue, a fuzzy energy management strategy based on long short-term memory neural network driving pattern recognition is proposed. The time-frequency characteristics of vehicle speed are obtained using the Hilbert–Huang transform method. The multi-dimensional features are composed of the time-frequency features of vehicle speed and the time-domain signals of the accelerator pedal and brake pedal. A novel driving pattern recognition approach is designed using a long short-term memory neural network. A dual-input and single-output fuzzy controller is proposed, which takes the required power of the vehicle and the state of charge of the battery as the input, and the comprehensive power of the range extender as the output. The parameters of the fuzzy controller are selected according to the category of driving pattern. The results show that the fuel consumption of the method proposed in this paper is 5.8% lower than that of the traditional fuzzy strategy, and 4.2% lower than the fuzzy strategy of the two-dimensional feature recognition model. In general, the proposed EMS can effectively improve the fuel consumption of extended-range electric vehicles.

Keywords: power management strategy; driving pattern recognition; fuzzy control; extended-range electric logistics vehicle

1. Introduction

The ongoing evolution of vehicle powertrain is primarily associated with the transition from conventional fossil energy to new energy sources such as hydrogen and electricity, impacting environmental considerations [1,2]. This transition necessitates innovative solutions to deal with the challenge arising from long-distance operations without frequent recharging. The growing concern for environmental sustainability and the need to reduce greenhouse gas emissions have led to the increased development and adoption of electric vehicles (EVs) [3]. In particular, extended-range electric logistics vehicles (ERELVs) have gained attention due to their potential to address the limitations of pure electric vehicles by incorporating an auxiliary power source, such as an internal combustion engine or a fuel cell. The integration of these power sources allows for an extended driving range and improved operational flexibility.

ERELVs offer increased flexibility by incorporating multiple power sources to meet their energy requirements. These sources include a combination of a diesel engine, an electric motor with a battery, and a fuel cell. Moreover, ERELVs utilize regenerative braking technology to generate electricity during deceleration, which can be stored in the battery. Consequently, ERELVs enhance fuel efficiency and minimize exhaust emissions [4,5]. However, due to the diverse range of energy sources in ERELVs, it is crucial to develop energy management strategies (EMSs) that enable the intelligent coordination and optimization of these sources. The objective is to achieve optimal fuel economy by monitoring the engine's operation within its optimal range, prolonging battery life, reducing CO_2 emissions, and effectively adapting to uncertainties and complex driving conditions [6,7]. To accomplish these objectives, the EMSs in ERELVs require access to real-time information about the driving conditions. This information can be obtained from various sources such as internet maps, global positioning systems (GPS), geographical information systems (GIS), or intelligent transportation systems (ITS) [8]. By integrating these data sources into the EMSs, ERELVs can make informed decisions regarding power allocation, optimizing the utilization of available energy sources, and adapting to the specific driving conditions in real time. This intelligent coordination and management of energy sources contribute to maximizing the overall efficiency and performance of ERELVs in terms of fuel economy and environmental sustainability.

1.1. Literature Review

Effective power management strategies are crucial for optimizing the performance and efficiency of ERELVs. Energy management strategies (EMSs) for vehicle power systems often involve the coordination and allocation of power among multiple power sources through manually set threshold values [9,10]. However, due to the fixed nature of these thresholds, it is difficult to adapt to the nonlinear characteristics of the vehicle system under changing operating conditions, resulting in suboptimal control performance [11]. To address this issue, various optimized energy management strategies have been proposed. Representative EMSs include rule-based energy management strategies (RB-EMSs) and optimization-based energy management strategies [12].

In rule-based energy management strategies, the parameter values in precise logical rules or fuzzy rules are mainly determined by empirical knowledge or the application of optimization methods. While an RB-EMS can provide real-time calculations, its control performance heavily relies on the engineer's experience. Minjae et al. [13] devised a hybrid thermostat energy management strategy for a series hybrid system. Engine output power and speed are determined by the state of charge (SOC) and required power while ensuring that the engine is operating at the optimal curve. Wassif et al. [14] put forward a rule-based real-time energy management strategy with variable thresholds. Three modes of the power system are defined: load balancing, load following, and load mixing. Economical driving control in different modes is achieved based on the driving status change threshold. To solve the problem of computational efficiency, Nicolas et al. [15] proposed a power-following strategy based on genetic algorithm. The results show that the genetic algorithm has a shorter calculation amount and the control effect is close to optimal under various working conditions. Deterministic rule-based energy management strategies cannot fully utilize hybrid powertrains. Many empirically determined parameters still have room for optimization [16]. In order to further improve the efficiency and performance of the deterministic rule energy management strategy, the results of the DP algorithm can be used to inform the formulation of the rule strategy [17]. In optimization-based energy management systems, the energy allocation in hybrid power systems aims to minimize fuel consumption locally or globally [18,19]. Optimization-based EMSs offer good fuel economy and adaptability to driving cycles, but their computational complexity often limits their implementation in whole vehicles. On the other hand, fixed threshold values in rule-based strategies make it challenging to adapt to varying vehicle operating conditions or maintain stable battery state of charge, resulting in suboptimal control [20].

The application of fuzzy intelligent control methods does not rely on high-precision mathematical models, and offers real-time performance and robustness [21]. Therefore, fuzzy logic-based energy management strategies have broad prospects for application in multi-power source systems [22]. However, their performance heavily relies on the formulation of fuzzy rules, which depends on expert knowledge. These rules can only be optimized for specific operating conditions during design, which may not yield satisfactory control performance in real-world complex and dynamic road conditions. Traditional approaches often rely on rule-based or optimization-based methods, which may not fully adapt to the dynamic nature of real-world driving conditions [23]. Moreover, these strategies often overlook the influence of driving pattern recognition on the power management system [24].

1.2. Motivation and Innovation

As previously discussed, a critical challenge in driving pattern recognition lies in seamlessly integrating the forecasted driving cycle with the power allocation strategy, all the while avoiding the burden of complex frameworks and extensive computational demands. This paper presents a novel approach to power management in ERELVs by incorporating driving pattern recognition and adaptive fuzzy logic techniques. The proposed strategy has the potential to significantly enhance the operational efficiency and sustainability of extended-range electric logistics vehicles, contributing to the advancement of electric transportation.

1.3. Article Overview

This paper is organized as follows: the system description and modeling of the ERELV are described in Section 2. An adaptive fuzzy power management strategy design is proposed in Section 3. Results and discussion are carried out in Section 4, and conclusions are given in Section 5.

2. ERELV Powertrain Modelling

2.1. Vehicle Structure and Longitudinal Dynamics Model

ERELV refers to a series PHEV designed for logistics and transportation purposes, known for its extended range capabilities [25]. Its powertrain system includes a three-cylinder engine, battery pack, electric motor, inverter, and generator, as shown in Figure 1a. The electrical energy in the battery comes from engine power generation, braking energy recovery, and external charging. As there is a decoupling of the mechanical connection between the three-cylinder engine and the wheel, the engine's high efficiency and low emissions region could be tracked properly.

The vehicle longitudinal dynamics model [26] is as shown in Figure 1b. The physical representation of the vehicle is illustrated in Figure 1c. The model parameters of the ERELV are listed in Table 1. The driving torque in motion as a function of the rolling resistance coefficient f and the coefficient of aerodynamic drag C_D is as follows:

$$\eta_T \cdot i_T \cdot T_P + T_b = (m\delta\frac{dv}{dt} + mg\cos\alpha \cdot f + mg\sin\alpha + \frac{1}{2}\rho_a v^2 C_D S) \cdot r_{roll} \tag{1}$$

where m is the overall mass of vehicle, S is the frontal area, and r_{roll} is the rolling radius.

Table 1. ERELV basic parameters.

Components	Description	
Final reduction drive	Final ratio	6.143
Engine	Maximum power (kW)	108
Electric machine	Maximum power (kW)	100
Battery	Rate capacity (Ah)	37
Vehicle	Vehicle mass (kg)	6000
	C_D	0.335
	S (m^2)	2
	Roll coefficient	0.0125
	Wheel radius (m)	0.376
	Rotational mass coefficient	1.06

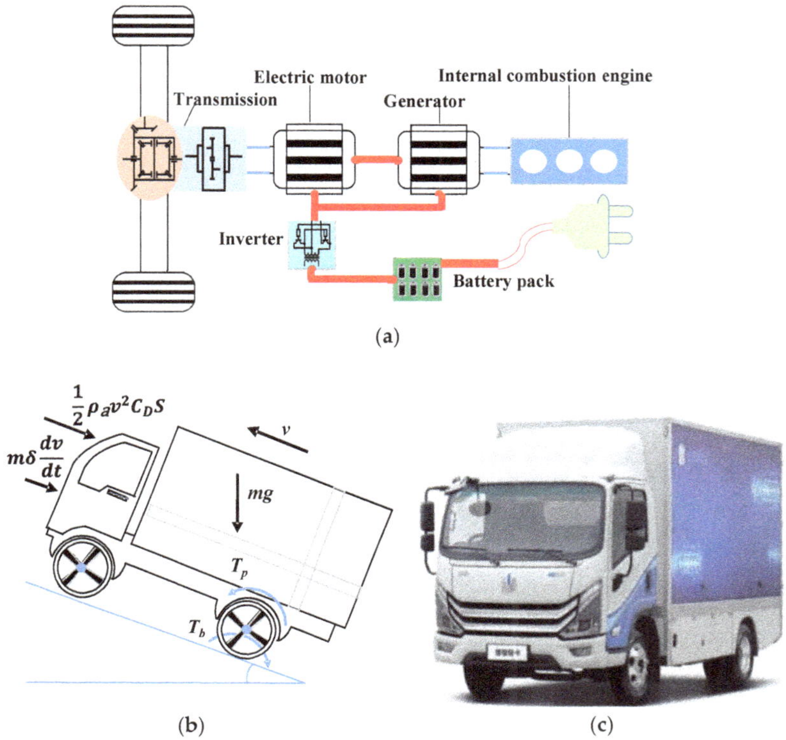

Figure 1. Schematic diagram of (**a**) hybrid powertrain structure; (**b**) forces on a vehicle in motion; and (**c**) vehicle.

2.2. Engine Model

Three-cylinder engines are considered as a promising power generator of the range extender, as a result of their excellent power and high efficiency performance and simple structure, which embedded in the investigated power system. The fuel rate can be defined as a function of the engine efficiency [27]. In addition, fuel consumption Q_e is derived from the following equation:

$$Q_e = \int_{t_1}^{t_2} \frac{T_E(t) \cdot \omega_E(t) \cdot b_E(t)}{1000\rho_f} dt \tag{2}$$

2.3. Electric Machine Model

In extended-range electric vehicles, the motor is the component that provides power. The efficiency of the drive motor is derived from a look-up table, with motor efficiency test data sourced from the literature [24]. The correlation among motor energy consumption Q_f, speed, and torque is described by the following equation:

$$Q_f = \int_{t_1}^{t_2} T_M(t) \cdot \omega_M(t) \cdot {\eta_M}^{\mathrm{sgn}(T_M)} dt \tag{3}$$

2.4. Battery Model

The electrochemical reaction of the power battery charge–discharge is an importantly complicated process, but the charge–discharge process is simplified to the Rint model including ideal voltage source and internal resistance [28,29]. The state of charge (SOC) is a function of initial SOC_{init}, open circuit voltage U_{VOC}, and battery internal resistance R_B, and can be defined by the following equation:

$$SOC(t) = SOC_{init} + \int_{t_0}^{t} \frac{U_{VOC} - \sqrt{U_{VOC}^2 - 4R_B P_B}}{2R_B Q_B} dt \quad (4)$$

3. Adaptive Fuzzy Power Management Strategy

3.1. Driving Pattern Recognition Based on Multi-Dimensional Features

3.1.1. Multi-Domain Feature Extraction Based on the Variational Mode Decomposition Algorithm

Driving pattern recognition is a technology that uses a vehicle-mounted driving state acquisition device to extract features of vehicle speed to identify and match working conditions. It is widely used in energy management systems, driving assistance systems, or intelligent driving systems, and is an important fulcrum for combining AI theory with actual industrial production.

The traditional driving pattern identification method uses manually extracted time-domain vehicle speed driving features for classification, such as average speed, maximum speed, average acceleration, and other time-domain features. Commonly used traditional pattern recognition methods include the Euclidean distance method, similarity distance method, and neural network method. A time-frequency feature extraction method of vehicle speed signals based on Hilbert transform is proposed. The time-frequency characteristic signal and the accelerator pedal and brake pedal signals are jointly constructed to produce multi-dimensional features. The extreme learning machine algorithm and the recurrent neural network are, respectively, integrated to form two multi-dimensional and multi-domain recognition methods. The effects of different parameters, different types of features, and different learning methods on the recognition effect are analyzed. The working condition identification framework based on multi-dimensional features is shown in Figure 2, and a driving database is constructed based on the collected vehicle speeds in typical cities. Multi-dimensional features in the time domain and time-frequency domain are extracted, and a working condition identification model of the sample is constructed under two learning methods.

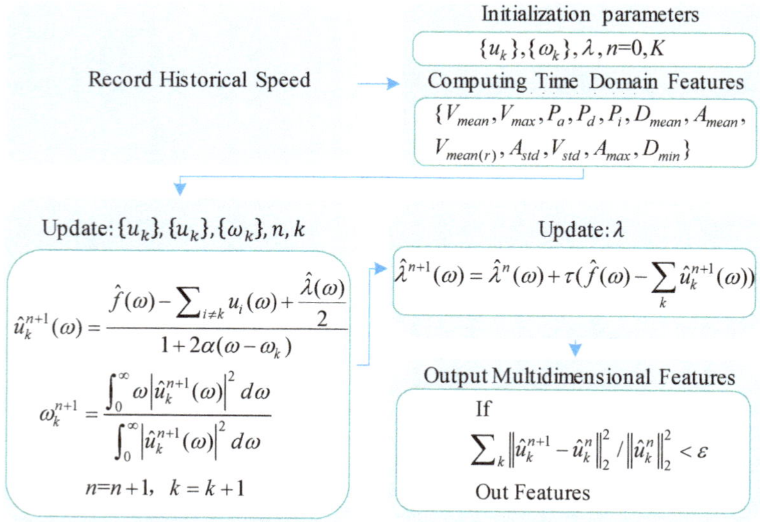

Figure 2. Multi-dimensional feature extraction method framework.

Assuming that the working mode will not change in a short period of time, the historical vehicle speed in a short period of time is selected to identify the current working mode, that is, the historical vehicle speed is selected as the acquisition signal of the working mode recognition method. Statistical methods are used to calculate three time-domain features: average speed, maximum speed, and maximum acceleration. Signal processing principles are used to transform the time-domain signal into a time-frequency-domain signal, and then the time-frequency-domain features are extracted. Hilbert transform proves advantageous in adaptively capturing the time-frequency characteristics of nonlinear historical vehicle speed signals. Its notable feature is the ability to achieve this without the need to select a specific basis function. Additionally, the transformed modal function components exhibit a notable degree of continuity in the frequency domain. Therefore, the variational mode decomposition algorithm (VMD) in Hilbert transform is selected to extract time-frequency features [30].

The multi-dimensional working condition feature extraction method based on VMD is mainly divided into two steps: historical vehicle speed sequence recording and time-frequency feature extraction.

(1) In the offline training recognition model, as many historical vehicle speed data as possible are recorded according to the capacity of the vehicle data storage platform to ensure the accuracy of the recognition model. During online identification, the history window time and update time need to be determined.

(2) The vehicle historical speed is decomposed into intrinsic mode functions (IMF) of different frequency bands using the VMD method. The characteristics of historical vehicle speed signals have a certain degree of randomness and complexity. The time-frequency analysis of vehicle speed is beneficial to achieve control performance. The input vector is the historical 150 s vehicle speed time series, and the elements of the output vector are the operating mode patterns of the next step size of 10 s. Each IMF is assumed to be a finite bandwidth with a center frequency. The variational problem can be described as seeking k modal functions to minimize the sum of the estimated bandwidths of each mode, with the constraint that the sum of each modal function is the original signal [31]. This constrained variation problem is expressed as

$$\begin{aligned} &\min_{\{u_k\}\{\omega_k\}} \left\{ \sum_k \left\| \partial_t \left[(\delta(t + \tfrac{j}{\pi t}) * u_k(t) \right] e^{-j\omega_k t} \right\|_2^2 \right\} \\ &s.t. \sum_k u_k(t) = f \end{aligned} \tag{5}$$

where ω_k is the center frequency of the k-th mode, and f is the original signal.

The quadratic penalty factor α and the Lagrangian multiplication algorithm λ are introduced to transform the constrained variation problem into an unconstrained variation problem and solve it.

$$\begin{aligned} &L(\{u_k\}, \{\omega_k\}.\lambda) \\ &\quad := \alpha \left\{ \sum_k \left\| \partial_t \left[(\delta(t + \tfrac{j}{\pi t}) * u_k(t) \right] e^{-j\omega_k t} \right\|_2^2 \right\} \\ &\quad + \left\| f(t) - \sum_k u_k(t) \right\|_2^2 + \left\langle \lambda(t), s(t) - \sum_k u_k(t) \right\rangle \end{aligned} \tag{6}$$

where k is the Lagrange multiplier.

Applying the multiplier, the alternating direction method iterative update can be expressed as

$$\hat{u}_k^{n+1}(\omega) = \frac{\hat{f}(\omega) - \sum_{i \neq k} u_i(\omega) + \frac{\hat{\lambda}(\omega)}{2}}{1 + 2\alpha(\omega - \omega_k)} \tag{7}$$

$$\omega_k^{n+1} = \frac{\int_0^\infty \omega \left| \hat{u}_k^{n+1}(\omega) \right|^2 d\omega}{\int_0^\infty \left| \hat{u}_k^{n+1}(\omega) \right|^2 d\omega} \tag{8}$$

where n is the number of iterations.

The Lagrange multiplier update formula is

$$\hat{\lambda}^{n+1}(\omega) = \hat{\lambda}^n(\omega) + \tau(\hat{f}(\omega) - \sum_k \hat{u}_k^{n+1}(\omega)) \tag{9}$$

where $\hat{u}_k^{n+1}(\omega)$ is the Fourier transform of the n + 1 iteration of k components. $\hat{\lambda}_k^{n+1}(\omega)$ is the (n + 1)-th Fourier transform of the Lagrange multiplier. The iteration termination condition is

$$\sum_k \left\| \hat{u}_k^{n+1} - \hat{u}_k^n \right\|_2^2 / \|\hat{u}_k^n\|_2^2 < \varepsilon \tag{10}$$

The process of using variational mode decomposition (VMD) to extract time-frequency features can be summarized in the following steps: (1) Start by initializing the modal functions, instantaneous frequencies, and Lagrangian operators based on the specified number of modal functions, denoted as "k". (2) Enter an iterative loop to update the modal functions, instantaneous frequencies, and Lagrangian operators for each component. The updates are performed according to a specific formula. (3) Continuously evaluate certain conditions to determine whether the convergence criteria have been met. If the conditions are satisfied, exit the iterative loop. (4) Once the convergence criteria are met, the algorithm returns the k modal function components obtained through the iterative process. The framework of the multi-dimensional feature extraction method using VMD is illustrated in Figure 2. The resulting modal components, which represent the time-frequency features extracted from the data, are depicted in Figure 3.

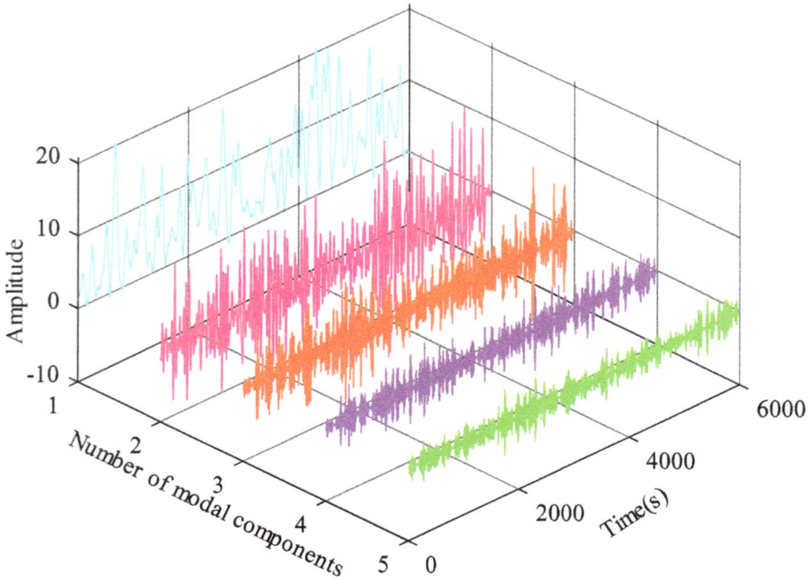

Figure 3. Time-frequency features extracted using the VMD algorithm.

3.1.2. Deep Learning Driving Pattern Identification Method

One of the most widely used methods for addressing time-series classification problems is the recurrent neural network (RNN) [32]. RNNs are designed to capture and store past information and current inputs by introducing state variables, enabling the prediction of current outputs. On the other hand, deep neural networks (DNNs) are an evolution of traditional neural networks, consisting of three main layers: input, hidden, and output. The strength of DNNs lies in their ability to effectively learn and represent complex data patterns. When it comes to working condition pattern recognition, deep neural networks excel at discovering deeper mapping relationships between driving data and patterns of working conditions. In comparison to conventional neural networks, the multi-layer structure of DNNs requires fewer parameters to describe intricate relationships. Consequently, training DNNs with a multi-layer structure tends to be faster and yields better training outcomes.

To balance the requirements of timing signals and computational efficiency while harnessing substantial amounts of data, long short-term memory (LSTM) networks introduce three essential control units: the input gate, the forget gate, and the output gate [33]. These units enhance the influence of historical timing information on current predictions. The network structure of LSTM is depicted in Figure 4. The mathematical expressions for these gate units are as follows:

$$\begin{cases} i_t = sigmoid(W_i \cdot [h_{t-1}, x_t] + b_i) \\ f_t = sigmoid(W_f \cdot [h_{t-1}, x_t] + b_f) \\ o_t = sigmoid(W_o \cdot [h_{t-1}, x_t] + b_o) \end{cases} \tag{11}$$

where the three logic gates are sigmoid. The coefficients are W_i, W_f, and W_a. h_{t-1} is the hidden unit information at the previous moment, x_{t-1} is the input vector, and b_i, b_f, and b_a are deviations.

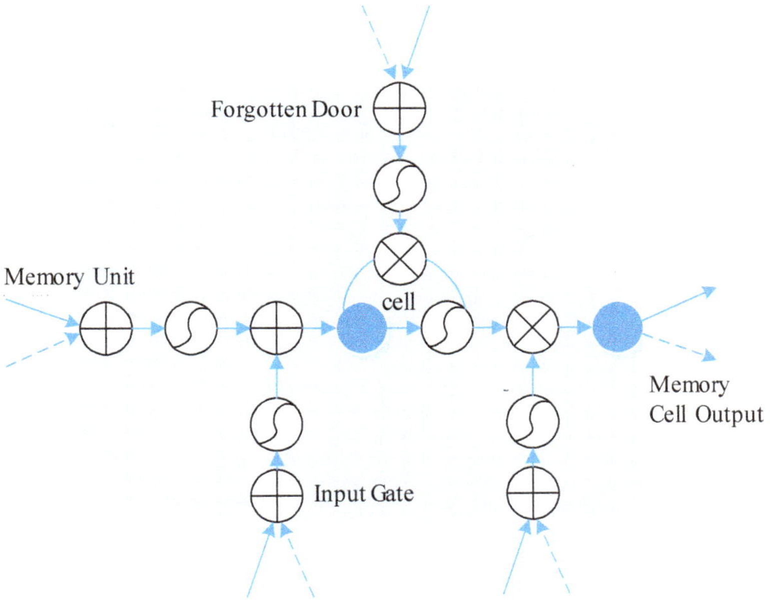

Figure 4. LSTM network structure.

The expression of the alternative information update process of the input gate is [34]

$$\begin{cases} \widetilde{C}_t = \tanh(W_e[h_{t-1}, x_t] + b_e) \\ f_t = f_t * C_{t-1} + i_t * \widetilde{C}_t \\ h_t = o_t * \tanh(C_t) \end{cases} \tag{12}$$

where the activation function is the tanh function, We is the weight matrix, be is the deviation of the updated information unit, and C_{t-1} is the information of the memory unit at the previous moment.

The output gate update formula is

$$h_t = o_t * \tanh(C_t) \tag{13}$$

where h_t is the hidden state information at the current moment.

During working mode recognition, the data volume of the input signal of the LSTM neural network will be different from that of the DNN recognition model, but the information included is the historical vehicle speed, accelerator pedal opening, and brake pedal opening, and the output is a different recognition step size. When using the LSTM-NN prediction model to identify the operating mode, the input information and output information are

$$\begin{aligned} Output(t) &= f_{LSTM-NN}(Input) \\ Input(t) &= \left\{ \begin{matrix} [v_{t-hh+1}, v_{t-hh+2}, \cdots, v_{t-1}, v_v] \\ [g_t, g_{t+1}, \cdots, g_{t+hp-1}, g_{t+hp}] \\ [a_t] \\ [s_t] \end{matrix} \right\} \\ Output(t) &= \left\{ v_{t+1}, v_{t+2}, \cdots, v_{t+hp} \right\} \end{aligned} \tag{14}$$

3.2. *Adaptive Fuzzy Energy Management Strategy*

In an effort to mitigate the energy consumption of extended-range electric vehicles, a dual-input single-output fuzzy controller has been devised [35,36]. As depicted in Figure 5, this fuzzy controller takes into account both the vehicle's power demand and the battery's state of charge (SOC) as inputs, yielding the comprehensive power output for the range extender. The purpose of employing the fuzzy rule-based controller is twofold: optimizing engine operation within the high-efficiency region while simultaneously ensuring the stability of the battery SOC.

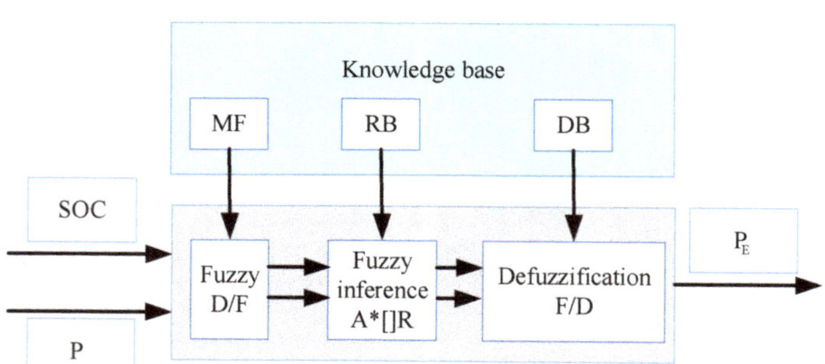

Figure 5. Fuzzy controller structure.

The control process begins by subjecting both the SOC and the required power to a process of fuzzification. Subsequently, the membership degrees of linguistic values are computed by utilizing the rule base and membership functions. Finally, the defuzzification module is employed to translate the fuzzy semantic values into precise control instructions, which are then conveyed to the vehicle's power system.

The fuzzy control rules utilized are presented in Table 2. The input variables SOC and demand power are each divided into three fuzzy subsets, denoted as {Low (L), Medium (M),

High (H)}. On the other hand, the output power is categorized into five fuzzy subsets: {Very Low (VL), Low (L), Medium (M), High (H), Very High (VH)}. The SOC range encompasses values from [0, 1], while the range for required power is [−150, 150].

Table 2. Fuzzy control rules.

Required Power	SOC		
	L	**M**	**H**
L	VL	L	M
M	L	M	H
H	M	H	VH

The calculation formula of the trapezoidal membership function is as follows:

$$\mu_F = \begin{cases} 0 & x < a \\ \frac{x-a}{b-a} & a \le x < b \\ 1 & b \le x < c \\ \frac{d-x}{d-c} & c \le x \le d \\ 0 & x > d \end{cases} \tag{15}$$

The method of defuzzification is the centroid removal method, that is, the output of the fuzzy controller can be obtained using the following formula:

$$u_f(k) = \frac{\sum_{j=1}^{r_1}\sum_{i=1}^{r_2} u_{I_1}^j(k) u_{I_2}^i(k) \Delta u_u^{ij}(k)}{\sum_{j=1}^{r_1}\sum_{i=1}^{r_2} u_{I_1}^j(k) u_{I_2}^i(k)} \tag{16}$$

where u_f is the fuzzy controller output, and $u_{I_1}^j(k)$ is the membership degree of SOC and demand power.

The membership function input to the fuzzy controller is shown in Figure 6. The performance of the fuzzy controller is mainly affected by the vertex coordinates of the input and output membership functions. Therefore, the 11 vertex coordinates are used as optimization variables, the fuel economy in the power maintenance stage is the optimization target, and the improved genetic algorithm is used as the optimization method.

The optimization variables are the vertex coordinates of the input and output membership functions, and the expression is

$$X = [x_1, x_2, x_3, x_4, x_5, x_6, x_7, x_8, x_9, x_{10}, x_{11}]^T \tag{17}$$

The coordinate position of the vertex of the optimization variable in the formula is shown in Figure 6. Each coordinate should satisfy the following:

$$\begin{cases} x_1 \le x_2 \le x_3 \\ x_4 \le x_5 \le x_6 \\ x_7 \le x_8 \le x_9 \le x_{10} \le x_{11} \end{cases} \tag{18}$$

To avoid the impact of electric power on fuel economy evaluation, the fuel consumption during the electric power maintenance phase is selected as the optimization target, and the expression is as follows:

$$J = \min\left(\int_{t_0}^{t_f} m_f(\omega_e(t), T_e(t))dt\right) \tag{19}$$

where $\omega_e(t)$ and $T_e(t)$ are the engine speed (rad/s) and torque (N·m) changing with time under test conditions; and m_f is the engine fuel consumption rate.

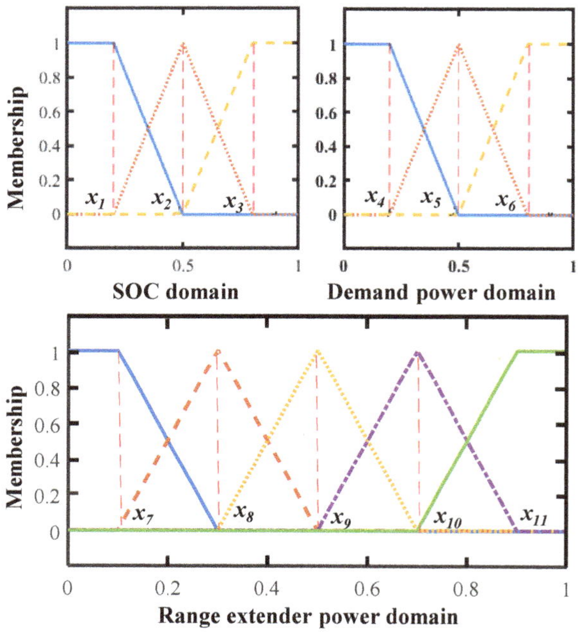

Figure 6. Membership function.

The designed framework for a fuzzy energy management strategy in neural network-based working condition identification is illustrated in Figure 7. It is divided into two key components: offline optimization and online fuzzy real-time control. Urban working conditions data are utilized as training samples to construct a recognition model. This model's purpose is to identify and classify three types of working conditions. During the power maintenance stage within these working condition cycles, the optimization objective is to minimize fuel consumption. To achieve this, an improved genetic algorithm is employed as the optimization method. The parameters of the fuzzy controller, operating under the three types of working conditions, are optimized offline using this approach. Within the fuzzy real-time control portion of the framework, the optimal parameters determined during the offline optimization stage are transferred to the controller. The controller takes several inputs, including the required power, state of charge (SOC), and the category of the working condition. After considering these inputs and ensuring compliance with various decision constraints, the fuzzy controller selects the optimal control parameters for real-time decision making. This ensures that the system operates optimally under different working conditions.

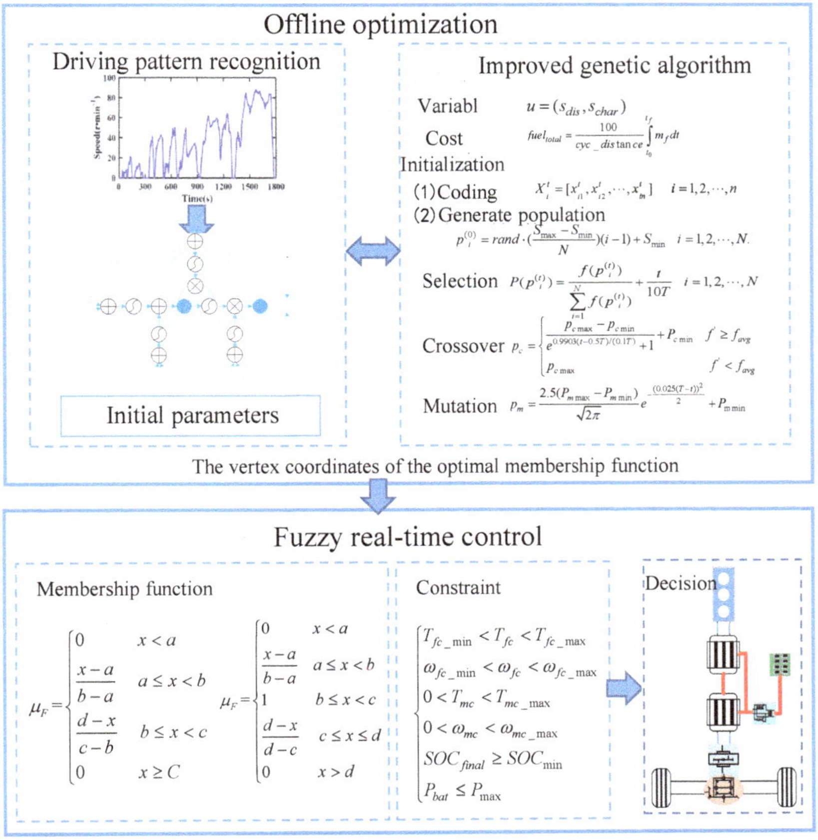

Figure 7. Fuzzy energy management strategy framework for working condition recognition using a neural network.

In summary, this framework combines offline optimization to ascertain optimal fuzzy controller parameters and online fuzzy real-time control to adapt and make optimal decisions based on the current working condition. It is designed to enhance energy management within a neural network-based system, ultimately aiming to minimize fuel consumption and improve overall performance.

4. Results and Discussion

4.1. Driving Pattern Recognition Performance Verification

To validate the efficacy of the method proposed in this article and investigate the influence of various factors and methodologies on its performance, we gathered vehicle speed data in Tianjin City from the literature [37]. Integrating the K-means clustering method, we segmented the data into three kinematic categories: urban, suburban, and expressway. Subsequently, we extracted 5000 s of historical data from each segment to train the model, as shown in Figures 8 and 9. Figure 10 depict the extracted accelerator pedal and brake pedal signals for their respective segments. While these three types of features are distinct in each kinematic segment, practical applications necessitate a fixed online identification window time.

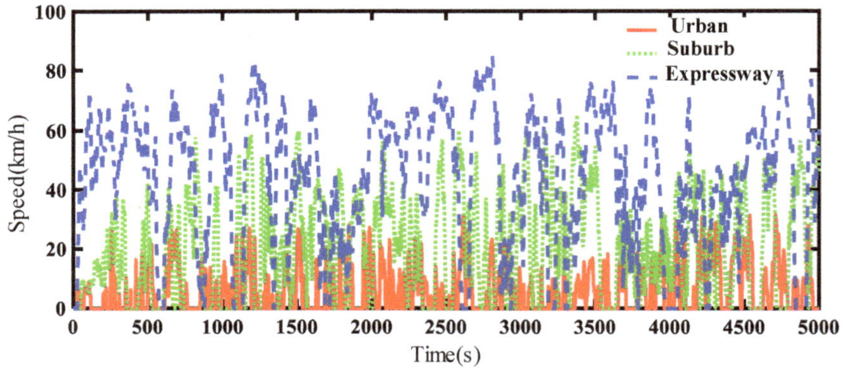

Figure 8. Training data.

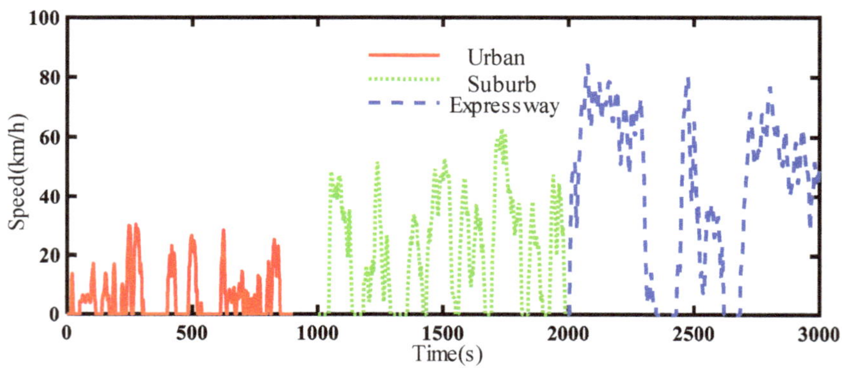

Figure 9. Test data.

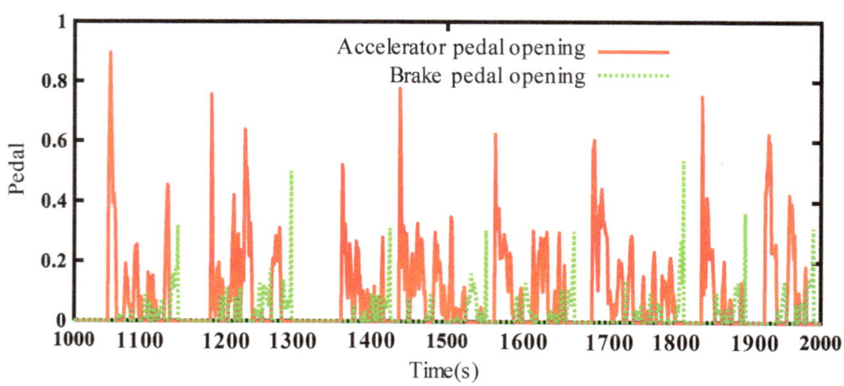

Figure 10. Acceleration and brake pedal opening at 1000 s–2000 s.

During the training phase, we utilized historical vehicle speed data to create training datasets based on the size of the data recording window. The term "update window" pertains to the interval for refreshing driving mode recognition, which is also called the recognition step. For instance, with a total of 5000 s of data and assuming a 150-s recording window and a 10-s update window, we generated 486 samples for the first data type. In total, we constructed 1458 samples across the three categories for training. Additionally, we selected 1000 s of each kinematic segment distinct from the training data to serve as test data. Employing the same 150-s recording window and 10-s update window, we created 258 samples for testing purposes.

Figures 8 and 9 display the velocity contour curves for both the training and test data. In the first type of training data, the sixth kinematic segment exhibits an average speed of 7.60 km/h, a maximum speed of 26.85 km/h, and an average acceleration of 0.65 m/s^2. Contrastingly, in the second type of training data, the same kinematic segment showcases an average speed of 11 km/h, a maximum speed of 41.60 km/h, and an average acceleration of 0.4988 m/s^2. These observations underscore the distinct kinematic characteristics between different types. In the third category, the sixth kinematic segment stands out with an average speed of 47.40 km/h, a maximum speed of 76.36 km/h, and an average acceleration of 0.2669 m/s^2. However, it is crucial to note that when using fixed historical time periods as input for the recognition model in practical applications, the differences in time-domain statistical characteristics may not be readily apparent, leading to lower recognition accuracy. Thus, it becomes imperative to incorporate multi-dimensional and multi-domain features such as time-frequency characteristics, accelerator pedal signals, and brake pedal signals.

Figure 10 illustrates the pedal opening curve ranging from 1000 s to 2000 s. In this graph, the solid line represents the accelerator pedal opening, while the dotted line represents the brake pedal opening. Notably, the average opening of the accelerator pedal surpasses that of the brake pedal. This disparity can be attributed to the distinct roles these pedals play during vehicle operation.

The accelerator pedal primarily controls the driving force of the vehicle and is tasked with overcoming various types of resistance such as air resistance, acceleration resistance, and slope resistance while ensuring smooth acceleration. On the other hand, the brake pedal is involved in the deceleration process, where it contends with two types of resistance: motor energy recuperation, which contributes to braking through energy recovery, and mechanical braking force.

4.1.1. The Impact of Time-Frequency Domain Feature Estimates on Recognition Performance

The term "number of time-frequency domain feature estimates" pertains to the quantity of time-frequency features derived from the time-domain signal through the variational mode decomposition algorithm. To investigate the influence of these estimates on recognition performance, we set the historical driving feature recording window time to 150 s. To assess the effectiveness of recognition, the evaluation relies on both the recognition rate and the overfitting degree. Their definitions are provided below:

$$P_{AC} = \frac{N_C}{N_T} \tag{20}$$

$$D = \frac{P_{AC\_train} - P_{AC\_test}}{P_{AC\_train}} \tag{21}$$

where P_{AC} is the recognition rate, N_C and N_T are the correct classification and the total number, respectively, D is the overfitting degree, and P_{AC_train} and P_{AC_test} are the training recognition rate and the test recognition rate, respectively.

Subsequently, we varied the update time for future working condition pattern recognition to 5 s, 10 s, 15 s, and 20 s, respectively, and assessed the degree of overfitting associated with different estimates ranging from three to nine. The outcomes are presented in Figure 11. Across the four update time intervals, the recognition rate initially ascends and then declines as the number of estimates increases, with the peak recognition rate occurring when the estimated number is seven. Specifically, the maximum recognition rates at 5 s, 10 s, 15 s, and 20 s are 0.980, 0.973, 0.957, and 0.955, respectively. Conversely, the overfitting degree initially decreases and then increases with an increasing number of estimates. The minimum overfitting degree is observed when there are seven estimates, with corresponding values of 0.020, 0.027, 0.029, and 0.032 under the four update times, respectively. Consequently, the optimal number of time-frequency domain feature estimates is determined to be seven. The recognition rate diminishes as the update time increases. This trend suggests that the longer the update time, the more pronounced the impact on the recognition rate, following an initial rise and fall as the estimated number increases. Additionally, the overfitting degree amplifies with longer update times, with the value initially decreasing and then increasing as the estimate number grows.

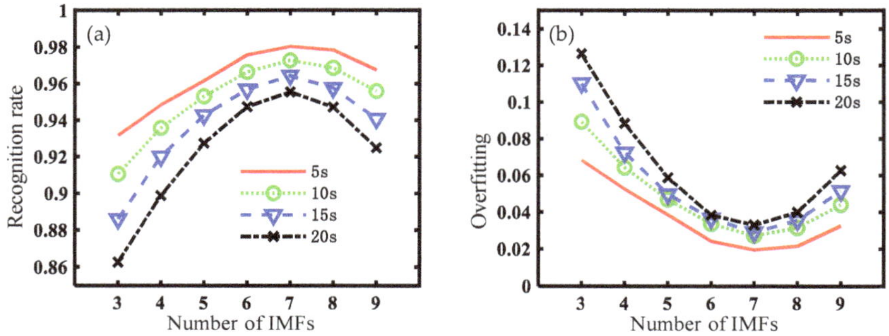

Figure 11. Result comparison under different update times: (**a**) recognition rate; (**b**) overfitting.

Moving on to Figure 12, it provides a comparison of recognition performance under various historical driving information recording times. Notably, under the same time-frequency feature estimates, the recognition rate is at its highest and the overfitting degree is at its lowest when the recording time is set to 150 s. This observation underscores that a recording time of 150 s yields the most favorable recognition results among the four time intervals considered.

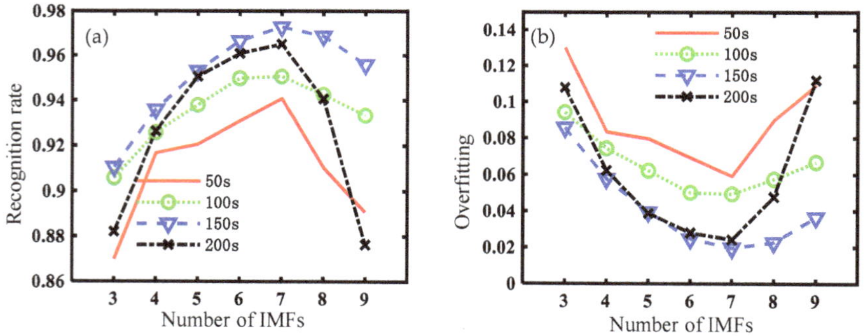

Figure 12. Comparison of results at different recording times: (**a**) recognition rate; (**b**) overfitting.

4.1.2. Comparison of the Two Identification Methods

To compare and analyze the performance of the two recognition methods, we calculate the overfitting degree of the recognition rate domain for both methods at recognition step lengths of 5 s, 10 s, 15 s, and 20 s. Table 3 presents the recognition effects of the two learning methods proposed in this article. The table indicates that when the recognition step size is consistent, LSTM exhibits a higher recognition rate than the extreme learning machine

(ELM), and its overfitting degree is lower compared to ELM. Consequently, the recognition effect of LSTM surpasses that of ELM.

Table 3. The recognition effect of different learning methods.

Recognition Step Size	LSTM Recognition Rate	ELM Recognition Rate	LSTM Overfitting Degree	ELM Overfitting Degree
5 s	0.980	0.960	0.020	0.040
10 s	0.973	0.953	0.027	0.047
15 s	0.964	0.940	0.032	0.051
20 s	0.955	0.920	0.031	0.068

4.1.3. The Impact of Different Types of Features on Recognition Performance

Table 4 provides a comprehensive overview of the working condition identification results, considering different feature types and recognition step sizes. These features encompass time-domain attributes (T, including average speed, maximum speed, and average acceleration), time-frequency-domain characteristics (F, derived from the signal post VMD decomposition), accelerator pedal signals (A), and brake pedal signals (B). The recognition steps of 5 s, 10 s, 15 s, and 20 s were utilized, with historical driving information recorded over a 150 s time period. When feature types remain constant, the recognition rate exhibits a decreasing trend as the recognition step size grows larger, accompanied by an escalation in overfitting. For instance, with only time-domain features and a 5 s recognition step, the recognition rate stands at 0.917, with an overfitting degree of 0.083. Increasing the recognition step to 10 s results in a recognition rate of 0.894 and an overfitting degree of 0.101. Extending the step to 20 s leads to a reduced recognition rate of 0.887, accompanied by an increased overfitting degree of 0.113.

Table 4. Recognition results under different types of features.

Feature Type	Recognition Step	Recognition Rate	Overfitting Degree
T	5 s	0.9171	0.0828
	10 s	0.8941	0.1014
	15 s	0.8872	0.1057
	20 s	0.8783	0.1129
T + F	5 s	0.9675	0.0323
	10 s	0.9566	0.0386
	15 s	0.9300	0.0624
	20 s	0.9235	0.0671
T + A	5 s	0.9226	0.0772
	10 s	0.9203	0.0751
	15 s	0.9002	0.0925
	20 s	0.8914	0.0996
T + B	5 s	0.9232	0.0766
	10 s	0.9047	0.0907
	15 s	0.8958	0.0969
	20 s	0.8894	0.1016
T + F + A + B	5 s	0.9803	0.0195
	10 s	0.9728	0.0223
	15 s	0.9644	0.0278
	20 s	0.9554	0.0348

However, maintaining a consistent recognition step size while introducing additional feature types can yield improvements in recognition rate and a reduction in overfitting. For instance, with a 15 s recognition step, the recognition rate for time-domain features alone is 0.887, with an overfitting degree of 0.106. When combining time-domain and time-frequency-domain features, the recognition rate increases to 0.930, with a reduced overfitting degree of 0.062. Similarly, incorporating time-domain features with accelerator pedal signals results in a recognition rate of 0.900 and an overfitting degree of 0.093. The recognition rate under time-domain features combined with brake pedal signals is 0.896, with an overfitting degree of 0.097. The most significant improvement is achieved when all four features are employed together, yielding a recognition rate of 0.964 and the lowest overfitting degree of 0.028.

Among these features, time-frequency-domain features have the most substantial impact on recognition rate, followed by accelerator pedal signal features, with brake pedal signal features having the smallest influence. Combining time-frequency-domain features, accelerator pedal signals, and brake pedal signals results in the most substantial enhancement in recognition efficiency. In summary, integrating diverse feature types for working condition identification can further elevate the recognition rate and optimize overall performance.

4.2. *Performance Comparison of Energy Management Strategies*

The fuzzy controller's state of charge (SOC) membership function and demand power vertex coordinate parameters have been optimized. The normalized SOC membership function parameters are [0.4497, 0.8644, 0.9540], while the demand power membership function parameters are [0.8165, 0.7201, 0.1220]. The output membership function parameters are [0.0830, 0.8391, 0.3500, 0.9056, 0.5779]. Table 5 displays the optimal fuzzy controller parameters for three different operating modes. It is evident from the table that the optimal control parameters vary for different operating modes. To assess the effectiveness of the adaptive fuzzy energy management strategy that incorporates working condition identification, a simulation comparison analysis was conducted between the traditional fuzzy energy management strategy and the fuzzy energy management strategy using various working condition identification methods.

Figure 13 illustrates the SOC curves under different strategies. The SOC fluctuations are relatively small in the traditional fuzzy strategy. The difference between the maximum SOC and the minimum SOC is 0.0134. In contrast, the SOC difference of the fuzzy strategy under T + F + A + B is 0.0198, under T + F it is 0.0212, under T it is 0.0237, under T + B it is 0.0218, and under T + A it is 0.0201. This difference is because the traditional fuzzy strategy lacks working condition identification during the power maintenance stage, making it unable to accurately determine the working condition. Consequently, the SOC must be balanced as much as possible at any vehicle speed.

Table 6 clearly demonstrates that the fuel consumption of the traditional fuzzy energy management strategy is 2408 g. When compared to this traditional approach, the fuzzy energy management strategy incorporating multi-dimensional characteristic working condition identification (T + F + A + B) achieves a fuel consumption of 2268 g, resulting in a fuel saving of 5.8%. Similarly, the fuzzy strategy using T + F characteristics achieves a fuel consumption of 2294 g, translating to a fuel saving of 4.7%. For the fuzzy strategy with T characteristics, fuel consumption is 2327 g, offering a fuel saving of 3.4%. When employing the fuzzy strategy with T+B characteristics, fuel consumption totals 2307 g, resulting in a fuel saving of 4.2%. Lastly, the fuzzy strategy with T + A characteristics records a fuel consumption of 2316 g, representing a fuel saving of 3.8%. In terms of engine start time, all of these strategies exhibit reductions of 5.8%, 4.7%, 3.4%, 4.2%, and 3.8%, respectively. Furthermore, the average rate of change of battery power decreases in each case. These results highlight that the fuzzy energy management strategy, when integrated with working condition recognition, can significantly enhance fuel economy during the power maintenance stage. Among the various working condition identification methods based on different types of characteristics, the strategy incorporating multi-dimensional characteristics, encompassing time-domain, time-frequency-domain, accelerator pedal signal, and brake pedal signal data, proves to be the most effective for improving fuel efficiency.

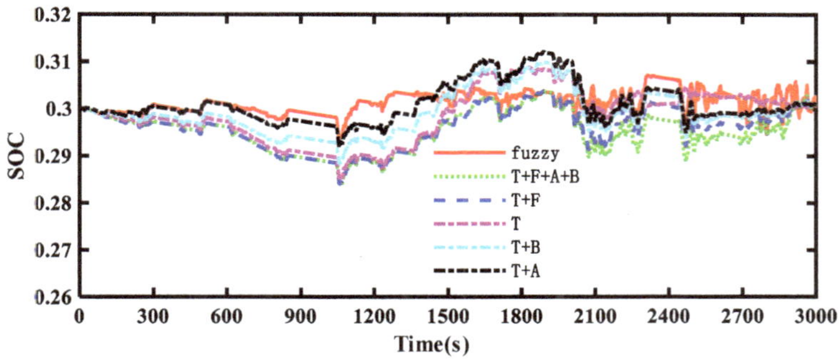

Figure 13. SOC with different strategies.

Table 5. Fuzzy controller parameters under different operating modes.

Abscissa of Vertex	Urban	Suburban	Expressways
x_1	0.7523	0.7336	0.4690
x_2	0.8949	0.3011	0.0873
x_3	0.8418	0.4956	0.8287
x_4	0.1309	0.2582	0.6859
x_5	0.1892	0.7329	0.2673
x_6	0.1536	0.1168	0.9695
x_7	0.0289	0.7460	0.1838
x_8	0.0091	0.8098	0.2999
x_9	0.5965	0.7452	0.4112
x_{10}	0.6090	0.3371	0.2365
x_{11}	0.9189	0.5843	0.1951

Table 6. Performance comparison under different strategies.

Different Strategies	Fuel Consumption (g)	SOC_{end}	Number of Engine Starts	Engine on Time	Average Rate of Change of Battery Power (kw/s)
Fuzzy	2408	0.3013	177	1057	6.67
T + F + A + B	2268	0.3010	165	973	6.48
T + F	2294	0.3014	165	974	6.53
T	2327	0.3007	166	971	6.62
T + B	2307	0.3006	170	1011	6.53
T + A	2316	0.3013	165	983	6.60

Figure 14 presents the distribution of engine operating points under different strategies. It is evident from the figure that all three strategies are operating along the optimal operating curve for range extender efficiency. Since the range extender only provides power according to the strategy, decoupling the range extender's rotational speed from the vehicle speed allows the operating point to align with the optimal operating curve. The operating point distribution of the fuzzy strategy under the T + F + A + B working condition identification method appears relatively narrow, while the traditional fuzzy power distribution exhibits a wider spread. This distinction arises because the fuzzy energy management strategy, after incorporating working condition identification, reduces the number of operating points in low-efficiency areas. This optimization is aimed at further enhancing fuel economy by avoiding operation in less efficient regions of the engine's performance curve.

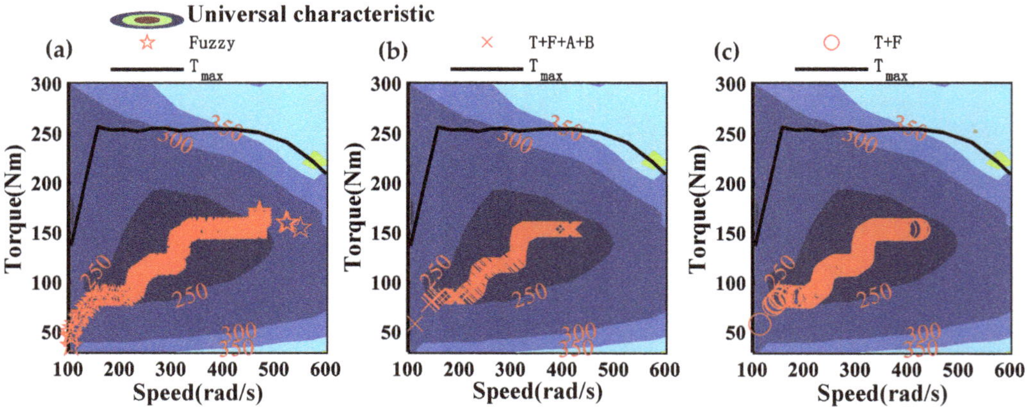

Figure 14. Engine operating point distribution: (**a**) fuzzy strategy; (**b**) fuzzy strategy under T + F + A + B features; and (**c**) fuzzy strategy under T + F features.

To provide a more detailed comparison of fuel economy among the three strategies, the proportion of engine fuel consumption within specific ranges was statistically calculated. The results are presented in Figures 15 and 16. The fuel consumption rate of the traditional fuzzy strategy falls within the range of [220, 240) grams per kilowatt-hour (g/kWh), accounting for 29.32% of the data. Moreover, within the range of [220, 260) g/kWh, the proportion is 68.52%. In contrast, the fuzzy strategy incorporating T + F + A + B working condition identification exhibits a fuel consumption rate within the [220, 240) g/kWh range, accounting for 42.20% of the data. This represents a 12.88% increase compared to the corresponding range of the traditional fuzzy strategy. Additionally, within the [220, 260) g/kWh range, the proportion is 71.82%, which is 3.3% higher than the corresponding range of the traditional fuzzy strategy. These results clearly demonstrate that the fuzzy energy management strategy incorporating T + F + A + B working condition identification achieves the optimal fuel economy among the three strategies, with significantly improved fuel efficiency in specific fuel consumption rate ranges.

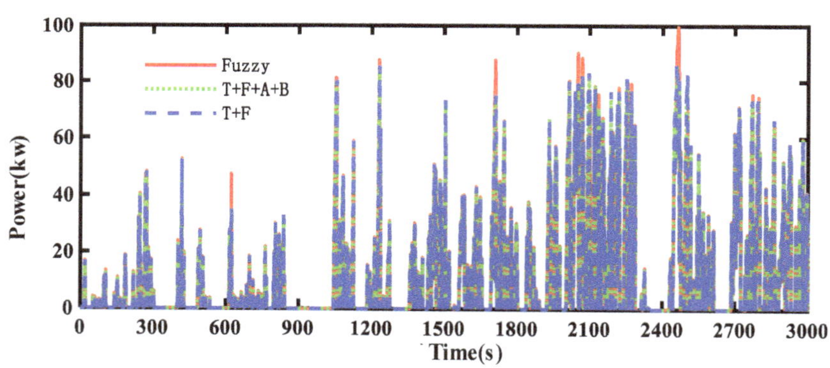

Figure 15. Engine power under three strategies.

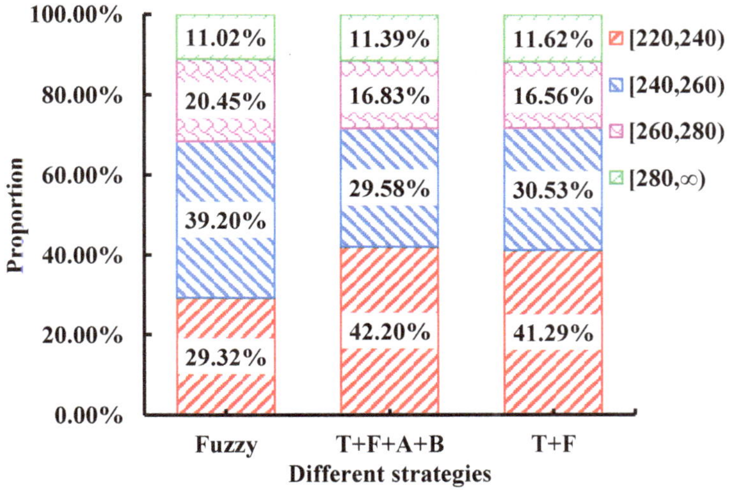

Figure 16. Engine fuel consumption rate distribution under different strategies.

5. Conclusions

The objective of the energy management strategy for hybrid energy storage systems in extended-range electric vehicles is to efficiently allocate energy to achieve optimal fuel economy. The energy storage control system effectively manages energy sources based on real-time traffic information and the current vehicle status obtained from the driving environment. However, vehicles often operate in uncertain and variable conditions, introducing challenges for energy management strategies to adapt quickly to sudden changes. To address these challenges, we introduce a working condition pattern recognition method. We combine this method with a practical fuzzy control strategy for real-time applications, allowing us to optimize parameters for different operating conditions. From our analysis, the following conclusions can be drawn:

(1) The optimal estimate for time-frequency-domain features is found to be seven. The recognition rate decreases as the update time increases. Longer update times show a trend where the recognition rate initially increases and then decreases as the estimated number increases. This suggests that the impact of update time on the recognition rate increases with the estimated number, with an initial increase, a subsequent decrease, and then a small increase. As update time increases, the degree of overfitting also increases, with the trend initially decreasing and then increasing with an increasing estimate number.

(2) When employing an identical recognition step size, long short-term memory (LSTM) surpasses the extreme learning machine (ELM) in recognition accuracy while concurrently demonstrating reduced levels of overfitting. Consequently, LSTM showcases superior recognition performance in comparison to ELM.

(3) The addition of time-frequency features, accelerator pedal signals, and brake pedal signals to the time-domain features has varying effects on the recognition rate. Time-frequency-domain features have the most significant impact, followed by accelerator pedal signals, while brake pedal signals have the smallest impact. Combining time-frequency-domain features, accelerator pedal signals, and brake pedal signals comprehensively yields the most substantial improvement in recognition efficiency. The fusion of diverse feature types for working condition identification can further enhance the recognition rate.

(4) Incorporating fuzzy energy management strategies after integrating working condition identification can further enhance fuel economy. Among the various working condition identification methods using different types of characteristics, the fuzzy energy management strategy utilizing multi-dimensional characteristics, including time domain, time-frequency domain, accelerator pedal signals, and brake pedal signals, performs the best.

Author Contributions: Conceptualization, C.W.; methodology, C.W. software, C.W. and X.W.; validation, Y.C. (Yong Chen), Y.C. (Yunxing Chen) and H.W. Writing of the manuscript: C.W. All authors have read and agreed to the published version of the manuscript.

Funding: This work was supported in part by the Ningbo Science and technology project of China (No. 2019B10111) and the National Key R&D Program of China (No. 2018YFB0106403).

Data Availability Statement: Not applicable.

Conflicts of Interest: The authors declare no conflict of interest.

Abbreviations

EMS	Energy Management Strategy	GIS	Geographical Information Systems
ERELV	Extended-Range Electric Logistics Vehicle	ELM	Extreme Learning Machine
GPS	Global Positioning Systems		
DP	Dynamic Programming	RB-EMS	Rule-basedEnergy Management Strategy
DPR	Driving Pattern Recognition	SOC	State of Charge
LSTM	Long Short-Term Memory Neural Network	PHEV	Plug-in Hybrid Electric Vehicle
ITS	Intelligent Transportation Systems	VMD	Variational Mode Decomposition Algorithm

Nomenclature

η_T	Transmission Efficiency	r_{roll}	Rolling Radius
i_T	Transmission Ratio	Q_e	Fuel Consumption
T_P	Driving Torque	$T_E(t), \omega_E(t)$	Engine Torque and Speed
T_b	Brake Torque	$b_E(t)$	Engine Fuel Consumption Rate
m	Overall Mass of Vehicle	ρ_f	Fuel Specific Gravity
δ	Rotational Mass Coefficient	η_M	Electrical Circuitry Efficiency
v	Vehicle Speed	U_{VOC}	Open Circuit Voltage
ρ_a	Air Density	ω_k	Mode Center Frequency
α	Road Slope	$\hat{u}_k^{n+1}(\omega)$	Fourier Transform Component
f	Coefficient of Rolling Resistance	W_i, W_f, W_a	Input Weight Vector
C_D	Air Resistance Coefficient	u_f	Fuzzy Controller Output
S	Wind Area	r_{roll}	Rolling Radius

References

1. Guo, J.; Wang, J.; Xu, Q.; Wang, B.; Li, K. Deep Reinforcement Learning-based Hierarchical Energy Control Strategy of a Platoon of Connected Hybrid Electric Vehicles through Cloud Platform. *IEEE Trans. Transp. Electrif.* **2023**. [CrossRef]
2. Wu, L.; Lyu, Z.; Huang, Z.; Zhang, C.; Wei, C. Physics-based battery SOC estimation methods: Recent advances and future perspectives. *J. Energy Chem.* **2023**. [CrossRef]
3. Li, Y.; Hu, Z.; Liu, H.; Xu, L.; Li, J.; Xu, L.; Ouyang, M. Comprehensive analysis of cathode air pressure of fuel cell powertrain system of aircraft: Performance, efficiency, and control. *Energy Convers. Manag.* **2023**, *283*, 116903. [CrossRef]
4. Wei, C.; Sun, X.; Chen, Y.; Zang, L.; Bai, S. Comparison of architecture and adaptive energy management strategy for plug-in hybrid electric logistics vehicle. *Energy* **2021**, *230*, 120858. [CrossRef]
5. Xiao, B.; Walker, P.D.; Zhou, S.; Yang, W.; Zhang, N. A power consumption and total cost of ownership analysis of extended range system for a logistics van. *IEEE Trans. Transp. Electrif.* **2021**, *8*, 72–81. [CrossRef]
6. Xiao, B.; Yang, W.; Wu, J.; Walker, P.D.; Zhang, N. Energy management strategy via maximum entropy reinforcement learning for an extended range logistics vehicle. *Energy* **2022**, *253*, 124105. [CrossRef]
7. Rahman, K.M.; Jurkovic, S.; Stancu, C.; Morgante, J.; Savagian, P.J. Design and performance of electrical propulsion system of extended range electric vehicle (EREV) chevrolet volt. *IEEE Trans. Ind. Appl.* **2014**, *51*, 2479–2488. [CrossRef]
8. Taherzadeh, E.; Hamid, R.; Ali, M.S. A comprehensive study of the parameters impacting the fuel economy of plug-in hybrid electric vehicles. *IEEE Trans. Intell. Veh.* **2020**, *5*, 596–615. [CrossRef]
9. Xiao, B.; Ruan, J.; Yang, W.; Walker, P.; Zhang, N. A review of pivotal energy management strategies for extended range electric vehicles. *Renew. Sustain. Energy Rev.* **2021**, *149*, 111194. [CrossRef]
10. Hu, D.; Xie, H.; Song, K.; Zhang, Y.; Yan, L. An apprenticeship-reinforcement learning scheme based on expert demonstrations for energy management strategy of hybrid electric vehicles. *Appl. Energy* **2023**, *342*, 121227. [CrossRef]

11. Yan, M.; Xu, H.; Li, M.; He, H.; Bai, Y. Hierarchical predictive energy management strategy for fuel cell buses entering bus stops scenario. *Green Energy Intell. Transp.* **2023**, *2*, 100095. [CrossRef]
12. Zhang, Y.; Gao, B.; Jiang, J.; Liu, C.; Zhao, D.; Zhou, Q.; Chen, Z.; Lei, Z. Cooperative power management for range extended electric vehicle based on internet of vehicles. *Energy* **2023**, *273*, 127238. [CrossRef]
13. Kim, M.; Jung, D.; Min, K. Hybrid thermostat strategy for enhancing fuel economy of series hybrid intracity bus. *IEEE Trans. Veh. Technol.* **2014**, *63*, 3569–3579. [CrossRef]
14. Shabbir, W.; Evangelou, S.A. Threshold-changing control strategy for series hybrid electric vehicles. *Appl. Energy* **2019**, *235*, 761–775. [CrossRef]
15. Denis, N.; Dubois, M.R.; Trovao, J.P.F.; Desrochers, A. Power split strategy optimization of a plug-in parallel hybrid electric vehicle. *IEEE Trans. Veh. Technol.* **2017**, *67*, 315–326. [CrossRef]
16. Chen, B.-C.; Wu, Y.-Y.; Tsai, H.-C. Design and analysis of power management strategy for range extended electric vehicle using dynamic programming. *Appl. Energy* **2014**, *113*, 1764–1774. [CrossRef]
17. Peng, J.; He, H.; Xiong, R. Rule based energy management strategy for a series–parallel plug-in hybrid electric bus optimized by dynamic programming. *Appl. Energy* **2017**, *185*, 1633–1643. [CrossRef]
18. Yu, X.; Lin, C.; Zhao, M.; Yi, J.; Su, Y.; Liu, H. Optimal energy management strategy of a novel hybrid dual-motor transmission system for electric vehicles. *Appl. Energy* **2022**, *321*, 119395. [CrossRef]
19. Saiteja, P.; Ashok, B. Critical review on structural architecture, energy control strategies and development process towards optimal energy management in hybrid vehicles. *Renew. Sustain. Energy Rev.* **2022**, *157*, 112038. [CrossRef]
20. Yang, Y.; Bremner, S.; Menictas, C.; Kay, M. Modelling and optimal energy management for battery energy storage systems in renewable energy systems: A review. *Renew. Sustain. Energy Rev.* **2022**, *167*, 112671. [CrossRef]
21. Yang, Y.; Bremner, S.; Menictas, C.; Kay, M. Numerical investigation on fuzzy logic control energy management strategy of parallel hybrid electric vehicle. *Energy Procedia* **2019**, *158*, 2643–2648.
22. Shen, Y.; Li, Y.; Tang, Y.; Sun, J.; Zhao, J.; Yang, X. An energy management strategy based on fuzzy logic for hybrid energy storage system in electric vehicles. *IEEJ Trans. Electr. Electron. Eng.* **2022**, *17*, 53–60. [CrossRef]
23. Marzougui, H.; Kadri, A.; Martin, J.-P.; Amari, M.; Pierfederici, S.; Bacha, F. Implementation of energy management strategy of hybrid power source for electrical vehicle. *Energy Convers. Manag.* **2019**, *195*, 830–843. [CrossRef]
24. Wei, C.; Chen, Y.; Li, X.; Lin, X. Integrating intelligent driving pattern recognition with adaptive energy management strategy for extender range electric logistics vehicle. *Energy* **2022**, *247*, 123478. [CrossRef]
25. Yang, Y.; Zhang, Y.; Tian, J.; Li, T. Adaptive real-time optimal energy management strategy for extender range electric vehicle. *Energy* **2020**, *197*, 117237. [CrossRef]
26. Vatanparvar, K.; Faezi, S.; Burago, I.; Levorato, M.; Al Faruque, M.A. Extended range electric vehicle with driving behavior estimation in energy management. *IEEE Trans. Smart Grid* **2018**, *10*, 2959–2968. [CrossRef]
27. Yao, M.; Zhu, B.; Zhang, N. Adaptive real-time optimal control for energy management strategy of extended range electric vehicle. *Energy Convers. Manag.* **2021**, *234*, 113874. [CrossRef]
28. Zhu, D.; Pritchard, E.; Dadam, S.R.; Kumar, V.; Xu, Y. Optimization of rule-based energy management strategies for hybrid vehicles using dynamic programming. *arXiv* **2022**, arXiv:2207.06450. [CrossRef]
29. Zeng, P.; Li, H.; He, H.; Li, S. Dynamic energy management of a microgrid using approximate dynamic programming and deep recurrent neural network learning. *IEEE Trans. Smart Grid* **2018**, *10*, 4435–4445. [CrossRef]
30. Li, J.; Song, Z.; Wang, X.; Wang, Y.; Jia, Y. A novel offshore wind farm typhoon wind speed prediction model based on PSO–Bi-LSTM improved by VMD. *Energy* **2022**, *251*, 123848. [CrossRef]
31. Zhao, L.; Li, Z.; Qu, L.; Zhang, J.; Teng, B. A hybrid VMD-LSTM/GRU model to predict non-stationary and irregular waves on the east coast of China. *Ocean. Eng.* **2023**, *276*, 114136. [CrossRef]
32. Cheng, Z.; Chen, B.; Lu, R.; Wang, Z.; Zhang, H.; Meng, Z.; Yuan, X. Recurrent neural networks for snapshot compressive imaging. *IEEE Trans. Pattern Anal. Mach. Intell.* **2022**, *45*, 2264–2281. [CrossRef] [PubMed]
33. Ewees, A.A.; Mohammed, A.A.; Laith, A.; Elaziz, M.A. HBO-LSTM: Optimized long short term memory with heap-based optimizer for wind power forecasting. *Energy Convers. Manag.* **2022**, *268*, 116022. [CrossRef]
34. Granata, F.; Fabio, D.N.; Giovanni, D.M. Stacked machine learning algorithms and bidirectional long short-term memory networks for multi-step ahead streamflow forecasting: A comparative study. *J. Hydrol.* **2022**, *613*, 128431. [CrossRef]
35. Pan, Y.G.; Wu, Y.M.; Lam, H.-K. Security-based fuzzy control for nonlinear networked control systems with DoS attacks via a resilient event-triggered scheme. *IEEE Trans. Fuzzy Syst.* **2022**, *30*, 4359–4368. [CrossRef]
36. Lin, H.; Han, Y.; Cai, W.; Jin, B. Traffic signal optimization based on fuzzy control and differential evolution algorithm. *IEEE Trans. Intell. Transp. Syst.* **2022**, *24*, 8555–8566. [CrossRef]
37. Liu, H.; Zhao, J.Y.; Qing, T.; Li, X. Energy consumption analysis of a parallel PHEV with different configurations based on a typical driving cycle. *Energy Rep.* **2021**, *7*, 254–265. [CrossRef]

actuators

Article

Vehicle Sideslip Angle Estimation Based on Radial Basis Neural Network and Unscented Kalman Filter Algorithm

Chuanwei Zhang *, Yansong Feng, Jianlong Wang, Peng Gao and Peilin Qin

College of Mechanical Engineering, Xi'an University of Science and Technology, Xi'an 710054, China; 21205224108@stu.xust.edu.cn (Y.F.); 20105016014@stu.xust.edu.cn (J.W.); 21205016042@stu.xust.edu.cn (P.G.); 20105016012@stu.xust.edu.cn (P.Q.)

* Correspondence: zhangcw@xust.edu.cn

Abstract: Most existing ESC (electronic stability control) and ADS (auto drive system) stability controls rely on the measurement of yaw rate and sideslip angle. However, the existing sensors are too expensive, which is one of the factors that makes it difficult to measure the side slip angle of vehicles directly. Therefore, the estimation of sideslip angle has been extensively discussed in the relevant literature. Accurate modeling is complicated by the fact that vehicles are highly nonlinear. This article combines a radial basis function neural network with an unscented Kalman filter to propose a new sideslip angle estimation method for controlling the dynamic behavior of vehicles. Considering the influence of input data type and sensor ease of measurement factors on the results, a two-degrees-of-freedom vehicle nonlinear dynamic model was established, and a radial basis function neural network estimation algorithm was designed. In order to reduce the impact of noise and improve the reliability of the algorithm, the neural network algorithm was combined with the Kalman filter. The information collected from low-cost sensors for actual vehicle operation (longitudinal vehicle speed, steering wheel angle, yaw rate, lateral acceleration) was trained using a radial basis function neural network to obtain a "pseudo slip angle". The "pseudo slip angle", yaw rate, and lateral acceleration are input as observations of the Kalman filter. The sideslip angle obtained from different observation methods was compared with the values provided by the Carsim 2020. The experiment shows that the sideslip angle estimator based on the radial basis function neural network and unscented Kalman filter achieves the optimal effect.

Keywords: sideslip angle estimation; neural network; Kalman filter; sensor fusion

1. Introduction

From electronic stability control (ESC) and advanced driver assistance system (ADAS) to partial or full automatic driving, vehicle stability is an essential consideration in developing reliable active safety and auto drive system (ADS) to enhance passenger safety and improve the reliability of these systems in intelligent transportation settings [1,2]. The accuracy and robustness of the estimation of sideslip angle and yaw rate as key information for evaluating vehicle stability are necessary [3]. Yaw rate can be directly measured through IMU, and sideslip angle can be directly measured by optical or GPS sensors [4]. However, the accuracy and reliability of the yaw rate and sideslip angle are related to cost. In the case of an actual production car, the expense of these direct measurement methods far outweighs the cost of using them.

Researchers have conducted extensive research on this issue. According to different research methods, there are mainly two types [5,6]: building vehicle models and data-driven. The estimation methods based on vehicle models can be divided into dynamic models, kinematic models [7], and a combination of the two [8]. The estimation method based on kinematic models mainly involves the numerical integration of sensors [9]. Due to the accumulation of errors in the long-term integration process, the accuracy of the sensor is required to be very high. The method based on dynamic models often involves constructing vehicle models with different degrees of freedom, such as the vehicle two-degrees-of-freedom model (2-dof) [10], the vehicle three-degrees-of-freedom model (3-dof) [11], and higher vehicle models [12,13]. Kalman filters have been widely used for their ability to effectively reduce the noise impact of sensors. When the tire is running in a linear region, linear Kalman filtering is sufficient, but its estimation effect deteriorates when the vehicle is driving in certain extreme conditions. In this case, most of the literature selects nonlinear tire models such as the brush model [14], Magic Formula of Pacejka [15], Arctangent tire model [16], and extended Kalman filter (EKF) and unscented Kalman filter (UKF) based on nonlinear vehicle models, which exhibit better estimation results [17–19]. A high degree of freedom vehicle dynamics model was established, taking into account lateral and longitudinal forces and road adhesion factors, using existing onboard sensors for real-time estimation. However, the estimation performance based on dynamic models strictly depends on the complexity of the model and the accuracy of the vehicle parameters and does not adapt well to the uncertainty of vehicle parameters or driving conditions. In response to the shortcomings of the above two methods, some scholars have adopted a combination of kinematics and dynamics. Reference [20] uses the Inertial Navigation System (R-INS) and global navigation satellite system (GNSS) sensors to jointly improve the heading angle error through kinematics and dynamics to estimate the sideslip angle. Reference [21] used Kalman filtering technology for vehicle state estimation, selecting the noise covariance of each Kalman filter as the weight, and data fusion was performed on the estimation results of the two. Experimental results showed that the fusion estimation results have better robustness to sensor bias and model parameter changes.

The data-driven approach mainly involves training input and output data to obtain a black box model. A related study found artificial intelligence (AI) algorithms that avoid the problems associated with identifying and tuning the parameters of the reference model [22]. In references [23–25], an adaptive fuzzy inference system (ANFIS) was used to estimate the sideslip angle based on artificial neural networks (ANN) and deep neural networks (DNN), respectively. In addition, reference [26] used the generalized regression neural network (GRNN) from radial basis function (RBF) for estimation, and the results showed that the trained GRNN model had high estimation accuracy and fast speed. The main problem with existing neural networks is that it is difficult to obtain all the driving condition training datasets, so it is difficult to generalize.

In recent years, multiple studies have shown a trend of data-driven and model fusion. Reference [27] proposes a principle component analysis (PAC) multivariate analysis, K-means improved RBF, and DRBF-EKF vehicle state estimation method for estimating road friction coefficient and vehicle speed using two RBFs. Reference [28] adopts the fusion method of ANFIS and Kalman filtering and uses the predicted sideslip angle of ANFIS as the input of Kalman filtering in the form of a "pseudo slip angle" to overcome the shortcomings of insufficient training data in existing ANFIS methods. However, this method only uses yaw rate and "pseudo slip angle" in the input of measured values, resulting in insufficient accuracy of the results. Reference [29] considers the impact of uncertainty in neural network results, using RNN to predict the estimated value of sideslip angle and its uncertainty. However, uncertainty learning based on neural networks is a difficult and complex process. Reference [30] only used the information obtained from IMU to fuse ANN and kinematic models to estimate the sideslip angle, but using IMU-measured data to generate neural networks alone cannot accurately estimate the magnitude of the sideslip angle.

Therefore, this article proposes a new observer based on the radial basis function neural network and Kalman filter to estimate vehicle sideslip angle, which effectively combines the advantages of the radial basis function neural network and Kalman filter. Considering the impact of input data types and sensor ease of measurement factors on the results, referring to the vehicle nonlinear dynamic model and the vehicle nonlinear model, a mapping relationship between system input and output can be established. Based on this mapping relationship, a radial basis function neural network estimator is used in order to reduce the impact of noise and improve the reliability of the algorithm, and the neural network algorithm is combined with Kalman filtering. The radial basis function neural network estimates the "pseudo slip angle" based on the vehicle parameters (longitudinal speed, lateral acceleration, yaw rate, steering wheel angle) that are easily measured by actual vehicle sensors and uses lateral acceleration, yaw rate, and "pseudo slip angle" as the Kalman filter observation values. This observer has good robustness, and due to the presence of lateral acceleration and yaw rate in the Kalman filter observation equation, it can reduce the impact of inaccurate "pseudo slip angle" values output by the radial basis function.

The structure of this article is as follows. In the Section 2, two-degrees-of-freedom vehicle dynamics models were established based on linear and Arctangent tire models. The Section 3 describes the basic structure of the observer, radial basis neural network algorithm, and neural network fusion with KF, EKF, and UKF. The Section 4 discusses the effect of multiple observers under different operating conditions. The Section 5 is the conclusion section.

2. Vehicle Dynamics Model

We consider a 2-DOF vehicle model, which is widely adopted to describe the vehicle's lateral and yaw motion. The vehicle two-degrees-of-freedom model is shown in Figure 1. The assumptions considered in this model include the following:

(1) Ignore the impact of the steering system and directly use the front wheel angle as input.
(2) Ignore the effect of suspension; it is considered that the car compartment only moves in a plane parallel to the ground.
(3) The forward speed of the vehicle along the *x*-axis is considered constant.
(4) The driving force is not large, and the impact of tangential ground forces on tire cornering characteristics is not considered.
(5) There is no aerodynamic effect, and the changes in tire cornering characteristics caused by load changes on the left and right tires and the effect of tire alignment torque are ignored.

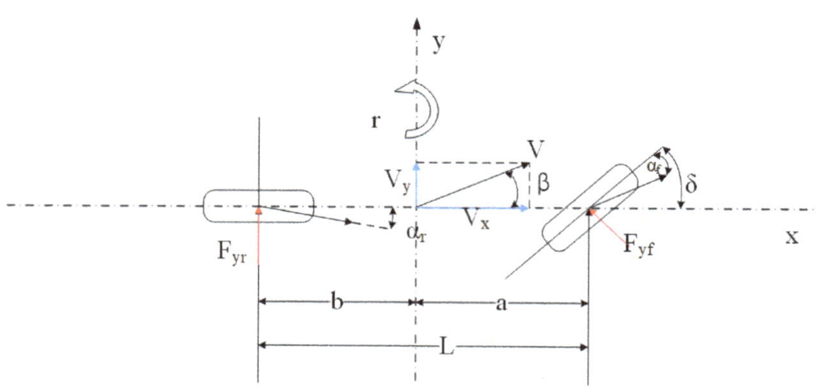

Figure 1. 2-DOF vehicle dynamics model.

The state space equations of the model are:
Lateral motion:

$$m\left(\dot{V}_y + rV_x\right) = F_{yf}\cos(\delta) + F_{yr} \quad (1)$$

Yaw motion:

$$I_z r = a\left(F_{yf}\cos(\delta)\right) - bF_{yr} \quad (2)$$

where m is the vehicle mass; V_y and V_x are longitudinal and lateral velocities of CG; F_{yf} is the lateral tire force of the front wheels; F_{yr} is the lateral tire force of the rear wheels; I_z is the moment of inertia of the vehicle; r the yaw rate velocity of the vehicle; δ is the front and steering angles; α_f, α_r are the sideslip angles of the front and rear wheels; L is the wheelbase of the front and rear axles; a is the distance from the center of mass to the front axle; b is the distance from the center of mass to the rear axle.

Two tire models are used to verify the effectiveness of the algorithm: the linear tire model and the nonlinear model.

The linear tire model is as follows:

$$F_y = C\alpha \tag{3}$$

where C is the tire cornering stiffness.

The front and rear wheel side slip angle are:

$$\alpha_f = -\beta + \delta - \frac{a}{V_x} r \tag{4}$$

$$\alpha_r = -\beta + \frac{b}{V_x} r \tag{5}$$

Equations (3)–(5) are combined with (1) and (2) to obtain a linear two-degrees-of-freedom model:

$$\dot{\beta} = \frac{C_f + C_r}{mV_x}\beta + \left(\frac{aC_f - bC_r}{mV_x^2} - 1\right)r - \frac{C_f}{mV_x}\delta \tag{6}$$

$$\dot{r} = \frac{aC_f - bC_r}{I_z}\beta + \frac{a^2C_f + b^2C_r}{I_zV_x}r - \frac{aC_f}{I_z}\delta \tag{7}$$

where $\dot{\beta}$ is the angular acceleration of the sideslip angle of the center of mass, and $\dot{r}$ is yaw angular acceleration. C_f and C_r represent the cornering stiffness of the front and rear axle.

The nonlinear tire model:

$$F_y = c_1 \arctan(c_2\alpha) \tag{8}$$

where c_1, c_2 are tire model parameters. The arctangent tire model is easy to design and solve for estimators and can also have good fitting accuracy in the area of sideslip angle before the tire reaches the adhesion limit.

$$\dot{\beta} = -r + \frac{c_{f1}\arctan\left(c_{f2}\left(-\beta + \delta - \frac{a}{V_x}r\right)\right)\cos(\delta) + c_{r1}\arctan\left(c_{r2}\left(-\beta + \frac{b}{V_x}r\right)\right)}{mV_x} \tag{9}$$

$$\dot{r} = \frac{1}{I_z}\left[ac_{f1}\arctan\left(c_{f2}\left(-\beta + \delta - \frac{a}{V_x}r\right)\right)\cos(\delta) - bc_{r1}\arctan\left(c_{r2}\left(-\beta + \frac{b}{V_x}r\right)\right)\right] \tag{10}$$

Among them, c_{f1}, c_{f2}, c_{r1}, c_{r2} are the front and rear wheel tangent model parameters, respectively.

3. Proposed Vehicle State Observer Based on RBF and Kalman Filter

The sideslip angle of a vehicle is a fundamental parameter, and its knowledge is the foundation of vehicle lateral stability control. The sideslip angle of a vehicle is the angle between the direction of travel and the direction of the vehicle, defined as:

$$\beta = \arctan\frac{V_x}{V_y} \tag{11}$$

This article introduces a new fusion estimation algorithm to estimate sideslip angles. The structure of this fusion estimation method is shown in Figure 2.

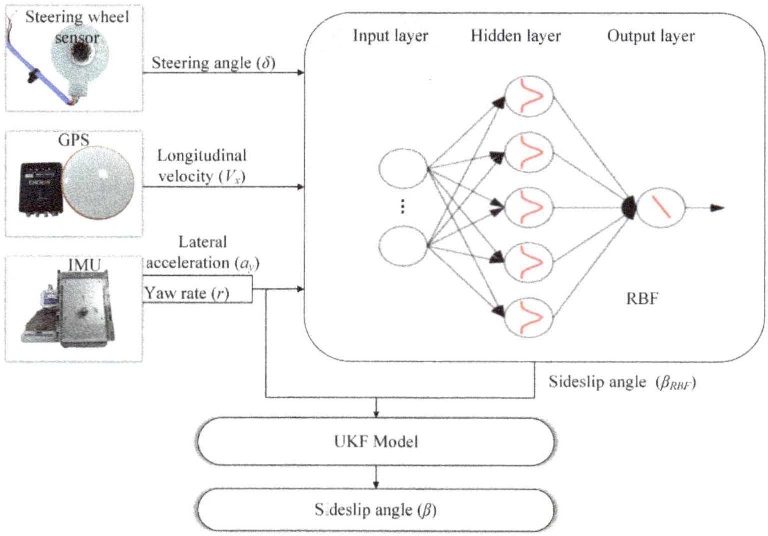

Figure 2. Estimator architecture.

The former is an estimator based on RBF neural networks, which can estimate the centroid sideslip angle (pseudo sideslip angle). The input for this section is the steering angle measured by the steering wheel angle sensor, as well as the longitudinal velocity measured using GPS and the longitudinal acceleration and yaw rate measured using the IMU inertial measurement unit. The advantage of this observer is to use existing sensors in the vehicle for measurement. The observer based on RBF outputs a "Pseudo sideslip angle", which serves as the input for the second part. Because this "Pseudo sideslip angle" is affected by noise, it cannot be directly used as a reference variable for vehicle lateral control. Therefore, we need the second part to filter out the impact of noise.

The second part is the Kalman filter. The module takes the lateral acceleration and yaw rate measured by IMU and the "pseudo sideslip angle" as the observations and combines the two-degrees-of-freedom model to obtain the final sideslip angle.

Finally, in the update stage of the Kalman filter, a new centroid sideslip angle is obtained to minimize the estimation error and obtain the optimal estimation value. To demonstrate the effectiveness of the proposed filtering algorithm, different Kalman filters (Kalman filter, extended Kalman filter, unscented Kalman filter) are considered.

3.1. *Radial Basis Function Neural Network*

RBF neural networks have attracted much attention due to their excellent generalization ability, simple network structure, and fast training speed. At the same time, relevant research also shows that RBF neural networks can approximate any nonlinear function with any precision [31,32].

The factors considered in selecting inputs for the RBF algorithm are the minimum type of input data; data that can be measured by the sensors included in the vehicle body are selected.

As well as from Equations (4) and (7), it can be seen that:

$$\beta = f\left(r, V_x, \dot{\beta}, \delta\right) \tag{12}$$

Assume V_x is a constant:

$$\frac{d\beta}{dt} = \dot{\beta} = \frac{a_y}{V_x} \tag{13}$$

We can think that:

$$\beta = f(r, V_x, a_y, \delta) \tag{14}$$

In summary, the following types of data are selected as inputs for the neural network: yaw rate (r), longitudinal velocity (V_x), steering angle (δ), lateral acceleration (a_y).

The RBF neural network has two layers: the hidden layer and the linear layer. Its structure is shown in Figure 3.

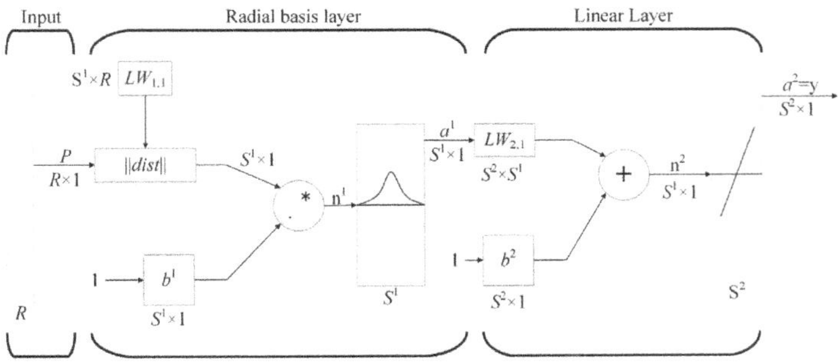

Figure 3. Radial basis function neural network architecture.

Where R denotes a single input vector and the matrix P denotes a matrix consisting of R input vectors.

$$P = \begin{bmatrix} p_1 \\ p_2 \\ p_3 \\ \cdots \\ p_{R-1} \\ p_R \end{bmatrix} \tag{15}$$

Hidden layer: The neuron activation function of the hidden layer is composed of radial basis functions. The array operation units composed of hidden layers are called hidden layer nodes. The hidden layer node contains a center vector $\boldsymbol{LW}_{1,1}$; $\boldsymbol{LW}_{1,1}$ and the input parameter vector P have the same dimension, and the Euclidean distance between them is defined as $\left|\left|\boldsymbol{LW}_{1,1(j)}(\boldsymbol{t}) - \boldsymbol{P}(\boldsymbol{t})\right|\right|$.

$$a^1{}_j(t) = exp\left(-\left|\left|\boldsymbol{LW}_{1,1(j)}(\boldsymbol{t}) - \boldsymbol{P}(\boldsymbol{t})\right|\right|^2 b_j{}^1\right),\ j = 1, \ldots, S^1 \tag{16}$$

Among them, b_j is a positive scalar representing the width of the Gaussian basis function; S^1 : S is the number of neurons in the hidden layer, 1 denotes the hidden layer, and a^1 is the results calculated by the hidden layer.

The output of the network is implemented using the following weighting function:

$$a^2{}_i(t) = \left(\sum_{j=1}^{m} \boldsymbol{LW}_{2,1(ji)}(\boldsymbol{t})\boldsymbol{h}_j(\boldsymbol{t})\right) + b_i{}^2 \quad i = 1, \ldots, S^2 \tag{17}$$

Among them, b_i is neuron thresholds of linear layers, $\boldsymbol{LW}_{2,1}$ is the weight of the linear layer, S^2: S is the number of neurons in the linear layer, 2 denotes the linear layer, and a^2 is the output of linear layers.

In this paper, Matlab's built-in radial basis neural network function was used to complete this work. The training process of the RBF neural network is shown in Figure 4. The four inputs mentioned in the above figure were selected for training to generate the required neural network model. At the same time, in order to prevent the algorithm from falling into local minima, a gradient descent method was used to obtain the number of hidden neurons, and finally, the optimal network was generated by testing the true values.

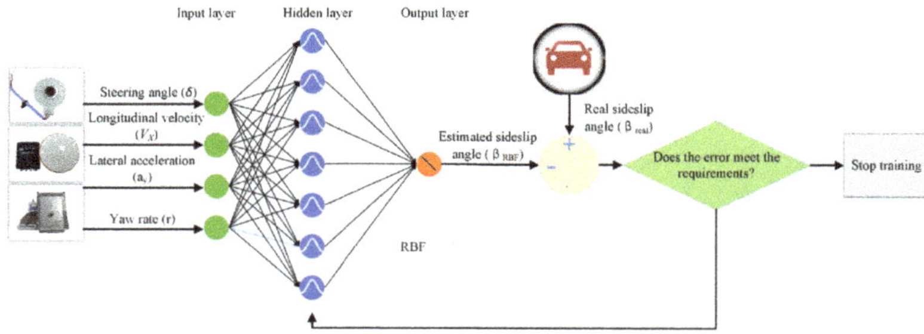

Figure 4. RBF learning process.

The selection of a dataset is also very important, as it directly affects the effectiveness of neural networks. Therefore, for data selection, we hope that it can simultaneously represent both linear and nonlinear features of the vehicle while also demonstrating its responsiveness to different vehicle speeds.

One hundred and twenty J-turn conditions with different vehicle speeds and steering wheel angles were selected. The training data are shown in Table 1.

Table 1. Set of simulation maneuvers for deep neural network training.

Road Friction Coefficient	Maneuver	Speed (km/h)	Steering (deg)	Frequency (Hz)
$\mu = 0.9$	Ramp steering	30/50/70/90/110	±0/±45/±75/±100/±125/±150	0.025
$\mu = 0.3$	Ramp steering	30/50/70/90/110	±0/±45/±75/±100/±125/±150	0.025

3.2. *Kalman Filter Incorporating Radial-Based Neural Networks*

3.2.1. Vehicle Dynamics Modeling for Estimator

The discrete nonlinear system of Kalman filtering can be represented as:

$$X(k+1) = f[X(k), U(k)] + W(k) \tag{18}$$

$$Z(k) = h[X(k), U(k)] + V(k) \tag{19}$$

where $f[\cdots]$ and $h[\cdots]$ are nonlinear equations; $X(k)$ represent the state vector; $U(k)$ is the input vector; $Y(k)$ is the measurement vector; $W(k)$ and $V(k)$ are assumed to be white noise, zero mean, and uncorrelated:

$$\begin{aligned} W(k) &\sim N(0, Q_k) \\ V(k) &\sim N(0, R_k) \end{aligned}$$

where Q_k and R_k are covariance matrices that describe the state and measurement noise.

By combining Equations (6) and (7), the state vector $X = \begin{bmatrix} \beta & r \end{bmatrix}^T$ of the state space equation of the dynamic linear model, the measurement vector $Y = \begin{bmatrix} a_y & r \end{bmatrix}^T$ can be represented as follows:

$$\dot{X} = \begin{bmatrix} \dot{\beta} \\ \dot{r} \end{bmatrix} = \begin{bmatrix} \frac{C_f+C_r}{mV_x}\beta + \left(\frac{aC_f-bC_r}{mV_x^2} - 1\right)r - \frac{C_f}{mV_x}\delta \\ \frac{aC_f-bC_r}{I_z}\beta + \frac{a^2C_f+b^2C_r}{I_zV_x}r - \frac{aC_f}{I_z}\delta \end{bmatrix} \tag{20}$$

$$Y = \begin{bmatrix} a_y \\ r \end{bmatrix} = \begin{bmatrix} \frac{C_f+C_r}{m}\beta + \left(\frac{aC_f-bC_r}{mV_x} - 1\right)r - \frac{C_f}{m}\delta \\ r \end{bmatrix} \tag{21}$$

By combining Equations (9) and (10), the state vector $X = \begin{bmatrix} \beta & r \end{bmatrix}^T$ of the state space equation of the dynamic nonlinear model, the measurement vector $Y = \begin{bmatrix} a_y & r \end{bmatrix}^T$ can be represented as follows:

$$\dot{X} = \begin{bmatrix} \dot{\beta} \\ \dot{r} \end{bmatrix} = \begin{bmatrix} -r + \frac{c_{f1}\arctan\left(c_{f2}\left(-\beta+\delta-\frac{a}{V_x}r\right)\right)\cos(\delta)+c_{r1}\arctan\left(c_{r2}\left(-\beta+\frac{b}{V_x}r\right)\right)}{mV_x} \\ \frac{1}{I_z}\left[ac_{f1}\arctan\left(c_{f2}\left(-\beta+\delta-\frac{a}{V_x}r\right)\right)\cos(\delta) - bc_{r1}\arctan\left(c_{r2}\left(-\beta+\frac{b}{V_x}r\right)\right)\right] \end{bmatrix} \tag{22}$$

$$Y = \begin{bmatrix} a_y \\ r \end{bmatrix} = \begin{bmatrix} -rV_x + \frac{c_{f1}\arctan\left(c_{f2}\left(-\beta+\delta-\frac{a}{V_x}r\right)\right)\cos(\delta)+c_{r1}\arctan\left(c_{r1}\left(-\beta+\frac{b}{V_x}r\right)\right)}{m} \\ r \end{bmatrix} \tag{23}$$

3.2.2. Radial Basis Neural Networks and Linear Kalman Filtering

When Equations (18) and (19) are linear, the state equation and measurement equation of the vehicle two degrees of freedom can be expressed as follows:

$$X(k+1) = AX(k) + BU(k) + W(k) \tag{24}$$

$$Y(k) = HX(k) + CU(k) + V(k) \tag{25}$$

where matrices $A = \begin{bmatrix} \frac{C_f+C_r}{mV_x} & \frac{aC_f-bC_r}{mV_x{}^2} - 1 \\ \frac{aC_f-bC_r}{I_z} & \frac{a^2C_f+b^2C_r}{I_zV_x} \end{bmatrix}$, $B = \begin{bmatrix} -\frac{C_f}{mV_x} \\ -\frac{aC_f}{I_z} \end{bmatrix}$.

To estimate the vehicle sideslip angle, Equations (20) and (21) are used as the state space equation and the measurement equation. The algorithmic process based on a radial basis neural network with Kalman filter is shown in Algorithm 1, where lines 6–10 are the standard Kalman filter process, $g(\cdot)$ is the neural network model, θ_k represents the sensor value sets for k second as explained in Section 3.1, and β_{rbf} is the output of the neural network model. The measurement vectors and covariance matrices (3–5 lines), matrices H and matrices C, are updated as follows:

$$\widetilde{Y}_k = \left[a_y, r, \beta_{rbf}\right]^T, \widetilde{R}_k = diag\left(\sigma_{ay}^2, \sigma_r^2, \sigma_{\beta rbf}^2\right), C = \begin{bmatrix} -\frac{C_f}{m} \\ 0 \\ 0 \end{bmatrix}, H = \begin{bmatrix} \frac{C_f+C_r}{m} & \frac{aC_f-bC_r}{mV_x} - V_x \\ 0 & 1 \\ 1 & 0 \end{bmatrix}$$

Algorithm 1 Radial Basis Neural Networks and Kalman Filtering.

1. **Function** RBF_KF($X_k, U_k, P_k, Y_k, \theta_k$)
2. Basic parameter input
3. $\beta_{rbf} = g(\theta_k)$
4. $\widetilde{Y}_k = \begin{pmatrix} Y_k \\ \beta_{rbf} \end{pmatrix}$
5. $\widetilde{R}_k = \begin{pmatrix} R_k & 0 \\ 0 & \sigma_{rbf}{}^2 \end{pmatrix}$
6. $X_{k+1|k} = AX_k + BU_k$
7. $Y_{k+1|k} = HX_{k+1|k} + CU_k$
8. $P_{k+1|k} = AP_kA^T + Q_k$
9. $K_{k+1} = P_{k+1|k}H^T\left[HP_{k+1|k}H^T + \widetilde{R}_k\right]^{-1}$
10. $X_{k+1} = X_{k+1|k} + K_{k+1}\left[\widetilde{Y}_k - Y_{k+1|k}\right]$
11. $P_{k+1} = [I - K_{k+1}]P_{k+1|k}$
12. **Return** X_{k+1} P_{k+1}

3.2.3. Radial Basis Neural Networks with Extended Kalman Filtering

Extended Kalman filtering involves first-order Taylor expansion of the nonlinear state equation around the $X(k)$. The Equations (18) and (19) can be described as:

$$X(k+1) = A(k)X(k) + B(k)U(k) + W(k) + \varnothing(k) \tag{26}$$

$$Y(k) = H(k)X(k) + C(k)U(k) + y(k) + R(k) \tag{27}$$

In the equations, $\varnothing(k)$, $R(k)$ are the second-order or higher-order terms of the Taylor series of the nonlinear state function at the filtered value $X(k)$, and $A(k)$, $H(k)$ are the Jacobian matrices of $f(\ldots)$ and $h(\ldots)$.

$$A(k) = \begin{bmatrix} \frac{\partial f_1}{\partial x_1} & \frac{\partial f_1}{\partial x_2} \\ \frac{\partial f_2}{\partial x_1} & \frac{\partial f_2}{\partial x_2} \end{bmatrix}$$

$$\frac{\partial f_1}{\partial x_1} = \frac{d_t}{mV_x}\left(-\frac{c_{f1}c_{f2}\cos(\delta)}{1+c_{f2}^2\left(\delta-\beta-\frac{a}{V_x}r\right)^2} - \frac{c_{r1}c_{r2}}{1+c_{f2}^2\left(-\beta+\frac{b}{V_x}r\right)^2}\right) + 1 \tag{28}$$

$$\frac{\partial f_1}{\partial x_2} = d_t\left[\frac{1}{mV_x}\left(-\frac{c_{f1}c_{f2}\frac{a}{V_x}\cos(\delta)}{1+c_{f2}^2\left(\delta-\beta-\frac{a}{V_x}r\right)^2} - \frac{c_{r1}c_{r2}\frac{b}{V_x}}{1+c_{f2}^2\left(-\beta+\frac{b}{V_x}r\right)^2}\right) - 1\right] \tag{29}$$

$$\frac{\partial f_2}{\partial x_1} = \frac{d_t}{I_z}\left(-\frac{ac_{f1}c_{f2}\cos(\delta)}{1+c_{f2}^2\left(\delta-\beta-\frac{a}{V_x}r\right)^2} + \frac{bc_{r1}c_{r2}}{1+c_{f2}^2\left(-\beta+\frac{b}{V_x}r\right)^2}\right) \tag{30}$$

$$\frac{\partial f_2}{\partial x_2} = \frac{d_t}{I_z}\left(-\frac{ac_{f1}c_{f2}\frac{a}{V_x}\cos(\delta)}{1+c_{f2}^2\left(\delta-\beta-\frac{a}{V_x}r\right)^2} - \frac{bc_{r1}c_{r2}\frac{b}{V_x}}{1+c_{f2}^2\left(-\beta+\frac{b}{V_x}r\right)^2}\right) + 1 \tag{31}$$

where d_t is the time interval between two data.

In order to estimate the vehicle sideslip angle, Equations (22) and (23) are used as state space equations and measurement equations. The algorithmic process based on a radial basis neural network with an extended Kalman filter is shown in Algorithm 2, where lines 6–10 are the standard extended Kalman filter process and lines 3–5 are the same as the same steps described in Algorithm 1. The new measurement vector and covariance are as in Algorithm 1. The new matrix $H(k)$ is shown below:

$$H(k) = \begin{bmatrix} \frac{\partial h_1}{\partial x_1} & \frac{\partial h_1}{\partial x_2} \\ \frac{\partial h_2}{\partial x_1} & \frac{\partial h_2}{\partial x_2} \\ \frac{\partial h_3}{\partial x_1} & \frac{\partial h_3}{\partial 2} \end{bmatrix}, \begin{bmatrix} \frac{\partial h_2}{\partial x_1} & \frac{\partial h_2}{\partial x_2} \\ \frac{\partial h_3}{\partial x_1} & \frac{\partial h_3}{\partial 2} \end{bmatrix} = \begin{bmatrix} 0 & 1 \\ 1 & 0 \end{bmatrix}$$

$$\frac{\partial h_1}{\partial x_1} = \frac{1}{m}\left(-\frac{c_{f1}c_{f2}\cos(\delta)}{1+c_{f2}^2\left(\delta-\beta-\frac{a}{V_x}r\right)^2} - \frac{c_{r1}c_{r2}}{1+c_{f2}^2\left(-\beta+\frac{b}{V_x}r\right)^2}\right) \tag{32}$$

$$\frac{\partial h_2}{\partial x_2} = \frac{1}{m}\left(-\frac{c_{f1}c_{f2}\frac{a}{V_x}\cos(\delta)}{1+c_{f2}^2\left(\delta-\beta-\frac{a}{V_x}r\right)^2} - \frac{c_{r1}c_{r2}\frac{b}{V_x}}{1+c_{f2}^2\left(-\beta+\frac{b}{V_x}r\right)^2}\right) - V_x \tag{33}$$

Algorithm 2 Radial Basis Neural Networks and Extended Kalman Filtering.

1. **Function** RBF_EKF$(X_k, U_k, P_k, Y_k, \theta_k)$
2. Basic parameter input
3. $\beta_{rbf} = g(\theta_k)$
4. $\widetilde{Y}_k = \begin{pmatrix} Y_k \\ \beta_{rbf} \end{pmatrix}$
5. $\widetilde{R}_k = \begin{pmatrix} R_k & 0 \\ 0 & {\sigma_{rbf}}^2 \end{pmatrix}$
6. $X_{k+1|k} = f(X_k, U_k)$
7. $Y_{k+1|k} = h\left(X_{k+1|k}, U_k\right)$
8. $P_{k+1|k} = AP_kA^T + Q_k$
9. $K_{k+1} = P_{k+1|k}H^T\left[HP_{k+1|k}H^T + \widetilde{R}_k\right]^{-1}$
10. $X_{k+1} = X_{k+1|k} + K_{k+1}\left[\widetilde{Y}_k - Y_{k+1|k}\right]$
11. $P_{k+1} = [I - K_{k+1}]P_{k+1|k}$
12. **Return** X_{k+1} P_{k+1}

3.2.4. Radial Basis Neural Networks with Unscented Kalman Filtering

Unscented Kalman filtering abandons the traditional practice of linearization of nonlinear functions and adopts a linear filtering framework. For the one-step prediction equation, the unscented transformation is used to deal with the nonlinear transfer problem of mean and covariance. The UKF approximates the probability density distribution of the nonlinear function and uses a series of determined samples to approximate the posterior probability density of the state.

The Equations (18) and (19) can be expressed as follows:

$$X(k+1) = f(X(k), U(k)) + W(k) \tag{34}$$

$$Y(k) = h(X(k), U(k)) + V(k) \tag{35}$$

The UKF first defines sigma points and then passes them to nonlinear functions. The mean and covariance of Gaussian are recovered by using the passed sigma points. The sigma points are defined symmetrically to the mean of the Gaussian as follows:

$$\begin{cases} X^{(0)} = \overline{X}, i = 0 \\ X^{(i)} = \overline{X} + \left(\sqrt{(n+\lambda)P}\right)_i, i = 1 \sim n \\ X^{(i)} = \overline{X} - \left(\sqrt{(n+\lambda)P}\right)_i, i = n+1 \sim 2n \end{cases} \tag{36}$$

where $\lambda = \left(\varepsilon^2(n+\kappa) - n\right)$, ε and κ are scaling parameters that determine how far the sigma points from the mean would be.

In order to recover the mean and the covariance of the Gaussian from the sigma points, each sigma point is given a weight as follows:

$$\begin{cases} \omega_m^{(0)} = \frac{\lambda}{n+\lambda} \\ \omega_m^{(0)} = \frac{\lambda}{n+\lambda} + \left(1 - \varepsilon^2 + \zeta\right) \\ \omega_m^{(i)} = \omega_c^{(i)} = \frac{1}{2(n+\lambda)}, i = 1 \sim 2n \end{cases} \tag{37}$$

where $\zeta \geq 0$ is the parameter containing prior information of the distribution.

Here, the mean and the covariance of the Gaussian distribution after the nonlinear function, $f(.)$, can be recovered from the passed sigma points, $\xi^i = f\left(X^{(i)}\right)$, and the weights, $\omega_m^{(i)}, \omega_c^{(i)}$, as shown in Equation (38).

$$\begin{gathered} \overline{X} = \Sigma_{i=0}^{2n} \omega_m^{(i)} \xi^i \\ P = \Sigma_{i=0}^{2n} \omega_c^{(i)} \left(\xi^i - \overline{X}\right)\left(\xi^i - \overline{X}\right)^T \end{gathered} \tag{38}$$

To estimate the vehicle sideslip angle, Equations (22) and (23) are used as the state space equation and the measurement equation. The algorithmic process based on a radial basis neural network with a traceless Kalman filter is shown in Algorithm 3, where lines 6–16 are the standard unscented Kalman filter process, and lines 3–5 are the same description of the same steps as in Algorithm 1. $g(\cdot)$ is the neural network model, and β_{rbf} is the output of the neural network model.

Algorithm 3 Radial Basis Neural Networks and Unscented Kalman Filtering.

1. **Function** RBF_UKF$(X_k, U_k, P_k, Y_k, \theta_k)$
2. Basic parameter input
3. $\beta_{rbf} = g(\theta_k)$
4. $\widetilde{Y}_k = \begin{pmatrix} Y_k \\ \beta_{rbf} \end{pmatrix}$
5. $\widetilde{R}_k = \begin{pmatrix} R_k & 0 \\ 0 & {\sigma_{rbf}}^2 \end{pmatrix}$
6. ${X_k}^i = \left[X_k, X_k + \sqrt{(n+\lambda)P_k}, X_k - \sqrt{(n+\lambda)P_k}\right]$
7. ${X_{k+1|k}}^i = f\left({X_k}^i, U_k\right)$
8. $X_{k+1|k} = \Sigma_{i=0}^{2n} \omega^{(i)} {X_{k+1|k}}^i$
9. $P_{k+1|k} = \Sigma_{i=0}^{2n} \omega^{(i)} \left[X_{k+1|k} - {X_{k+1|k}}^i\right]\left[X_{k+1|k} - {X_{k+1|k}}^i\right]^T + Q_t$
10. ${X_{k+1|k}}^i = \left[X_{k+1|k}, X_{k+1|k} + \sqrt{(n+\lambda)P_k}, X_{k+1|k} - \sqrt{(n+\lambda)P_k}\right]$
11. ${Y_{k+1|k}}^{(i)} = h\left[{X_{k+1|k}}^{(i)}, U_k\right]$
12. $Y_{k+1|k} = \Sigma_{i=0}^{2n} \omega^{(i)} {Y_{k+1|k}}^{(i)}$
13. $P_{Z_k Z_k} = \Sigma_{i=0}^{2n} \omega^{(i)} \left[{Y_{k+1|k}}^{(i)} - Z_{k+1|k}\right]\left[{Y_{k+1|k}}^{(i)} - Y_{k+1|k}\right]^T + \widetilde{R}_k$
    $P_{X_k Z_k} = \Sigma_{i=0}^{2n} \omega^{(i)} \left[{X_{k+1|k}}^{(i)} - Y_{k+1|k}\right]\left[{Y_{k+1|k}}^{(i)} - Y_{k+1|k}\right]^T$
14. $K_{k+1} = P_{X_k Z_k} P_{Z_k Z_k}^{-1}$
15. $X_{k+1} = X_{k+1|k} + K_{k+1}\left[\widetilde{Y}_k - Y_{k+1|k}\right]$
16. $P_{k+1} = P_{k+1|k} - K_{k+1} P_{Z_k Z_k} {K_{k+1}}^T$
17. **Return** X_{k+1} P_{k+1}

4. Results

Carsim 2020 software has been proven to be used to verify the effectiveness of the algorithm. Therefore, Carsim 2020 software was used to obtain relevant experimental data to prove the proposed RBF joint Kalman filter algorithm.

Since the real experimental data always have noise interference, we added Gaussian noise with a mean of zero and a variance of 0.01°, 0.01°/s, 0.01 m/s^2, and 0.01 km/h to the steering wheel angle, yaw rate, lateral acceleration, and longitudinal velocity obtained by Carsim 2020.

The vehicle model is a C-class hatchback car, and the tire model is 205/45 R17. The interval between the two data is 0.025 s. The hardware configuration we used is CPU 2.5 G HZ, no GPU, and an 8 GB memory module. Table 2 shows the parameters of the vehicle.

The effectiveness and strong robustness of the designed estimator are demonstrated by estimating the sideslip angle of vehicles under different operating conditions, as shown in Table 3. Firstly, there are four types of double lane changing conditions, and the 1.2 conditions, respectively, demonstrate the effectiveness of the designed sideslip angle estimator in high speed with high adhesion and high speed with low adhesion states, and the 3.4 conditions show the effectiveness of the designed sideslip angle estimator in high and medium adhesion states at lower vehicle speeds. The 5.6 conditions are the high attachment, middle attachment, high-speed J-turn maneuver. The above vehicle speeds are fixed, and the test conditions are not included in the training process.

Table 2. Vehicle of parameters for the C-Class hatchback car.

Vehicle Configuration Parameters	Value	Unit
Sprung mass (m_s)	1270	kg
Unsprung mass (m_u)	144	kg
Yaw inertia (I_z)	153.6	kg·m^2
Distance from front tire to COG (a)	1.015	m
Distance from rear tire to COG (b)	1.895	m
Effective rolling radius (R_w)	0.308	m
Height of COG (H)	0.540	m
Wheel track (L)	0.1675	m

Table 3. Experimental conditions for test dataset.

Condition	Condition Description	Speed (km/h)
Con.1	DLC (double lane change) and friction coefficient ($\mu = 0.85$)	100
Con.2	DLC (double lane change) and friction coefficient ($\mu = 0.3$)	100
Con.3	DLC (double lane change) and friction coefficient ($\mu = 0.85$)	50
Con.4	DLC (double lane change) and friction coefficient ($\mu = 0.5$)	50
Con.5	J-turn (75 deg) and friction coefficient ($\mu = 0.85$)	100
Con.6	J-turn (75 deg) and friction coefficient ($\mu = 0.5$)	100

Figure 5 shows the comparison overall estimation results of the RBF-based observer, the UKF-based observer, and the RBF-UKF-based observer with the double lane changing maneuver on the road with a friction coefficient of 0.85 at a speed of 100 km/h. It is seen that the three observers of RBF, UKF, and RBF-UKF can better reflect the changing trend of the actual vehicle sideslip angle, but their estimation accuracy is very different. RBF, compared with UKF and RBF-UKF, always presents the characteristics of larger and earlier changes to the expected results. The estimated value of the UKF observer is greater than the reference value in the range of 2.0~2.6 s and 5.0~5.5 s, and the estimated value is less than the reference value in the range of 3.5~4.5 s. The observation results of the RBF-UKF observer in the limit state are closer to the reference value. From a numerical point of view, the maximum errors of the observer based on RBF, UKF, and RBF-UKF observers are 0.0080, 0.0069, and 0.0061, respectively. The root mean square errors of the three observers are 0.0027, 0.0023, and 0.0016, respectively. The root mean square error and maximum error of the RBF-UKF observer are the smallest.

Figure 6 shows a double lane changing operation at a speed of 100 km/h on a road with a friction coefficient of 0.3. This condition shows that the proposed observer cannot provide the required force of the vehicle, that is, the observation effect under the out-of-control state. We can see that the RBF observer can roughly reflect the changing trend of the actual sideslip angle of the vehicle, but the estimation accuracy of the sideslip angle in −0.025~−0.01 and 0.015~0.03 (large sideslip angle or limit state) is still insufficient. Based on UKF, the estimation results of RBF-UKF observer have no reference value. The maximum errors of the observer based on RBF, UKF, and RBF-UKF observers are 0.0189, 0.0696, and 0.0587, respectively. The root mean square errors of the three observers are 0.0086, 0.0273, and 0.0237, respectively.

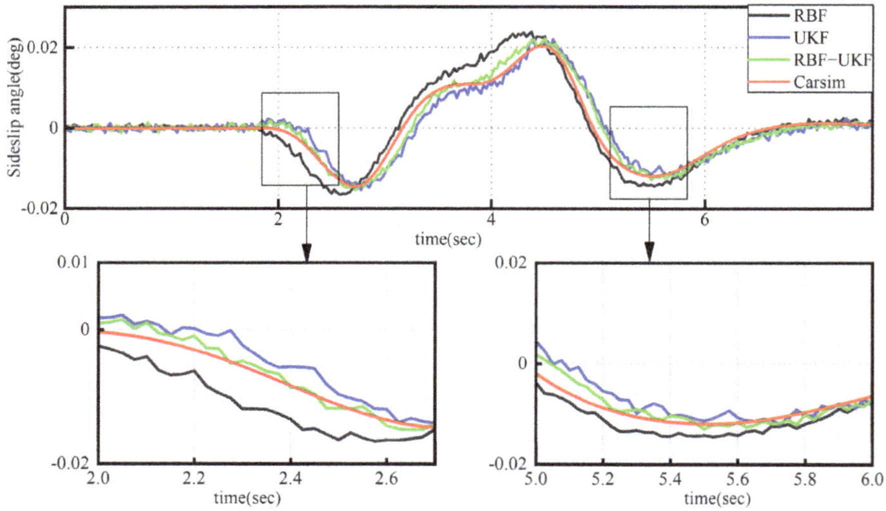

Figure 5. Results for sideslip angle estimation for condition 1.

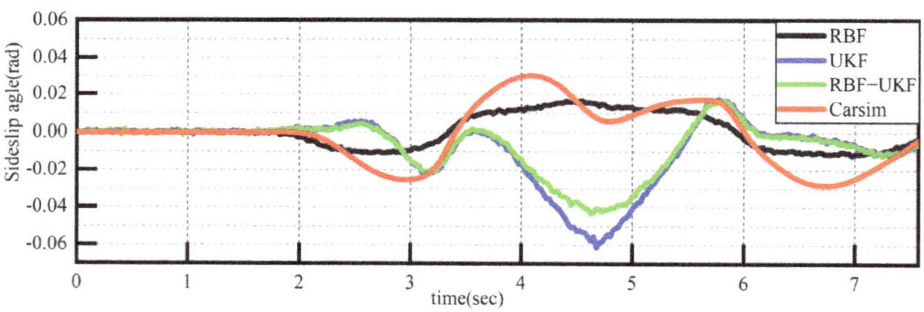

Figure 6. Results for sideslip angle estimation for condition 2.

Figure 7 shows the results for a vehicle traveling a double line changing at 50 km/h on a road surface with a friction coefficient of 0.85. The figure shows that the observation accuracy of all three observers exhibits good observation accuracy at medium speed and high attachment; when there is noise input, the UKF and RBF-UKF observers using the filtering algorithms show significantly less fluctuation in the estimation compared to the RBF-based observer to obtain the sideslip angle. The root-mean-square errors based on RBF, UKF, and RBF-UKF are 0.0011, 0.0016, and 0.0012, respectively. The maximum errors are 0.0041, 0.0050, and 0.0048, respectively.

Figure 8 shows that the vehicle performs a J-turn at a speed of 100 km/h on a road with a road adhesion coefficient of 0.85. The diagram shows that the three observers have good estimation results for the sideslip angle of the vehicle after entering the steady state (2–10 s), but the estimation results of the RBF observer are more consistent with the reference value; when the steering wheel suddenly changes and is fixed to a specific value, the transient change in the vehicle sideslip angle (0–1 s), the three observers do not show good estimation results; the UKF-RBF observer has a better observation effect during 1~2 s. The root of mean square errors of RBF, UKF, and RBF-UKF are 0.0055, 0.00106, and 0.0052, respectively, and the maximum errors are 0.0248, 0.0287, and 0.0248, respectively. According to these data, RBF-UKF fits the data better for the whole working condition.

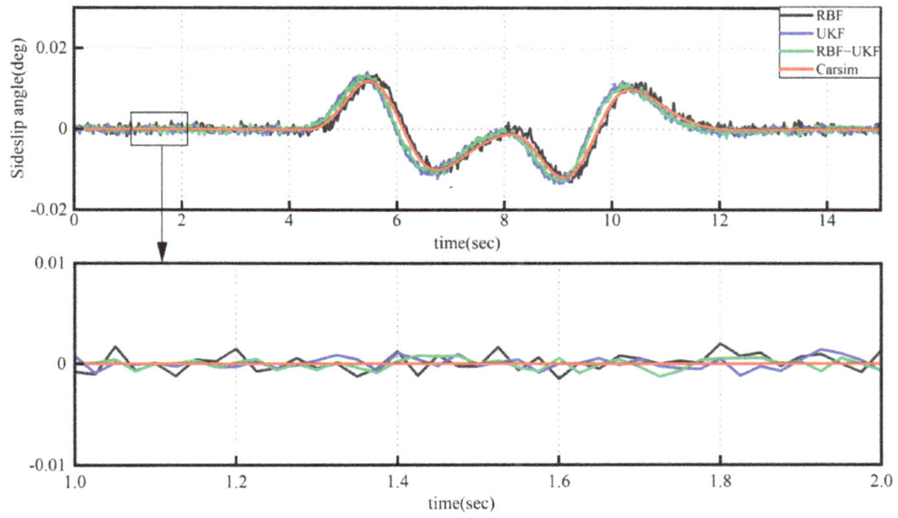

Figure 7. Results for sideslip angle estimation for condition 3.

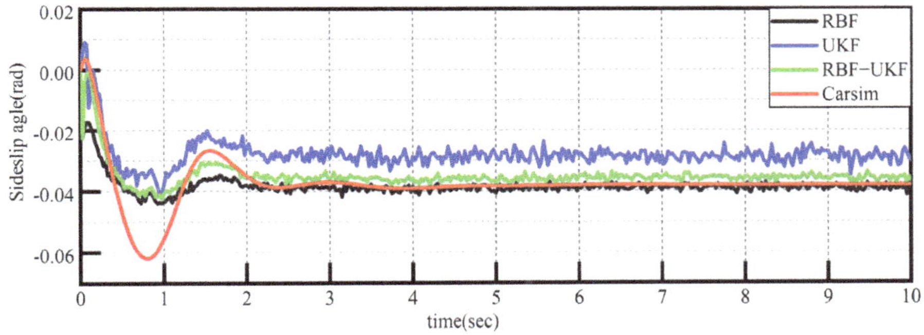

Figure 8. Results for sideslip angle estimation for condition 5.

Figure 9 shows that the vehicle performs a J-turn at a speed of 100 km/h on a road with a road adhesion coefficient of 0.5, and the estimated results of the RBF-UKF observer are more consistent with the reference values. RBF observer shows poor estimation of performance. The root of mean square errors of RBF, UKF, and RBF-UKF are 0.0139, 0.0150, and 0.0090, respectively, and the maximum errors are 0.0255, 0.0423, and 0.0343, respectively.

From the results, we can conclude that the proposed observer based on RBF-UKF achieves a better estimation of the sideslip angle.

In order to compare the performance, the root mean square error (RMSE) and the maximum error (Emax) are chosen. It has been shown that Kalman filtering is necessary to reduce noise measurements. From Table 4, it can be concluded that the root means square error of RBF-UKF is reduced by 0.0009 (Con.1), 0.001 (Con.4), 0003 (Con.5), and 0.0049 (Con.6) in comparison to RBF. The root means square error of RBF-UKF is reduced by 0.0007 (Con.1), 0.0004 (Con.3), 0.0003 (Con.4), 0.0054 (Con.5), and 0.006 (Con.6), respectively, as compared to UKF. Meanwhile, the root means square error of RBF-EKF is reduced by 0.0001 (Con.3) compared to that of RBF-UKF. Table 5 shows that RBF-EKF reduces the maximum error by 0.0006 (Con.1) and 0.0013 (Con.3) compared to RBF-UKF. In the whole test scenario, all observers fusing neural networks with Kalman filtering algorithms show better performance compared to considering RBF or Kalman algorithms alone. The RBF-

UKF-based observer provides better or the same performance compared to other Kalman-based observers (RBF-KF, RBF-EKF). In some cases, the RBF-EKF observer achieves better results than RBF-UKF, but it does not achieve a significant advantage in terms of overall working conditions.

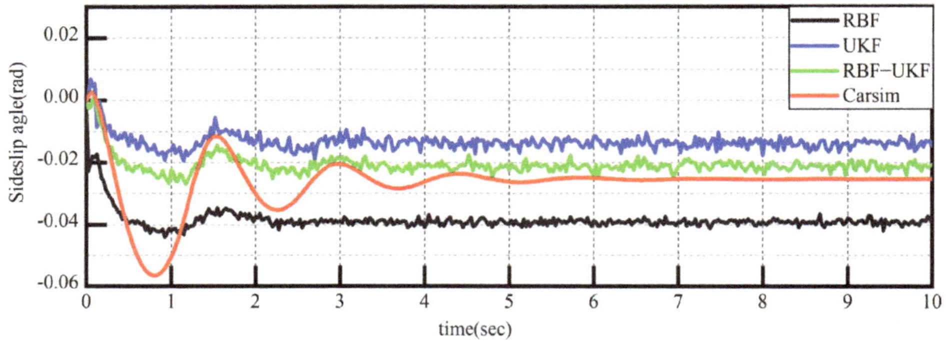

Figure 9. Results for sideslip angle estimation for condition 6.

Table 4. Root means square error of different test conditions (RMSE).

	Con.1	Con.2	Con.3	Con.4	Con.5	Con.6
RBF	0.0027	**0.0086**	0.0011	0.0024	0.0055	0.0139
KF + RBF	0.0017	0.0243	0.0011	0.0015	0.0055	0.0103
EKF	0.002	0.0273	0.0012	0.0014	0.011	0.0155
EKF + RBF	0.0019	0.0244	**0.0011**	0.0014	0.0083	0.0107
UKF	0.0023	0.0273	0.0016	0.0017	0.0106	0.0150
UKF + RBF	**0.0016**	0.0237	0.0012	**0.0014**	**0.0052**	**0.0090**

Table 5. Maximum error for different test conditions (Emax).

	Con.1	Con.2	Con.3	Con.4	Con.5	Con.6
RBF	0.008	**0.0189**	0.0041	0.0102	0.0248	**0.0255**
KF + RBF	0.0071	0.0701	0.0049	0.0057	**0.0239**	0.0393
EKF	0.0064	0.0696	0.0044	**0.0052**	0.0281	0.0442
EKF + RBF	**0.0055**	0.0587	**0.0035**	0.0056	0.0278	0.0373
UKF	0.0069	0.0696	0.005	0.0057	0.0287	0.0423
UKF + RBF	0.0061	0.0587	0.0048	0.0053	0.0248	0.0343

5. Conclusions

In this paper, a new RBF-UKF estimator is used to estimate the sideslip angle. The value of the "pseudo sideslip angle" is obtained by the neural network model, and the information is combined with the Kalman filter based on the dynamic model. The estimator fully integrates the advantages of the neural network and vehicle two-degrees-of-freedom model, which solves the generalization difficulty caused by the lack of training data of the neural network and also solves the accuracy problem of the two-degrees-of-freedom model under certain working conditions. At the same time, the Kalman filter algorithm can also effectively reduce the interference of the estimation process noise. The model trained by the neural network is still biased toward the effect of the vehicle dynamics model. In the state of the vehicle out of control, the RBF-based filtering algorithm cannot accurately estimate the magnitude of the sideslip angle under extreme operating conditions, and the RBF-UKF estimator cannot effectively estimate the vehicle sideslip angle. The model has been verified by a set of working conditions representing different test conditions.

Compared with the observer based on RBF and different Kalman filters, the RBF-UKF observer can more accurately reflect the change in vehicle sideslip angle.

6. Prospect

For neural network models, a high-precision vehicle simulation model can be constructed to generate ideal network training signals suitable for various driving situations to compensate for data deficiencies; in order to improve the robustness of the estimator, future research can attempt to estimate the real-time state of the vehicle or reduce the credibility of the dynamic state equation by incorporating kinematics in the state of runaway. At the same time, future research can attempt to use real-world vehicles for experiments.

Author Contributions: Conceptualization, C.Z., Y.F. and P.Q.; methodology, C.Z.; software, Y.F.; validation, C.Z., Y.F. and J.W.; formal analysis, P.G.; investigation, C.Z.; resources, C.Z.; data curation, C.Z. and P.Q.; writing—original draft preparation, Y.F.; writing—review and editing, P.G.; visualization, J.W.; supervision, J.W.; project administration, C.Z.; funding acquisition, C.Z. All authors have read and agreed to the published version of the manuscript.

Funding: This research was funded by the Shaanxi Innovation Talent Promotion Plan—Science and Technology Innovation Team (2021TD-27) and the 2022 Youth Innovation Team Construction Scientific Research Program of Shaanxi Provincial Education Department (22JP045).

Data Availability Statement: Not applicable.

Conflicts of Interest: The authors declare no conflict of interest.

References

1. Suzuki, Y.; Takeda, M. An overview on vehicle lateral dynamics and yaw stability control systems. *J. Adv. Vehic. Eng.* **2016**, *2*, 182–190.
2. Tang, X.; Yan, Y.; Wang, B.; Xu, X.; Zhang, L. Analysis of Intrinsic Mechanistic of Stability-Tracking Control for Distributed Drive Autonomous Electric Vehicle. *Electronics* **2021**, *10*, 3010. [CrossRef]
3. Bengler, K.; Dietmayer, K.; Farber, B.; Maurer, M.; Stiller, C.; Winner, H. Three decades of driver assistance systems: Review and future perspectives. *IEEE Intell. Transp. Syst. Mag.* **2014**, *6*, 6–22. [CrossRef]
4. Sun, B.; Zhang, T.; Gao, S.; Ge, W.; Li, B. Design of brake force dis-tribution model for front-and-rear-motor-drive electric vehicle based on radial basis function. *Arch Transp.* **2018**, *48*, 87–98. [CrossRef]
5. Guo, H.; Cao, D.; Chen, H.; Lv, C.; Wang, H.; Yang, S. Vehicle dynamic state estimation: State of the art schemes and perspectives. *IEEE/CAA J. Autom. Sin.* **2018**, *5*, 418–431. [CrossRef]
6. Chindamo, D.; Lenzo, B.; Gadola, M. On the vehicle sideslip angle estimation: A literature review of methods, models, and innovations. *Appl. Sci.* **2018**, *8*, 355. [CrossRef]
7. Xiong, L.; Xia, X.; Lu, Y.; Liu, W.; Gao, L.; Song, S.; Yu, Z. IMU-based automated vehicle body sideslip angle and attitude estimation aided by GNSS using parallel adaptive Kalman filters. *IEEE Trans. Veh. Technol.* **2020**, *69*, 10668–10680. [CrossRef]
8. Liu, W.; Xia, X.; Xiong, L.; Lu, Y.; Gao, L.; Yu, Z. Automated vehicle sideslip angle estimation considering signal measurement characteristic. *IEEE Sens. J.* **2021**, *21*, 21675–21687. [CrossRef]
9. Xia, X.; Xiong, L.; Lu, Y.; Gao, L.; Yu, Z. Vehicle sideslip angle estimation by fusing inertial measurement unit and global navigation satellite system with heading alignment. *Mech. Syst. Signal Process.* **2021**, *150*, 107290. [CrossRef]
10. Du, H.; Zhang, N.; Dong, G. Stabilizing vehicle lateral dynamics with considerations of parameter uncertainties and control saturation through robustyaw control. *IEEE Trans. Veh. Technol.* **2010**, *59*, 2593–2597.
11. Guo, H.; Chen, H.; Cao, D.; Jin, W. Design of a reduced-order non-linear observer for vehicle velocities estimation. *IET Control. Theory Appl.* **2013**, *7*, 2056–2068. [CrossRef]
12. Kim, J. Effect of vehicle model on the estimation of lateral vehicle dynamics. *Int. J. Automot. Technol.* **2010**, *11*, 331–337. [CrossRef]
13. Wenzel, T.A.; Burnham, K.; Blundell, M.; Williams, R. Dual extended kalman filter for vehicle state and parameter estimation. *Veh. Syst. Dyn.* **2006**, *44*, 153–171. [CrossRef]
14. Huang, X.; Wang, J. Robust sideslip angle estimation for lightweight vehicles using smooth variable structure filter. In Proceedings of the 2013 ASME Dynamic Systems and Control Conference, Palo Alto, CA, USA, 21–23 October 2013.
15. Davoodabadi, I.; Ramezani, A.A.; Mahmoodi-k, M.; Ahmadizadeh, P. Identification of tire forces using Dual Unscented Kalman Filter algorithm. *Nonlinear Dyn.* **2014**, *78*, 1907–1919. [CrossRef]
16. Guo, K.H.; Lei, R. *A Unified Semi-Empirical Tire Model with Higher Accuracy and Less Parameters*; SAE: Warrendale, PA, USA, 1999.
17. Li, L.; Jia, G.; Ran, X.; Song, J.; Wu, K. A variable structure extended kalman filter for vehicle sideslip angle estimation on a low friction road. *Veh. Syst. Dyn.* **2014**, *52*, 280–308. [CrossRef]

18. Bechtoff, J.; Isermann, R. Cornering stiffness and sideslip angle estimation for integrated vehicle dynamics control. *IFAC-PapersOnLine* **2016**, *49*, 297–304. [CrossRef]
19. Mosconi, L.; Farroni, F.; Sakhnevych, A.; Timpone, F.; Gerbino, F.S. Adaptive vehicle dynamics state estimator for onboard automotive applications and performance analysis. *Veh. Syst. Dyn.* **2022**, 1–25. [CrossRef]
20. Xia, X.; Hashemi, E.; Xiong, L.; Khajepour, A. Autonomous vehicle kinematics and dynamics synthesis for sideslip angle estimation based on consensus kalman filter. *IEEE Trans. Control. Syst. Technol.* **2022**, *31*, 179–192. [CrossRef]
21. Lee, H. Reliability indexed sensor fusion and its application to vehicle velocity estimation. *J. Dyn. Sys. Meas. Control.* **2006**, *128*, 236–243. [CrossRef]
22. Saadeddina, K.; Abdel-Hafezd, M.F.; Jaradatb, M.A.; Jarrahd, M.A. Performance enhancement of low-cost, high-accuracy, state estimation for vehicle collision prevention system using ANFIS. *Mech. Syst. Signal Process.* **2013**, *41*, 239–253. [CrossRef]
23. Du, H.; Zhang, N. Robust Vehicle Stability Control Based on Sideslip Angle Estimation. In *Robust Control Theory and Applications*; Bartos-zewicz, A., Ed.; InTech: Rijeka, Croatia, 2011; ISBN 978-953-307-229-6.
24. Melzi, S.; Sabbioni, E. On the vehicle sideslip angle estimation through neural networks: Numerical and experimental results. *Mech. Syst. Signal Process.* **2011**, *25*, 2005–2019. [CrossRef]
25. Boada, B.L.; Boada, M.J.L.; Gauchía, A.; Olmeda, E.; Díaz, V. Sideslip angle estimator based on ANFIS for vehicle handling and stability. *J. Mech. Sci. Technol.* **2015**, *29*, 1473–1481. [CrossRef]
26. Wei, W.; Shaoyi, B.; Lanchun, Z.; Kai, Z.; Yongzhi, W.; Weixing, H. Vehicle sideslip angle estimation based on general regression neural network. *Math. Probl. Eng.* **2016**, *2016*, 3107910. [CrossRef]
27. Shaohua, L.; Guiyang, W.; Zekun, Y.; Xuewei, W. Dynamic joint estimation of vehicle sideslip angle and road adhesion coefficient based on DRBF-EKF algorithm. *Li Xue Xue Bao* **2022**, *54*, 1853–1865.
28. Boada, B.; Boada, M.; Diaz, V. Vehicle sideslip angle measurement based on sensor data fusion using an integrated anfis and an unscented kalman filter algorithm. *Mech. Syst. Signal Process.* **2016**, *72*, 832–845. [CrossRef]
29. Kim, D.; Min, K.; Kim, H.; Huh, K. Vehicle sideslip angle estimation using deep ensemble-based adaptive Kalman filter-ScienceDirect. *Mech. Syst. Signal Process.* **2020**, *144*, 106862. [CrossRef]
30. Tommaso, N.; Renzo, C.; Claudio, A. An integrated artificial neural network–unscented Kalman filter vehicle sideslip angle estimation based on inertial measurement unit measurements. *Proc. Inst. Mech. Eng. Part D J. Automob. Eng.* **2018**, *233*, 095440701879064.
31. Hartman, E.J.; Keeler, J.D.; Kowalski, J.M. Layered neural networks with Gaussian hidden units asuniversal approximations. *Neural Comput.* **1990**, *2*, 210–215. [CrossRef]
32. Park, J.; Sandberg, L.W. Universal approximation using radial-basis-function networks. *Neural Comput.* **1991**, *3*, 246–257. [CrossRef]

actuators

MDPI

Review

A Survey of Vehicle System and Energy Models

Lingyun Hua, Jian Tang and Guoming Zhu *

Department of Mechanical Engineering, Michigan State University, East Lansing, MI 48824, USA; hualingy@msu.edu (L.H.); tangji11@msu.edu (J.T.)
* Correspondence: zhug@msu.edu

Abstract: Vehicle system models can be roughly divided into two categories, dynamic and steady-state (or quasi-steady-state) models, and can be applied to evaluate vehicle transient performance such as vehicle longitudinal and lateral dynamics, as well as energy economies like fuel or electricity consumption. This paper reviews various energy consumption models for automotive systems, focusing on component- and vehicle-level models. As the foundation to calculate the energy consumption, powertrain component models of three main vehicle types (internal combustion engine (ICE) vehicles, electric vehicles (EVs), and hybrid vehicles) are reviewed with their key components, including internal combustion engines, electric motors, and batteries. Three types of vehicle energy consumption models are explored according to their interpretability: white-box, black-box, and grey-box models. Optimizing vehicle energy usage based upon a vehicle energy consumption model is reviewed from the aspects of eco-driving and eco-routing problems at the end of the paper. Eco-driving research primarily selects models focusing on transient performance; whereas eco-routing focuses on steady-state or quasi-steady-state conditions to balance the needs of model accuracy and calculation efficiency for real-time applications. This review aims to guide model selection and inspire future applications of energy consumption models for advancing sustainable automotive technologies.

Keywords: vehicle powertrain model; powertrain component model; vehicle energy consumption model; economy route optimization; economy route–speed optimization

1. Introduction

The study of vehicle dynamics and energy consumption models is essential for developing vehicle control systems, such as speed regulation, powertrain calibration, and vehicle route optimization, especially as transportation efficiency and environmental sustainability become increasingly critical along with the rapid evolution of automotive technologies [1] in electric vehicles (EVs), hybrid electric vehicles (HEVs), and other propulsion systems. Vehicle energy consumption models integrated with energy source models are fundamental in estimating energy consumption under various driving conditions, ensuring that powertrain systems (conventional internal combustion engines (ICEs), hybrid electric systems, or fully battery–electric systems) are optimized for the best possible efficiency while meeting regulatory standards and consumer expectations. In recent years, significant research has focused on developing and refining powertrain and vehicle energy consumption models. These models have been widely used to enhance vehicle energy efficiency by optimizing vehicle routes and/or speed, minimizing fuel consumption, and improving overall vehicle performance. With the rise of advanced driver assistance systems (ADASs) and automated driving technologies, the integration of vehicle dynamics and energy models is even more critical, as these systems rely on precise predictions of vehicle dynamics

https://doi.org/10.3390/act14010010

behaviors and energy usage to make intelligent decisions in real time. This paper aims to provide a comprehensive review of existing energy consumption models, along with their component models, and their applications to different energy sources to optimize the vehicle route and speed simultaneously with reduced energy consumption. This review intends to allow researchers and engineers to compare methodologies, highlight the best practices, uncover limitations in existing models, and assess the effectiveness of physical, data-driven, or hybrid approaches to predicting real-world vehicle energy consumption performance, which could inspire the development of more robust, scalable, and adaptable approaches to support the improvement of vehicle energy consumption in a rapidly changing transportation landscape.

A comprehensive discussion of vehicle energy consumption models needs to be extended to their integration with different energy sources, such as fuel for internal combustion engine vehicles (ICEVs), batteries for electric vehicles (EVs), and hybrid energy (a combination of fuel and electric energy) used in hybrid electric vehicles (HEVs). Each of these powertrain systems can be modeled using its physical principles to describe the dynamics of energy flow and consumption. The study of energy consumption models must account for the distinct energy sources associated with individual types that impact vehicle performance differently. Among them, ICE powertrains have dominated the automotive industry for over a century, relying on fossil fuels to generate power. With growing environmental concerns, HEVs emerged, combining ICEs with electric motor(s) to improve fuel efficiency. More recently, EVs, relying entirely on electric motor(s), have gained popularity due to their zero-emission capabilities and advancement in battery technology. The performance of these powertrain systems can be modeled using fundamental physical principles [2], trained by extensive experimental data [3], and/or identified through system identification algorithms [4].

The powertrain layout and its power flow for ICEVs, EVs, and HEVs are illustrated in Figure 1. As mentioned, ICEVs are powered by an internal combustion engine that converts the fossil fuel to mechanical energy; see the diagram in the blue box of Figure 1. Then, the engine torque is transferred to the final drive and tires through the transmission, differential gears, and tires. The red box in Figure 1 describes an EV powertrain powered by electric motor(s) through the battery package. The DC/DC (or DC/AC) inverter supplies the electricity from the battery to the electric motor(s). Note that since the electric motor can also perform as a generator during regenerative braking, the motor torque output can be both positive and negative depending on its operational mode (driving or braking), and as a result, the motor operation could lead to battery discharge (providing driving power) or charge (regenerative braking), respectively. These EVs are usually equipped with a single-gear transmission (or without transmission). In addition, the EV also allows flexible powertrain architectures such as single-motor, dual-motor, and even triple-motor. The powertrain of an HEV, equipped with both an internal combustion (IC) engine and electric motor(s), is shown in the green box of Figure 1. The powertrain architecture of HEVs depends on how the engine and motor(s) are coupled, which can be divided into series and parallel HEVs. The IC engine can directly supply torque to drive the vehicle for parallel HEVs, or only be used to charge the battery through a generator for series HEVs.

There are three main approaches to vehicle system modeling in terms of the level of model transparency or the level of physical system knowledge used: physics-based, hybrid, and data-driven models, which are also known as white-box, grey-box, and black-box models, respectively [5,6]. Physical-based models are based on physical principles, such as the first law of thermodynamics and Newton's laws, to describe the behaviors of internal combustion engines, electric motors, and vehicle dynamics. Black-box models (or non-parametric models) encompass a wide range of techniques and approaches that rely on experimental data to represent system behavior without needing detailed knowledge of the system's internal structure. Black-box models can further be divided based on the methodologies or algorithms used to create them. There are also linear (or polynomial regression) and Gaussian process models, which use statistical techniques to understand relationships between variables based on historical data [7]; and decision trees (DTs), random forests (RFs), support vector machines (SVMs), and k-nearest neighbors (KNN); and machine learning (ML) methods can be utilized to find patterns in data without explicitly programming the relationships between inputs and outputs [8]. Furthermore, there is a subset of machine learning models called neural network models, inspired by biological neural networks and ideal for complex nonlinear relationships, such as artificial neural networks (ANNs), convolutional neural networks (CNNs), and recurrent neural networks (RNNs) [9–11]. Among them, machine learning and neural network-based methods are particularly popular due to their ability to model complex and nonlinear systems without the need for explicit knowledge of underlying physical processes. Grey-box models combine elements of both physical-based and data-driven approaches, balancing theoretical understanding with real-world data to improve model accuracy [12].

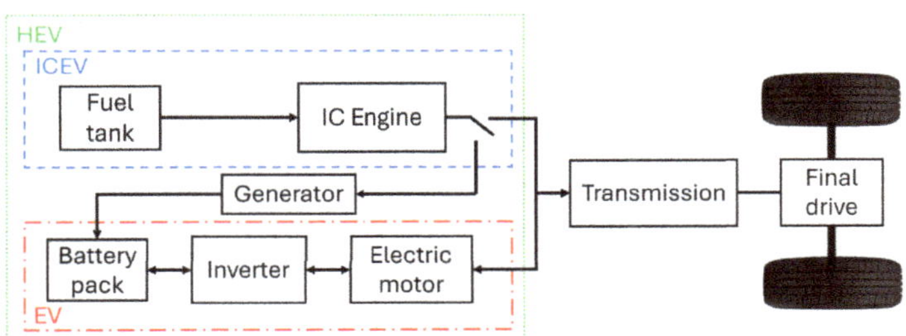

Figure 1. Example structure of ICEV, EV, and HEV.

In addition to the transparency of the model, there are two common modeling approaches used to represent vehicle characteristics: transient dynamics models and steady-state (or quasi-steady-state) models based on details of the system dynamics. Full transient dynamics models include detailed system dynamics which are mainly used for capturing the time-dependent behavior of vehicles and accounting for rapid transient changes in displacement, velocity, and power demand. Quasi-steady-state models, in contrast, assume that the vehicle operates in a close-to-steady-state mode, with slow variations in dynamic parameters over time.

In this paper, we review various models to estimate vehicle energy consumption. From the perspective of model interpretability, the models reviewed are divided into three categories: white-box, black-box, and grey-box models. From a power balance perspective, models can either follow a forward approach, where known powertrain characteristics are used to compute the energy required by the driver, or a backward approach, where kinematic quantities are used to calculate the energy generated by the powertrain [13]. The main contribution of this review lies in presenting the newly developed vehicle energy consumption models and introducing their architecture, and the advantages and disadvantages of using these models. It also provides application examples and explains how to choose and use these models in research and development.

This paper is organized as follows. Section 2 reviews three types of energy source models with associated applications to lay a foundation for vehicle model review. Section 3 describes the vehicle dynamics and energy consumption models with their disadvantages and advantages. Note that in Section 3, the vehicle energy consumption models will be discussed based on the models discussed in Section 2. The important applications of vehicle energy models are presented in Section 4 to address vehicle economy routes (eco-routes) and economy route–speed optimization problems. A summary is drawn in Section 5.

2. Powertrain Models of ICEVs, HEVs, and EVs

As the powertrain is a vehicle's power source, powertrain component models play a significant role in developing a vehicle energy consumption model. This section introduces the powertrain models from the aspect of powertrain types regarding three main vehicle types: internal combustion engine vehicles (ICEVs), electric vehicles (EVs), and hybrid vehicles. Vehicle models, based on powertrain dynamics, fall into three categories: transient, quasi-steady-state, and steady-state. Transient models capture rapid dynamic changes like braking and cornering, while quasi-steady-state models represent close-to-steady-state behavior with slow transients, suitable for emission analysis, such as in the quasi-steady NEDC (new European driving cycle [14]).

2.1. Internal Combustion Engine Vehicle (ICEV) Powertrain Model

The internal combustion engine plays a pivotal role in vehicle dynamics, serving as the heart that converts fossil fuels into mechanical power to propel the vehicle forward and backward [15]. IC engines are mainly divided into gasoline and diesel types. Gasoline engines offer high RPM and smooth operation, while diesel engines provide high torque and fuel efficiency. Both can be turbocharged, enhancing power by increasing air intake for better fuel–air mixtures and optimized boost control. Transmissions, such as continuously variable transmissions (CVTs) or geared types, optimize the power path.

Several modeling approaches exist to simulate and analyze engine internal dynamic performance with varying levels of complexity. For control purposes, at the micro-level [6], these models focus on the detailed combustion processes that contribute to torque generation. In contrast, macro-level models, such as map-based and data-driven models, are used for vehicle emission control and energy consumption analysis. In IC engines with fixed-geometry hardware and limited flexibility (e.g., fixed-geometry turbine, mechanical camshaft with fixed timing), performance tradeoffs are evident between high- and low-speed/load operations, particularly in engines with broad speed ranges, like gasoline engines. System integration often involves mapping engine speed–load characteristics to component maps, such as compressor maps, to optimize performance for specific applications, including vehicle or marine load cycles.

Map-based models are classified as steady-state, quasi-steady-state, and transient. Quasi-steady-state models, for instance, estimate emissions over predefined driving cycles, while steady-state models, like brake-specific fuel consumption (BSFC), torque, and power output maps, ignore transient effects like turbo lag. To account for these dynamics, transient elements are sometimes integrated with map-based models. From an energy consumption perspective, steady-state models (such as BSFC [15] maps showing fuel efficiency as fuel consumption per unit of power (e.g., g/kWh)), are valuable for highlighting optimal operating regions. An example of a BSFC map is shown in Figure 2, where the contours of constant BSFC on the torque–speed map show the locations at which the engine operates most efficiently. However, these maps require an extensive number of data points from experiments [16] or simulations (e.g., GT-Power models [17,18]), which are both time-consuming to generate and less suited for analytical combustion optimization and control.

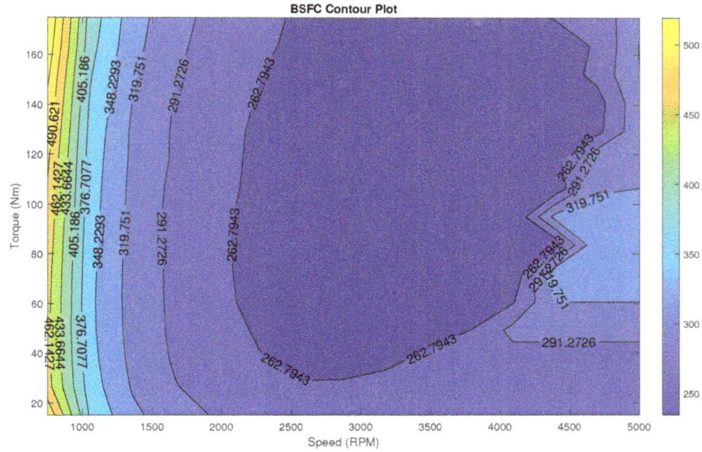

Figure 2. Example of BSFC contour map.

Researchers have developed various mathematical models to generate engine maps. Early models, such as those by Goering [16], mapped fuel consumption over engine speed and torque using fourth-order polynomials calibrated for a specific engine, though this process is time consuming. The BSFC model, which connects fuel consumption with brake thermal efficiency (BTE) and the fuel heating value, was later refined, including BTE as a function of engine speed and torque [19]. Data-driven approaches like ANNs and RVM further improve BSFC estimation accuracy [17], while methods like natural neighbor interpolation (NNI) account for operational altitude effects on efficiency [20]. Several other map-based models aim at assessing steady-state performance. Thermal efficiency maps [15] indicate the engine's fuel-to-power conversion efficiency at various operating points, critical for regulatory compliance evaluations [21]. Turbocharger boost maps outline boost pressure relative to engine RPM and throttle, essential for optimizing fuel consumption and emissions [22]. Torque and power curves show engine output across RPM, essential for understanding performance characteristics and tuning for applications like acceleration and towing. While these maps provide insights into steady-state performance, they do not capture transient dynamics, such as turbo lag or delays in air charge and fuel systems. Turbocharger dynamics can be modeled through differential equations that account for heat transfer and rotational inertia, and detailed CFD (computational fluid dynamics) models offer insights into flow and heat transfer under transient conditions. Emission maps, typically quasi-steady state, detail emissions (e.g., CO_2, NO_x) across various operating points and are critical for regulatory compliance [23]. Standards such as the new European driving cycle (NEDC) [14], the Federal Test Procedure (FTP) [24], and Worldwide Harmonized Light Vehicles Test Cycle (WLTC) [25] offer frameworks for assessing emission levels across different regions, with NEDC focusing on European standards and FTP addressing the US EPA (Environmental Protection Agency) standards with transient driving phases. In summary, these reviewed ICEV models are summarized in Table 1 below.

Table 1. Map-based macro-level models for IC engines.

Model Type	Steady-State Status	Model Name	Reference
Map-based model	Steady state	BSFC	Numerical-based map [15]
			Polynomial equation-based map [16,19]
			ANN-based map [17]
		Thermal efficiency map	Numerical-based map [15,21]
		Turbocharger boost map	Numerical-based map [22,26]
		Torque curve map	Numerical-based map [15,22]
		Power curve map	Numerical-based map [15,22]
	Transient state	Turbo lag model	Physical-based model [27]
	Quasi-steady state	Emission map model	NEDC and FTP [14,23–25]

For vehicle propulsion systems, transmissions are vital in ICE vehicles for optimal power delivery and fuel efficiency [28]. Key types include continuously variable transmission (CVT), which offers smooth gear changes but lower reliability and durability, and geared transmissions (such as manual transmission (MT), automated transmission (AT), and dual-clutch transmission (DCT)) known for their robustness under various operational conditions. A complete physical model for geared transmissions may include components like a torque converter, planetary gearbox, and differential [29]. For energy consumption modeling at the vehicle level, efficiency maps [30] are typically used for each gear. In AT systems, a torque converter map (steady-state) can be coupled with an observer to address transient errors [31]. Power losses, including gear windage [32] and shift losses [33], are calibrated experimentally and integrated into the efficiency map. Dynamic efficiency models for MT systems also generate gear-specific efficiency maps, which can be adjusted for system losses, including hydraulic and mechanical losses [34]. these reviewed transmission models are summarized in Table 2.

Table 2. Map-based transmission models.

Model Type	Steady-State Status	Model Application	Reference
Map-based model	Steady state	Efficiency map for each gear	Numerical-based map [30]
		Torque map of the torque converter for AT	Numerical-based map [31–33]
		Dynamic transmission efficiency model	Numerical-based map [30]
		Efficiency map for CVT	Numerical-based map [34]

2.2. Electric Vehicle (EV) Powertrain Model

Electric vehicle (EV) powertrain systems, including single-, dual-/triple-, and in-wheel motor configurations, are different in architecture and motor count [35]. EVs are purely powered by electric motor(s), known for high efficiency, zero emissions, instant torque, and regenerative braking capabilities [35]. Commonly used EV motor types include DC motors, induction motors, and permanent magnet (PM) motors [36]. A typical EV powertrain connects the battery (power source) to the motor via inverter(s), with transmissions featuring fewer gears than those in ICE vehicles due to the broader speed range of EV motors [37]. The architecture of a typical EV can be viewed in Figure 1 in Section 1, with the components in the box highlighted with red dash-dot lines. Note that some EVs use direct (hub wheel) drive instead of connecting to a transmission [38].

Electric motor models are categorized into micro- and macro-levels. Micro-level models, such as analytical electro-magnetic models and numerical models, consider transient motor performance. For the macro-level models, motor equivalent circuit models (ECMs) are simple and efficient for control applications, assuming ideal conditions, with a lack of accuracy for nonlinear transient behaviors [39,40]; analytical models (e.g., differential equation models, "d-q" transformation models [41,42], lumped parameter models) belong to the family of macro-models and use differential equations to describe electrical and mechanical dynamics, making them suitable for steady-state analysis. Among them, the lumped parameter model shares the concept of electric components with the ECMs or "d-q" transformation models and also describes the mechanical motor behaviors [42]. These models focus mainly on steady-state behaviors with limited descriptions of transient responses and nonlinear behaviors. Micro-level models, such as finite element analysis (FEA) models, though highly accurate for transient behaviors, are computationally intensive and not ideal for real-time applications [43].

For vehicle-level evaluations, steady-state behaviors are often modeled using ideal torque–speed models, assuming constant torque or power across speed ranges with zero loss [44]. To improve accuracy, efficiency maps [45], which incorporate experimental data on power losses, provide a realistic view of motor performance under various operating conditions, as shown in Figure 3.

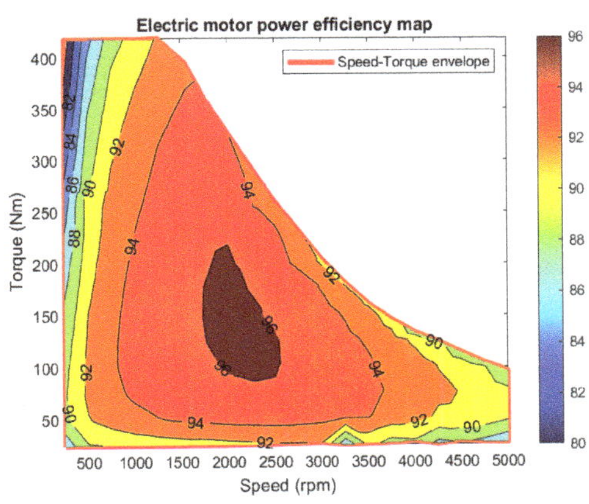

Figure 3. Example of motor speed–torque map with power efficiency.

The upstream components of electric motors used in EVs are the inverter and the battery, as shown in Figure 1. EV inverters convert DC power from the battery to AC (or pulse-width-modulated) power during driving and enable AC-to-DC conversion to charge the battery during regenerative braking. Inverters are modeled in detail at the micro-level based on inverter circuits or at a simplified macro-level (e.g., using efficiency maps). However, a constant-efficiency map is often used for vehicle-level control and energy analysis. The inverter efficiency model is frequently combined with the motor map for streamlined analysis [46].

As the EV's power source, the battery pack is typically modeled at the vehicle level using electro-mechanical, degradation, point-wise, and equivalent circuit models (ECMs) [47]. Electro-mechanical models capture detailed chemical reactions for high accuracy but are computationally intensive, making them unsuitable for real-time applications. Battery degradation models predict batteries' capacity loss due to aging, temperature, and usage, aiding battery management but requiring substantial processing time [48,49].

The point-wise inverter model is widely used in research due to its high accuracy in predicting battery state of charge (SOC) [50] and efficiency during charging and discharging processes. As a result, it can also be used to improve the charging efficiency [51]. For real-time vehicle-level controls and evaluations, equivalent circuit models (ECMs) are preferred due to their capability of estimating state of charge (SOC) and open-circuit voltage (V_{oc}), based on collected data, with computational efficiency. These models use simple electrical circuits (resistors, capacitors, and inductors), with additional RC (resistor and capacitor) branches enhancing accuracy at the cost of complexity [52]. ECMs can also account for battery aging by adjusting circuit component values over time, supporting control strategies for battery management and motor control [53]. A typical ECM with one parallel RC branch is given by the following equations (as shown in Figure 4):

$$SOC(t) = SOC(0) - \frac{1}{C_{max}} \int_0^t I(\tau)\, d\tau \tag{1}$$

$$V_t = V_{oc} - V_1 - R_0 I \tag{2}$$

$$\dot{V}_1 = -\frac{V_1}{R_1 C_1} + \frac{I}{C_1} \tag{3}$$

where R_0 is the internal resistance, I is the load current, and V_t is the terminal voltage that can be measured.

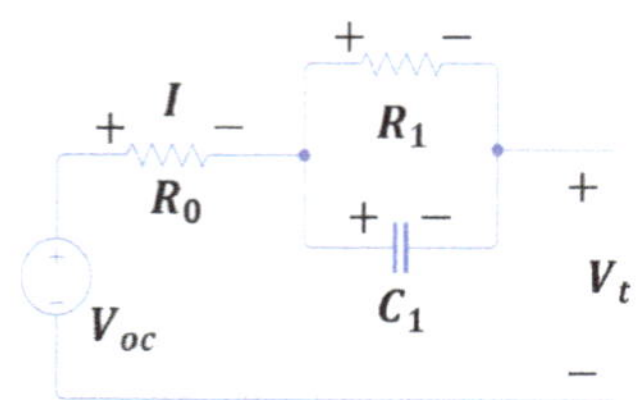

Figure 4. An example of a battery ECM with one parallel RC branch.

High-capacity EV battery packs are sensitive to temperature, challenging their thermal management in battery management systems (BMSs), especially during fast charging and high-power discharging. For optimal energy efficiency, the BMS must also manage charging and discharging efficiency to maximize driving range, reduce energy consumption, and extend battery life, which is crucial for route planning and reducing range anxiety. For instance, routes with more highway driving are preferred when the battery is nearly full to avoid overcharging from regenerative braking; and otherwise, routes with more local segments are chosen to utilize regenerative braking.

For calculating energy consumption, the electric motor power flows downstream to powertrain components (clutch, transmission, and final drive gears (if any)) and to wheels (see Figure 1). Normally, the downstream components have fixed transfer (gear) ratios that can be modeled by efficiency coefficients or maps.

2.3. Hybrid Electric Vehicle (HEV) Powertrain Model

Hybrid electric vehicles (HEVs) are valued for improved fuel economy, lower emissions, and better performance compared with ICE vehicles. Among HEV types (series, parallel, and split), the series hybrid electric vehicle (SHEV) is especially suited for urban transportation due to its strong transient performance, low cost, and simple architecture. Fuel economy and emissions in HEVs depend heavily on their supervisory control strategy, which dynamically allocates power between the internal combustion engine and the battery to meet power demands while minimizing fuel consumption and emissions [54]. Various control strategies have been developed based on series and parallel/power-split HEV powertrain models.

A general HEV powertrain model architecture is illustrated in Figure 1, where setting the engine power output switch downward to connect to the electrical power generator forms series HEVs, and engine power connecting to both the generator and transmission gives parallel/power-split HEVs. A sample serial hybrid powertrain model architecture can be found in Figure 5 from reference [55]. Similar to many other series hybrid electric vehicles, this sample series HEV utilizes a combination of two energy sources, fuel and electricity, as shown in Figure 5. There are two different power sources, where the main power source (the engine–generator set) provides the main power by converting the fuel chemical energy to mechanical energy using an internal combustion engine and then to electric energy using a generator; the assistant power source (the battery pack) acts as an electric power equalizer to provide the rest of the required power (positive or negative) through its charging and discharging processes.

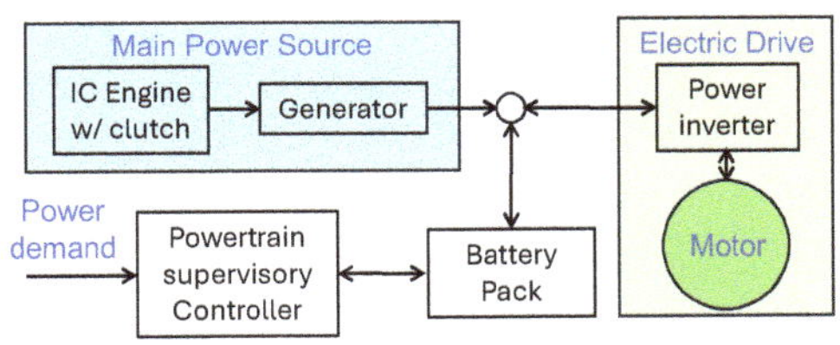

Figure 5. An example of a serial HEV powertrain system.

Note that most of these component models for ICEV and EV powertrain systems are discussed earlier in this section. The only new item is the power generator model. Most often, the generator model is modeled in a similar way to that of an electric motor; see the map in Figure 3. Note that given the motor-applied mechanical torque and speed, the associated efficiency on that operational condition can be found, and consequently the generator efficiency can be calculated and used for calculating power generation efficiency. This is an important modeling method, especially when the electric motor is used both as a motor and a generator.

The second HEV powertrain type is called parallel, one whose system architecture can be found in Figure 1, created by connecting the engine power output to the generator and also supplying the power to the drivetrain. Figure 6 shows a sample parallel HEV architecture [56], with all component models described earlier in this section.

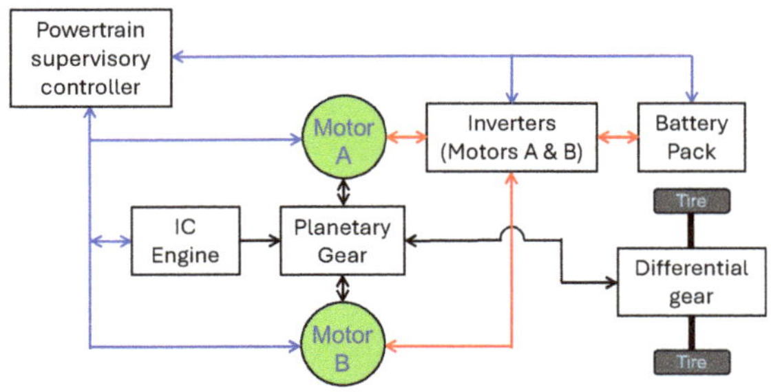

Figure 6. An example of parallel HEV powertrain system.

3. Vehicle Dynamics Model and Energy Consumption Model

Both vehicle dynamics (transient) and vehicle energy consumption (quasi-steady-state) models are widely used in many applications. Vehicle dynamics is a critical technical area in automotive engineering, encompassing the study of how vehicles respond to driver inputs and external forces. Dynamics models are essential for analyzing and predicting vehicle behaviors under various conditions such as vehicle stability and driver ability; while vehicle energy consumption models are designed to predict and analyze the vehicle energy usage under various operating conditions. The vehicle energy consumption models are crucial for optimizing vehicle/powertrain system performance, understanding vehicle fuel efficiency, and developing strategies for reducing energy consumption and emissions. They differ from vehicle dynamics models that focus primarily on the vehicle's transient motion and handling characteristics. Considering there are many different types of models for various design purposes in the literature, this section provides a review of some typical vehicle dynamic and energy consumption models.

3.1. Vehicle Dynamics Model (VDM)

Vehicle dynamics models range from simple linear representations to complex nonlinear systems, each serving different purposes in the research, development, and testing of automotive systems, particularly in developing advanced driver assistance systems (ADASs) and autonomous vehicles (AVs).

The simplest VDM is linear, such as the vehicle bicycle model that simplifies the vehicle dynamics down to a two-wheel bicycle, which is widely used since it represents foundational vehicle dynamics and captures essential dynamics with fewer model parameters. They are particularly useful for initial design and control studies due to their simplicity and analytical tractability. This model assumes the vehicle as a single track with combined front and rear axles, focusing on simplified lateral dynamics and handling characteristics. This is beneficial for understanding basic steering and stability concepts [57], and focusing on longitudinal motions. Nonlinear models provide a more accurate representation of vehicle dynamics by incorporating complex interactions and behaviors that linear models cannot. For example, these models include nonlinear tire–road interaction dynamics [58], nonlinear aerodynamics [57], nonlinear suspension dynamics, etc. These models, such as the two-track model (also called the four-wheel model) and the full-vehicle models, are crucial for understanding vehicle behaviors in extreme operational conditions. The two-track model includes individual wheel dynamics, accounting for longitudinal, lateral, and vertical forces, which is essential for studying detailed vehicle handling, ride comfort, and tire dynamics [59]. The full-vehicle model is more comprehensive, including detailed chassis, suspension, tires, and sometimes even the powertrain dynamics. It captures the complete vehicle behaviors, including pitch, roll, and yaw dynamics, and is used in advanced simulations and control system development [60]. Software packages such as

CarSim [61] and CarMaker [62] can be used to obtain nonlinear models by only providing the required vehicle parameters.

The vehicle handling and stability performances are pivotal in the vehicle model and are directly affected by the tire and suspension performance. Various tire models have been developed to simulate the tire–road interaction accurately [59]. Among those, the magic formula tire model was developed by Pacejka in reference [59]. This empirical model is widely used for its ability to accurately predict tire behavior under various operational conditions using a set of calibration coefficients. Another popular model, the brush tire model [63], provides a more physical understanding of tire forces by considering the deformation of the tire contact patch. The ride comfort and vehicle handling are also strongly related to the suspension. Suspension models range from simple lumped-mass representations to complex multi-body systems. The quarter-car model (QCM) represents one corner of the vehicle, focusing on vertical dynamics to study ride comfort and suspension characteristics [60]. A detailed model that includes all suspension components provides insights into vehicle dynamics during maneuvers over varying road surfaces [64]. When planning a vehicle driving trajectory, especially in ADAS applications, it is vital to predict the vehicle's future position and orientation with high accuracy. Tire models provide mathematical frameworks to predict tire behavior under different operational conditions. By integrating these models into a vehicle dynamics simulation model, ADAS systems can precisely estimate and predict a vehicle's future trajectory.

With the advent of ADASs and autonomous vehicles, advanced dynamics models, integrating vehicle control systems, have become crucial. These models incorporate real-time control algorithms and sensor data for simulations and validations. ADASs and the associated autonomous vehicle (AV) models are combined to simulate the interaction between vehicle dynamics and control systems such as adaptive cruise control, lane-keeping assist, and emergency braking. They are critical for developing and testing automated driving technologies [57].

Vehicle dynamics models in simulation software packages such as CarSim [61], CarMaker [62], and MATLAB play a critical role in the development and validation of advanced driver assistance systems (ADASs) and autonomous driving technologies. These simulators provide highly accurate and detailed representations of vehicle behaviors by integrating complex mathematical models, accounting for various aspects of vehicle dynamics, including tire–road interaction forces and motions, suspension characteristics, aerodynamics, etc., with the road and its surface conditions. CarSim [61] and CarMaker [62], for instance, offer comprehensive driving environments for simulating real-world driving scenarios, allowing engineers to test and refine control algorithms under diverse conditions.

3.2. Vehicle Energy Consumption Model

In contrast to the dynamics models that focus on the transient performance aspects of a vehicle, the energy consumption model is a computational tool used to estimate energy usage during vehicle operation. Energy consumption models take into account various factors such as driving conditions, terrain types, and the energy demands of auxiliary systems like air conditioning and heating [65]. By considering these factors, the energy consumption model can help identify strategies to reduce fuel consumption, extend the range of electric vehicles, and minimize environmental impact. Additionally, energy consumption models are essential in designing more efficient vehicles, planning eco-friendly routes, and developing effective battery management systems [45], etc. They play a crucial role in advancing sustainable transportation by enhancing energy efficiency and supporting regulatory compliance for emissions standards. To clearly illustrate the differences between the vehicle dynamics model and the energy consumption model, we have provided comparisons in Table 3.

Table 3. Comparisons of vehicle dynamics and energy consumption models.

	Dynamics Models	Energy Consumption Models
Main focuses	Vehicle handling and stability. Tire behaviors. Suspension dynamics. Aerodynamics, etc.	Fuel/battery usage. Driving conditions. Powertrain configuration. Auxiliary systems, etc.
Drawbacks	High complexity in model structure. Focus on performance, not time efficient. Require intensive calibration data, etc.	Simplified vehicle dynamics. Limited performance and handling ability. Assumptions on driving conditions, etc.
Applications	Transient performance optimization. Vehicle safety and stability improvement. Simulations of real-world driving, etc.	Route planning and eco-driving strategies. Battery management and optimization. Environmental impact assessments, etc.

In general, the vehicle energy consumption is the time integration of power generated from the power source: fuel consumed by the IC engine for ICE vehicles, battery power consumption by electric motor(s) for EVs, or a combination of fuel consumed by the IC engine and electric power consumption by electric motor(s) for hybrid vehicles. Whereas the energy consumption of these three powertrains is evaluated differently due to their power generation methods. For ICEVs, the fuel consumed due to driving through a unit distance can be measured as energy consumption, thus the engine combustion process needs to be modeled. The battery on the EV can supply the required electrical power instantaneously and maintain constant power output; as a result, the energy consumption is the total power provided by the battery [66]. The energy consumption of an HEV is a combination of fuel and battery power consumption since the hybrid system contains both the IC engine and the electric motor(s) with a battery. In addition, the HEV energy consumption is calculated based on the power-split method.

3.2.1. White-Box (Physics-Based) Vehicle Model

Generally, the vehicle energy consumption model can be categorized into forward and backward models [46]. The forward model is a micro-level model that uses the driver commands as inputs, such as throttle and brake pedals, to follow the speed profile from a driving cycle and calculate the amount of energy consumed by the powertrain through a combination of the physical equations, energy consumption models, and efficiencies of the vehicle system–component interactions [55,67]. A reminder that various modeling methods of vehicle components are introduced in Section 2. To match the speed profile and the driver's demands, acceleration and braking need to be properly controlled so that the required driving or braking torque can be supplied to overcome the driving resistance force or slow down the vehicle, respectively. As a result, the estimated energy consumption should be accurate under any driving behaviors and associated powertrain dynamics. Note that the driving behavior is one of the main factors that affects the vehicle energy consumption [68], thus the energy consumption can be sensitive to the driver or driver model.

The backward model estimates the energy usage based on a speed profile from a driving cycle and the associated resistance forces required for the vehicle to follow the target speed to calculate the powertrain energy consumption. The backward model is a quasi-steady-state model since it focuses on energy consumption caused by outside influences. The main sources of resistance are the wheel rolling resistance force F_r, gravity force by road grade F_g, and air resistance force F_a. Sample equations of the resistance forces are given below:

$$F_r = mgf\cos(\theta) \tag{4}$$

$$F_g = mg\sin(\theta) \tag{5}$$

$$F_a = 0.5C_D\rho Av^2 \tag{6}$$

where m, g, and θ are the vehicle mass, gravity acceleration, and road grade, respectively; and f is the rolling resistance coefficient considering the interaction between tires and road. Note that f has several formats (such as constant value and polynomial functions of vehicle speed with different orders) and is determined according to the required model accuracy. C_D, ρ, and A are the air resistance coefficient, air density, and front area of the vehicle, respectively; and v is the vehicle longitudinal speed. Then, with the vehicle longitudinal acceleration a, the vehicle driving force F_d can be calculated using the dynamics model below, where the energy consumption EC (kJ) is computed through Equation (7) based on the driving force F_d with the transfer ratio of the powertrain η_p.

$$\begin{aligned} ma &= F_d - F_r - F_g - F_a \\ EC &= \int_0^{t_{trip}} \eta_p F_d v(t)dt \end{aligned} \tag{7}$$

To expand the application field of the model above, Chang et al. modeled the energy consumption EC and propulsive power P_p based on the resistance forces represented by a third-order polynomial function and extended the usage to heavy duty vehicles, such as trucks and buses. The model is shown below [69].

$$EC = \int_0^{t_{trip}} P_p dt \tag{8}$$

s.t.

$$\begin{aligned} P_p &= (R + H + m\sin(\theta) + m\dot{v})v \\ R &= b_1 + b_2v + b_3v^2 \end{aligned} \tag{9}$$

where b_1, b_2, and b_3 are the parameters to be trained or calibrated for specific vehicles; H stands for the curvature resistance that compensates for the extra force for the vehicle to round a curve, and its value is almost zero for motor vehicles or a function of weight and curve radius for a train. Therefore, this model has the potential for application in different types of transportation applications [69].

The above are the ideal models developed with fixed parameters for a target vehicle under a given driving environment, which is convenient to use. The simple model structure with a single differential equation allows a relatively large simulation step size, leading to faster simulations suitable for prototyping or early estimation applications. The drawbacks of using this model come from the ignorance of the factors of powertrain component characteristics and the assumption that the torque or power calculated from the driving trajectory can always be met by the powertrain. This is not always true, for example, when the required motor torque for the desired acceleration exceeds the drivetrain capability. Thus, the simple model is not suitable for performance evaluations.

In reality, when the vehicle is driven under quasi-steady-state operation, the vehicle energy consumption could still be influenced by, but not limited to, the following factors that could change during driving: road terrain type, tire conditions (inflation pressure and wear), weather (wind and road surface condition), and uses of air conditioning and other electronics. Note that the terrain type can result in different rolling resistance coefficients. Meanwhile, this coefficient is affected by tire conditions such as tire pressure and wear. The weather can be a main factor in rolling resistance, too. Rain increases the tire slip, and the ambient temperature and driving speed may change the tire pressure [70]. In addition, this coefficient can be a function of vehicle speed that could introduce estimation errors, for example, if the test data used for calibration contains mostly high-speed data. At the same time, the driving speed under longitudinal gusts of wind will affect the air resistance force for the vehicle to overcome [71]. The gusting wind can either pull or push the vehicle under extreme weather. Thus, using the relative longitudinal speed (compensated by front vehicle wind speed) to calculate the air drag force can help improve the model accuracy in windy conditions. When predicting the vehicle cruising range based on energy consumption, the energy usage of electrical devices should be considered, especially for the heating, ventilation, and air conditioning (HVAC) system [72]. Note that people will turn on the air conditioner when the cabin temperature is high and turn on the heater when it is cold inside, so that a function of ambient temperature can represent the energy consumption of the HVAC. A typical energy usage plot for the HVAC system on the Nissan Leaf vehicle is presented in Figure 7 and reference [73]. The accuracy of an energy consumption model can be improved by considering these factors.

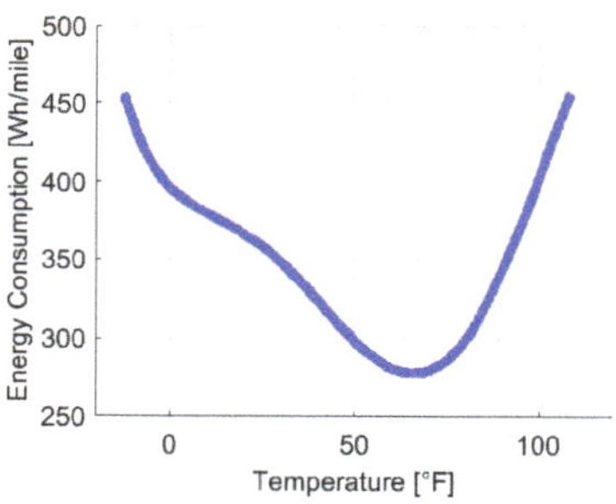

Figure 7. Energy consumption of HVAC for Nissan Leaf.

The forward vehicle model is at the micro-level and describes vehicle dynamics, subsystem dynamics and their interactions, and powertrain energy consumption for given driving behaviors. The forward model usually includes a driver model (usually a PI controller) that translates target speed trajectory into throttle and brake paddle commands, and a powertrain model consisting of an engine and/or electric motor(s) (map-based or mathematical model introduced in Section 2) to generate driving torque according to the throttle position, so that the energy (fuel or electricity) usage can be calculated. Then, the generated torque is transferred through the vehicle components shown in Figure 1 to the driving wheels with these subsystem models addressed in Section 2. A sample forward model diagram in Simulink is shown in Figure 8. The energy consumption model in the Autonomie simulation tool (a state-of-the-art vehicle system simulation tool developed by the U.S. Department of Energy's Argonne National Laboratory [74]) is a typical forward model that uses physics-based mathematical equations, maps of the engine and transmission, and explicit driving cycles [75]. This tool provides vehicle models component by component with associated control strategies, and it supports co-simulations with MATLAB/Simulink. Going through the powertrain model, the dynamics changes in the system are monitored and the transient energy usage can be captured to ensure a high prediction accuracy on energy usage. Meanwhile, using a driver model that provides throttle and brake commands, the power, and torque provided by the engine and/or motor(s) can match realistic powertrain characteristics and efficiency maps. Whereas a complicated micro-level model requires a high computational load. Therefore, implementing this type of model requires balancing model accuracy and computational load based on the application's specific requirements. In addition, forward models are mainly used to evaluate transient behaviors to provide higher estimation accuracy than backward models when estimating energy usage with a given driving profile. Note that forward models are widely used, for instance, to study driver behavior or certain powertrain components to improve energy consumption. However, forward models require a high computational load due to the multiple component equations, such as differential equations and point-wise models. On the other hand, more information is required to calibrate a micro-level forward model compared to a backward model, which is critical for the number of sensors required to be installed on an experimental vehicle for recording the required vehicle data. On the contrary, backward models are easier to establish due to less calibration information being needed and having a faster computational speed than forward models, which is beneficial for developing real-time control strategies and collecting data with reduced model accuracy.

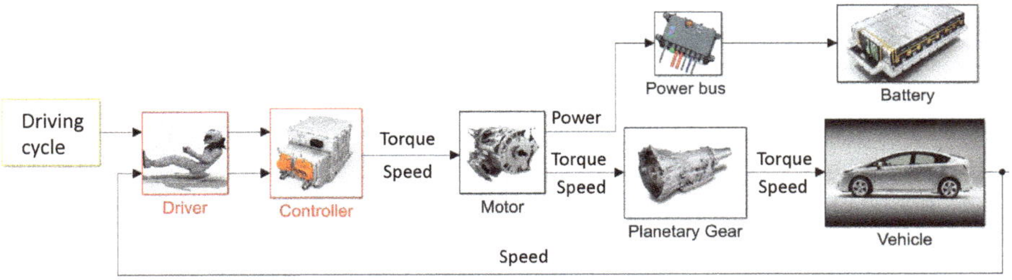

Figure 8. An example forward model diagram for an EV.

The ADVISOR (advanced vehicle simulator) is a vehicle model collecting the advantages of both backward and forward models that can be calculated efficiently and limit the torque or power request based on the powertrain capabilities [76]. Based on the assumption that the vehicle always operates below its powertrain limitations, the ADVISOR model first calculates the required values of these variables using the backward model, such as power and torque [76]. Then, it uses the forward model to constrain the required values from the last step and calculates the available values to replace them. By combining backward and forward models, the ADVISOR model can predict the energy usage for a specific vehicle with an error bound of 1.9%, and the simulation speed is up to 8.0 times faster than a comparable forward-facing vehicle model built in Simulink [76]. The disadvantage of using this model is still the extensive calibration information needed for the vehicle model, without which it will lose prediction accuracy.

3.2.2. Black-Box (Data-Based) Vehicle Model

The above energy consumption models are developed fully based on the knowledge of physics or chemical reaction processes, and can be recognized as white-box models. In addition to white-box models, according to the knowledge level of the vehicle system, energy consumption models can also be categorized as grey-box or black-box models. A black-box model is opposite to a white-box one: it is not based on physics and is purely established on experimental data. A grey-box model, developed based on combined physics and experimental data, is between white-box and black-box models. Black-box vehicle energy consumption models can be developed to combine multiple black-box energy models of vehicle components or for the entire vehicle directly. Black-box models are usually developed based on data regression. Due to the component complexity and multiple factors that affect energy consumption, individual black-box energy models are built for IC engines. Zhou et al. discovered that engine black-box models usually take the requested engine torque, engine speed, engine load, brake torque, and power demand as model inputs and use the fuel mass flow rate as the model output [12].

A polynomial function is often utilized to describe the power demand based on the vehicle speed demand [77] and a first-order equation of power is used to calculate the fuel consumption. Note that a polynomial function with a single input makes it hard to model the transient performance, thus it is more suitable for estimating the fuel consumption at steady state. By adding more factors (i.e., increasing the dimension of the polynomial function), the model can handle certain nonlinearity in fuel consumption [78]. Passenberg et al. adopted a two-dimensional second-order polynomial function to represent the fuel consumption according to the vehicle speed and demanding control inputs in the form as follows [79]:

$$FC = c_0 + c_1 v + c_2 v^2 + c_3 u + c_4 u^2 + c_5 v u \tag{10}$$

where FC denotes the fuel consumption per time step; v is the vehicle speed; u represents the demanded inputs to the engine, which could be reference torque or rotational speed; and c_i ($i = 0, 1, 2..., 5$) are constant parameters to be trained based on test data using machine learning methods. Aside from fuel consumption during driving, in some scenarios, the engine idle operation can also consume a lot of fuel. The VT-CPFM (Virginia Tech comprehensive power-based fuel consumption) model [80,81] is adopted to model the fuel consumption rate based on engine speed and propulsive power demand. Both Equations (11) and (12) are types of VT-CPFM models. Their difference lies in whether the engine speed (for driving and idling) is required or not.

$$FC = \begin{cases} \alpha_0 + \alpha_1 P_p + \alpha_2 P_p^2 & \text{if } P_p \geq 0 \\ \alpha_0 & \text{if } P_p < 0 \end{cases} \tag{11}$$

$$FC = \begin{cases} \alpha_0 \omega + \alpha_1 P_p + \alpha_2 P_p^2 & \text{if } P_p \geq 0 \\ \alpha_0 \omega_i & \text{if } P_p < 0 \end{cases} \tag{12}$$

$$P_p = \frac{F_r + F_g + F_a + (1 + \lambda + \gamma) m a}{3600 \eta_d} v \tag{13}$$

$$\gamma = 0.0025 \epsilon v^2 \tag{14}$$

where α_0, α_1, and α_2 are model parameters to be calibrated based on the propulsive power and force data from both highway and urban driving experiments, and the calibration equations for each model are listed in reference [80]; P_p is the propulsive power calculated based on the vehicle resistance forces, whose parameter λ and expression of γ are introduced by Wang et al. to match the characteristics of heavy duty vehicles, and λ with $\gamma = 0$ is used for light duty vehicles [81]; ω and ω_i denote the engine speeds during driving and idling, respectively. For improving model accuracy, Hung et al. discovered that the engine's instantaneous fuel consumption changes if a certain time threshold T_i is met [82].

Then, the engine fuel rate model was established as an exponential equation as follows [82]:

$$FC = \begin{cases} a_1 e^{-a_2 t_i} & \text{if } t_i < T_i \\ a_3 & \text{if } t_i \geq T_i \end{cases} \tag{15}$$

where a_1, a_2, and a_3 are the parameters to be trained; t_i is the engine idle time; and T_i is a threshold time for engine idling. This model can be added to the engine fuel consumption model for driving. As a result, the model will be accurate time-wise since it considers fuel consumption during waiting, which is important when performing route planning and considering traffic lights and stop signs. Once the engine black-box model is established, the vehicle energy consumption with the ICE powertrain can be calculated by converting vehicle speed to engine speed with added resistance torques. For EVs, due to the linear motor characteristics, they can be modeled linearly with associated efficiency maps and it is not necessary to use a black-box model.

At the vehicle level, the energy consumption model mostly focuses on the resistance force models for ICEVs, HEVs, and EVs. The vehicle specific power (VSP) model [83] was developed to calculate the instantaneous power consumption based on the same factors described by Equation (8). The VSP model structure is shown below:

$$VSP = v(d_1 \dot{v} + g\theta_{pct} + d_2) + d_3 v^3 \tag{16}$$

where d_1, d_2, and d_3 are the parameters to be trained or calibrated; and θ_{pct} is the road grade in percent. These generic parameters are often used by VSP models for typical light duty vehicles and the values for the parameters are $d_1 = 1.1, d_2 = 0.132$, and $d_3 = 0.000302$ [83].

The models introduced above are designed with explicit purposes at the engine or vehicle level. One may need to dive deep into the vehicle system and discover how energy consumption is affected by factors other than vehicle speed. To handle multiple factors, a detailed vehicle model is required. The forward energy consumption model in the Autonomie simulation tool, presented in Section 3.2.1, can meet the above requirements. However, the model development requires a lot of experimental data to calibrate the vehicle dynamics equations, leading to high development costs, and the developed model is only accurate to the specific vehicle. To propagate the built model to different vehicles with different powertrain configurations, vehicle-level controls, etc., the Argonne National Laboratory developed a large-scale simulation process (LSSP) utilizing machine learning algorithms, such as nonlinear regression with a feed-forward neural network model to improve the Autonomie [75]. As a result, this model is able to provide multiple estimations (such as energy consumption, emissions, vehicle mass, etc.) based on inputs like engine type, transmission type, battery peak power, etc. The prediction accuracy of the trained model in Autonomie is very high, with its normalized root mean square error less than 0.8% for model outputs [75]. Similar to the forward white-box model, this trained model still needs intensive calibration data.

Since black-box energy consumption models do not contain physical information, these models are developed purely based on experimental data, with certain data processing methods including system identification and machine learning. The advantage of using a black-box model is its flexible structure, making it simple to calculate the energy consumption and having the potential to propagate the model to different vehicles with various powertrain configurations. The drawbacks of using a black-box model are the need for a large number of experiments of a specific powertrain and an extended time to train the model, which lead to high costs (labor and time) for constructing the black-box model.

3.2.3. Grey-Box (Physical- and Data-Based) Vehicle Model

To reduce the development cost of black-box models and improve modeling accuracy, grey-box models are often utilized to model vehicle energy consumption. A grey-box model, combining system physical information and experimental data, sits between white- and black-box models. The model structure of a grey-box model contains some physical and experimental equation structures and is mainly calibrated through machine learning or parameter identification. Due to utilizing the physical model architecture, the model outputs have similar trajectories to those of the white-box model, being capable of avoiding over-fitting caused by noise or artifacts in data during the training process. Meanwhile, guided by the physics in the model structure during training, training a grey-box model is less time consuming and expensive than a black-box model because the reliance on a large experimental dataset for training a grey-box model is reduced. In addition, a grey-box model has the potential to consider additional factors affecting the model outputs (such as energy consumption) to improve model accuracy, depending on the model details included in the model structure. The comprehensive modal emission model (CMEM) was introduced by Barth et al. [84], where engine fuel consumption is modeled based on the demanded engine power P_e. As a result, engine fuel consumption can be calculated separately under the driving and idling conditions below:

$$FC = \begin{cases} \phi(KNE_d + P_e/\eta)/44 & \text{if } P_e > 0 \\ K_i NE_d & \text{if } P_e = 0 \end{cases} \tag{17}$$

where ϕ is the fuel–air equivalence ratio; K and K_i are the engine friction factors during driving and idling, respectively; N is the engine speed; E_d denotes the engine displacement; P_e is the engine output power; and η is a constant efficiency. Note that 44 kJ/g is the lower heating value of typical gasoline fuel [84]. Additional nonlinear factors (based on the physics) used in the model make it hard to form a grey-box model structure. Meanwhile, adding the factors to the model may need extra sensors or experiments to calibrate the model, which increases the expense of model training. Whereas the number of required experimental data can still be lower than that for a black-box model with the same capabilities due to the grey-box model utilizing physics-based equations.

Another grey-box model, handling the nonlinearity in engine fuel consumption, is the fuel consumption model introduced by Lindgren, which separates the model into transient and quasi-steady-state parts; see below [85].

$$\begin{aligned} FC &= FC_s(\omega, T) \times (1 + R_t(\omega, T, \dot{\omega}, \dot{T})) \\ R_t &= q_\omega(T, \dot{\omega}) + q_T(\omega, \dot{T}) - q_c q_\omega(T, \dot{\omega}) q_T(\omega, \dot{T}) \end{aligned} \tag{18}$$

where ω and T are the engine speed and torque, respectively; FC_s stands for the fuel consumption rate at steady state; and R_t is a dimensionless transient correction function that derives the energy consumption in transience for each step based on the engine status. In Equation (18), for R_t, q_ω is a dimensionless correction factor based on engine speed; q_T is a dimensionless correction factor based on engine torque; and q_c is a coefficient to be calibrated [85]. This model achieves high model accuracy for fuel consumption by including transient vehicle information and using engine torque, speed, and their rates as model inputs.

Including transient features in the energy consumption models helps improve estimation accuracy when the driving cycle is known but proves difficult for real-time applications, such as solving route and speed planning problems, without driving information such as acceleration or braking. A grey-box model was developed specifically for route planning problems and it estimates the steady-state energy usage for driving through a unit distance [86]. This model studies the energy consumption with constant vehicle speed, along with the effects of the environment, such as temperature, road grade, and wind speed with direction, to improve the estimation accuracy for driving at constant speed in various environmental conditions. The model structure is designed in two layers and given below:

$$EC_u(v, T_a, \theta, w) = \beta_{-1}v^{-1} + \beta_0 + \beta_1 v + \beta_2 v^2 + \beta_3(v+w)^3$$
$$\begin{cases} \beta_{-1} = \beta_{-11} + \beta_{-12}T_a + \beta_{-13}T_a^2 + \beta_{-14}T_a^3 + \beta_{-15}T_a^4 + \beta_{-16}T_a^5, \\ \beta_0 = m(\beta_{01} + \beta_{02}T_a + \beta_{03}\sin\theta) + \beta_{04} + \beta_{05}w + \beta_{06}w^2, \\ \beta_1 = \beta_{11} + \beta_{12}w + \beta_{13}w^2, \\ \beta_2 = m(\beta_{21} + \beta_{22}T) + \beta_{23} + \beta_{24}w + \beta_{25}w^2, \\ \beta_3 = \beta_{31} \end{cases} \quad (19)$$

where EC_u is the energy consumption for a unit driving distance d_u; β with subscriptions are parameters to be trained or calibrated; and T_a, θ, and w are the ambient temperature, road grade, and wind speed, respectively. This model is constructed in two layers and is trained layer by layer. The first layer is a polynomial function of vehicle speed and the parameters β_{-1}, β_0, β_1, β_2, and β_3 are those to be trained. Then, the trained parameters in the first-layer model are used as the outputs to train the parameters of the second-layer model in the curly bracket. The rest of the second-layer βs (with dual sub-indexes) are obtained. The model structures of the second layer are established based on a typical power plot P_d for HVAC devices (e.g., AC and heating) and dynamic models of resistance forces in Equations (4)–(6), introduced in Section 3.2.1. The base model of the grey-box energy consumption model is shown in Equation (20) below:

$$\begin{aligned} EC_u(v, T_a, \theta, w) &= [mg(f cos\theta + sin\theta) + 0.5C_D\rho A(v+w)^2 + P_d/v]d_u \\ &= [mg(c_2 - c_2c_3T_a)v^2 + mg(c_1 - c_1c_3T_a + \sin\theta) \\ &\quad + 0.5C_D\rho A(v^2 + w^2 + 2wv) + P_d v^{-1}]d_u \end{aligned} \quad (20)$$

where the rolling resistance is considered as a second-order equation of speed and is affected by temperature [86]. The constants c_1, c_2, and c_3 are the parameters to be trained. Then, these like terms in Equation (20) are combined based on the order of speed v and allocated to the equations in the second layer model of Equation (19). The first layer of the grey-box model (19) is trained by the stochastic Kriging training method, and the parameters of the second-layer model are obtained through the recursive least squares (RLS) method. The grey-box model can estimate the energy consumption with an error less than 2.5% for ideal scenarios and can provide good accuracy when comparing to real driving data; see Figure 9 [86]. Moreover, the applied training methods make it possible to update the model with improved accuracy utilizing newly collected data under real-world driving conditions [86].

In addition to the route planning solution using the grey-box model, the first layer of this model is further developed to find a speed at which the vehicle uses minimal energy by including energy consumption from necessary acceleration and braking due to traffic lights or stop signs [87]. In addition, since the trained equation of β_{-1} in (19) models the energy consumption from HVAC against ambient temperature, it is also used to estimate the HVAC usage while the vehicle is idling [87]. Due to physics being contained in the structures of grey-box models, this model is easier to train based on limited experimental data obtained using the design of experiment (DOE) method, which uses much less data and time than for training a black-box model [7].

In this subsection (Section 3.2), the vehicle energy consumption models are categorized into backward white-box, forward white-box, black-box, and grey-box models. For each category, several representative models are introduced and discussed. To clearly illustrate the differences between these models, Table 4 provides a summary of the models introduced in this subsection according to their types, advantages, and disadvantages. Note that it is difficult to find an energy consumption model that has high model accuracy with a simple model structure and real-time applicability simultaneously. Thus, it is essential to pick the right model with the desired accuracy or calculation speed according to the design requirement.

Table 4. Summary of vehicle energy consumption models.

Model Type	Advantages	Disadvantages
Backward white-box	1. Simple model structure. 2. High model interpretability.	1. Low estimation accuracy. 2. Limited information. 3. Assumption based.
Forward white-box	1. High estimation accuracy. 2. Better transient performance. 3. Components information available. 4. High model interpretability.	1. Complex model structure. 2. Slow calculation speed. 3. More information needed to calibrate.
Black-box	1. Flexible and simple model structure. 2. High estimation accuracy for certain design purposes. 3. Potential for online model update. 4. Low model interpretability.	1. More information needed to calibrate. 2. Time-consuming and expensive to train.
Grey-box	1. Structure simpler than forward white-box model. 2. High estimation accuracy for certain applications. 3. Easier to calibrate than black-box model. 4. Potential for online model update. 5. High model interpretability.	1. Structure more complex than black-box model. 2. Less accurate than forward white-box model. 3. More development cost than forward white-box model.

To evaluate the discussed energy consumption models and explore their potential usage, we conducted a study on the estimation accuracy of energy consumption for the selected energy models. The energy was calculated based on the FTP driving cycle, with a sampling time at 1 s; see speed profile in Figure 10. Considering the forward white-box models have the highest estimation accuracy according to Table 4 and require plenty of information that is not available for this study, as do the black-box models, we picked the backward white-box model (7) with the environmental factors shown in (20) as the baseline model and compared with the VSP model (16) and grey-box model (19), using the following vehicle parameters: $m = 1200$ kg, $f = 0.013$, $C_D = 0.37$, and $A = 0.5$ m^2. The driving condition was set at an environmental temperature of $T = 20$ °C (68 °F) with the HVAC running at low power, road grade 0%, and wind speed $w = 0$ m/s. Another driving scenario was set at $T = 30$ °C (86 °F) with more HVAC power, $g = 0$%, and $w = 0$ m/s. The results are presented in Figure 11 and Table 5. Regenerative braking is not calculated for a fair comparison.

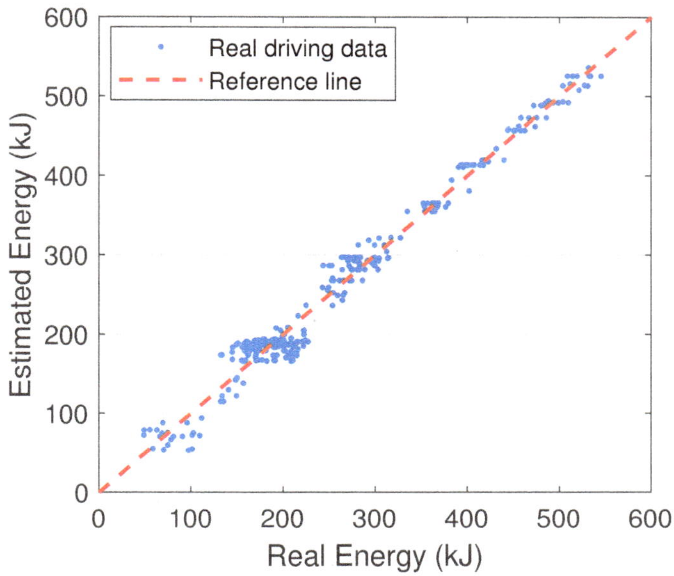

Figure 9. Grey-box model validation.

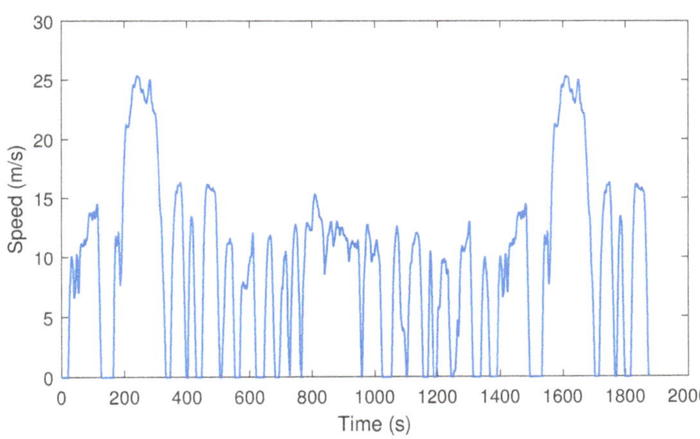

Figure 10. FTP driving cycle.

Table 5. Route selection using simple and grey-box models.

Model	Energy at 20 °C	Difference	Energy at 30 °C	Difference
Base model	6075.8 kJ	0%	9446.7 kJ	0%
Grey-box model	5018.3 kJ	−17.41%	8969.6 kJ	−5.05%
VSP model	5808.1 kJ	−4.41%	5808.1 kJ	−38.52%

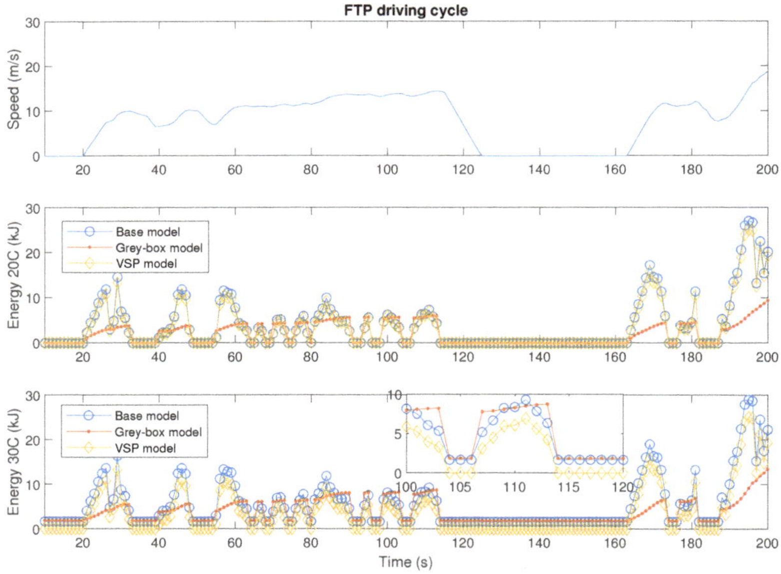

Figure 11. Model validation for grey-box model.

Figure 11 partially shows the estimated energy consumption of the FTP cycle for the three models. The chosen period includes rapid accelerations (20–60 s and 160 to 200 s), light acceleration (60–115 s), and idling (0–20 s and 115–160 s). As shown by the Energy at 20 °C column, the estimation of the VSP model matches the base model well for the whole driving period, with an error of 4.41%. However, the grey-box model mismatches a lot during the rapid accelerations, resulting in a large error when comparing with the base model, which indicates that it is necessary to consider the acceleration in energy consumption. On the other hand, the Energy at 30 °C column shows that the VSP model can catch up with the trend in energy instantly with some gaps, and the grey-box model still cannot follow the rapid acceleration period but matches the base model during light acceleration and idling periods. The total energy usage estimated by the grey-box model is 5.05% off the base model and the VSP model has a very large error.

This study shows that the VSP model is fit for calculating the instant energy consumption by considering acceleration and the grey-box model is good at handling driving scenarios with environmental effects or cruising. This study indicates that the VSP model can be used in micro-scale applications with driving cycle-based energy estimation and the grey-box model is designed for applications at the macro-scale, such as route selection, where the requirement for estimation accuracy is macroscopic (not microscopic). As a result, the selection of the form of energy consumption model should focus on the research purposes.

4. Applications of Vehicle Energy Consumption Models

This section provides some insights into the main applications of vehicle energy consumption models and discusses how different energy consumption models contribute to research/development and improve the outcomes. Since vehicle energy consumption models focus on describing the relationship between vehicle energy (electricity or fuel) usage and various system inputs, they are used to fulfill the demands for developing new control technologies to optimize energy consumption. According to the aspects of driving, these technologies can be categorized into two categories: eco-driving (economy-driving) and eco-routing (economy-routing). To develop eco-driving and eco-routing strategies, the proper energy consumption models are required for different applications, especially for real-time applications. This section reviews some application examples of energy consumption models, such as developing eco-driving and eco-routing strategies, and outlines our rationales for model selection.

4.1. Eco-Driving

Eco-driving focuses on optimizing powertrain performance and driving behaviors to reduce trip energy consumption. According to the characteristics of different powertrains for ICEVs, EVs, and HEVs, and their energy (fuel for ICEVs and HEVs and electricity for EVs and HEVs), consumption models are utilized for the optimization process.

For ICEVs, an engine fuel consumption model, typically in the form of a map-based model due to its system complexity and nonlinearity, is used for the optimization process. For example, to develop a real-time supervisory controller for power distribution in a series hybrid vehicle, an ADVISOR model can be adopted, since a high prediction accuracy of power is required [55]. Within the model, the BSFC map, generator efficiency map, and equivalent circuit models of the battery (see Section 2.1 for details) are utilized due to their simplicity and fast simulation capability. Meanwhile, the BSFC map illustrates the engine conversion efficiency from fuel to mechanical energy at the given engine operational range in terms of its speed and torque, making it possible for engineers to pinpoint the desired engine operational region with the best fuel conversion efficiency (see reference [55] for details). Similarly, fuel efficiency maps (e.g., BSFC map), which are speed and torque-related, are easy to use for identifying the optimal shifting points of transmission that minimize the fuel consumption [88]. Also, the engine speed is considered to be at steady state during the gear shifting, thus no transient response needs to be modeled, and using a BSFC map can produce high accuracy. In this reference, the brake-specific fuel consumption is reduced by 7.5% for certain engine speeds, with a six-speed transmission and a gear-shifting strategy developed based on the BSFC map. Moreover, acknowledging the BSFC map, specifications of transmission, and final drives allows for developing optimized strategies for vehicle driving speed. Saerens et al. developed an optimal strategy for controlling the throttle valve angle based on a simplified fuel consumption model that represents the relation between fuel usage and the throttle valve angle, which was identified on a universal engine dynamometer [89] and reduced fuel usage by up to 30.8%. Engine fuel consumption models have various applications and are not limited to the mentioned examples.

For EVs, the efficiency map of the electric motor, shown in Figure 3, for example, can be utilized to optimize the transmission shift points as well as the battery operational efficiency. Meanwhile, the motor power output has an almost linear relationship with motor torque and speed according to the speed–torque envelope in Figure 3, assuming no power loss in the motor, making it possible to optimize control strategies in real time based on vehicle speed information. In addition, since the motor can also perform as a generator and its torque–speed envelope is almost the same as the motor when assuming no power loss, to recover mechanical energy from vehicle deceleration kinetic energy to the battery, an optimal control algorithm (e.g., Pontryagin's maximum principle) can be used to maximize the kinetic energy recovered by controlling the regenerative braking process. By assuming 100% motor efficiency, the energy consumption model for acceleration is almost linear, except for the air drag force, which makes it possible to apply real-time linear controls. With the developed method in the reference, eco-acceleration can save 6.8%

energy over rapid acceleration, and eco-deceleration can recover at least 50.8% more kinetic energy compared to results without optimization [90].

Considering that the electric motor has high energy consumption during starting up or changing the torque significantly due to fast changes in the motor flux and current vectors, PI (proportional–integral) controllers are designed to utilize variable references of motor speed and flux instead of using a constant reference [91]. The energy consumption of an induction motor is calculated using a "d-q" model based on magnetic flux estimator and mechanical load models. The selection of this model is due to the strict requirement of high model accuracy and transient response. With the explored variable references, the energy consumption for motor operation can be reduced by 20–45%. (see [91] for details).

For HEVs, in addition to the above applications to the engine, electric motor, and battery, the power management system has the potential to be optimized. For a series HEV, based on the battery SOC, engine fuel consumption, and motor equivalent circuit models, power-split control strategies were proposed to control the engine–generator set to minimize the fuel consumption according to the commanded vehicle speed in real time [55]. According to the powertrain type of a parallel HEV with driving power supplied by both engine and motor(s), the torque and speed control of the engine and motor(s) are involved in the supervisory controls for parallel HEVs, capable of reducing fuel consumption by up to 50% under urban driving conditions [92]. In this reference, the torque and speed information of the engine and motor(s) are provided by map-based models for real-time computations. Plug-in HEVs that mostly drive on a battery have a similar power management system to EVs when driving purely relying on the battery, and the power management system migrates to the operation of HEVs with both engine and motor(s) providing power to the vehicle [93].

4.2. Eco-Routing

Eco-routing optimization concentrates on finding an optimal route with minimal energy consumption, considering that the fastest or shortest routes might not be the ones with the greatest energy-savings [94]. Thus, evaluating energy consumption is necessary when searching for the eco-route and the energy consumption estimation accuracy should be high to improve the precision of optimization results. Another challenge in finding the eco-route is that the optimal route may change due to the rapidly shifting traffic conditions, especially during the busy commuting time. Benefiting from developing vehicle-to-everything (V2X) technology, traffic data (such as dynamic map info, vehicle speeds, traffic lights, etc.) are available and can be updated in real time [95]. But still, it is impossible to predict future traffic exactly. This means that a global optimal eco-route is impossible to achieve but can be approached. Therefore, the challenges for eco-routing optimization methods are the capabilities of finding the current optimal route and updating the eco-route online according to the changing traffic conditions. To accomplish this target, a vehicle energy consumption model with a certain complexity can play a vital role in developing eco-routing optimization methods.

The energy consumption model's accuracy will affect the eco-route optimality determined by the eco-routing algorithm. Whereas calculating the energy consumption from a high-fidelity model can be time consuming, which undermines the more important real-time performance of the route planning methods. Thus, there is a tradeoff between the model accuracy and simplicity when developing and selecting the right energy consumption model for solving eco-routing problems. A backward white-box vehicle energy consumption model developed based on the vehicle resistance force in Equation (7) is used to estimate the energy consumption for the controlled vehicle, and the classic global route planning method, the Dijkstra algorithm [96], is adopted to generate the routes [97].

The Dijkstra algorithm is known to run very fast and can be used for real-time applications if the weighting matrix containing information on all available road segments is available. Aljohani et al. adopted deep reinforcement learning with a simple energy consumption model in (7) to make route decisions to reduce the energy consumption by up to 11.82% [98]. Going through the literature, it can be found that the calculation speed, a real-time performance measure, has a higher priority than the model accuracy when selecting models for developing eco-routing strategies.

Even though the model prediction accuracy has lower priority, and backward models are usually selected, researchers still seek ways to improve the model accuracy. Ahn et al. found that even a small portion of the route with a large powertrain torque, caused by the weather, could significantly affect total energy consumption, which indicates the need to consider environmental factors in economy route–speed optimization problems [94]. The accuracy of this simple model is improved by considering the passive energy usage, such as auxiliary or active energy consumed by encountering wind [99]. In addition, the energy consumption is considered to change with the time duration of acceleration and deceleration operations; so, the integration of time for computing energy usage in Equation (7) is separated for the times of acceleration, deceleration, cruise, and stop [99]. The developed model keeps the estimation error below 2.28%. With the integrated two-stage approach proposed in this reference, energy savings of at least 24% can be achieved for electric vehicles by considering the extra energy usage in addition to the resistance forces [99]. To improve accuracy, a complex energy consumption model, including differential equations for acceleration and deceleration, and transmission shifting strategies were utilized by Miao et al. [100]. This model was used to estimate the energy consumption for feasible roads and the Genetic Algorithm is adopted to generate the route, making it almost impossible for real-time optimization. Meanwhile, Miao mentioned that vehicle speed can also affect energy consumption significantly and needs to be optimized together with the route selection. Miao used the dual-loop Genetic Algorithms to optimize the route and speed simultaneously and achieved up to a 15% improvement in reducing energy usage [100]. To optimize the route and speed simultaneously for real-time applications, a grey-box model in (19), considering environmental factors with a given form of polynomial function, is used for real-time gradient descent optimization of all speed values on the route, and the energy usage values for acceleration, deceleration, and vehicle idling are added as correction values that are computed based on target speeds and idling time; see reference [87] for details. In that reference, Hua et al. used the Dijkstra algorithm to generate the optimal route but mentioned that the global optimum is impossible to achieve due to the requirement of future traffic data caused by rapidly changing traffic conditions. Thus, the calculated optimal route based on current traffic information is quasi-optimal and needs to be updated frequently to approach the global optimal solution. Also, it is mentioned that real-time traffic data are more valuable in areas close to the vehicle than in areas far from the vehicle, so optimization focusing on the close area is more efficient. An example of an economy route–speed strategy is given in Figure 12, which couples route and speed optimizations based on the vehicle energy consumption model [9], and the method can achieve energy savings of up to 23.83% [87].

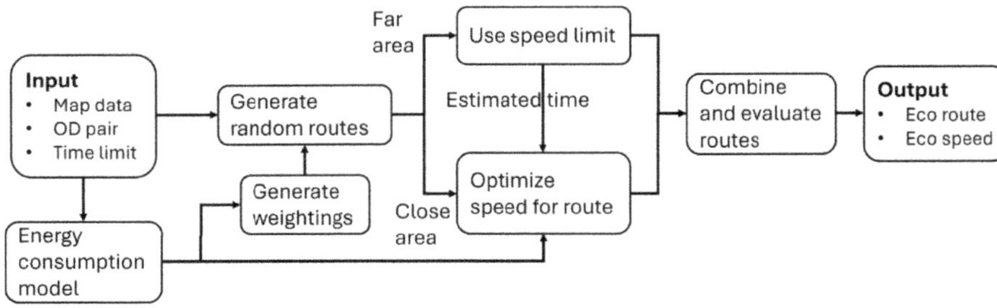

Figure 12. An example of an eco-route and speed optimization strategy [87].

5. Conclusions

This paper provides a review of various vehicle powertrain component models, vehicle energy consumption models, and their applications. Section 2 reviews the powertrain component models used to develop vehicle energy consumption models for three types of vehicles: internal combustion (IC) engine vehicles, electrical vehicles, and hybrid vehicles. The main components discussed are the IC engine, electric motor, and battery.

The energy consumption models discussed in this paper are divided into three main categories: white-box, black-box, and grey-box models, according to their level of interpretability. The white-box models are physics-based and usually built through backward and forward calculations depending on the required level of model accuracy and available knowledge of components. The backward models often concern only the external resistance forces, and the forward models are built on the components models introduced in Section 2 with the resistance forces. Black-box models benefit from their flexible model structure but require extensive data to train the model for acceptable accuracy. Grey-box models have predetermined physics-based structures to reduce the data needed for model training and have better interpretability than black-box models. The model structures, advantages, and disadvantages of these three model types are presented to be used for model selection dependent on the application.

In addition to the discussed model structures, pros, and cons of the energy consumption models above, examples of energy consumption model applications are introduced as an inspiration for potential applications. Among these applications of energy consumption models, vehicle energy consumption models play a vital role in developing eco-driving and eco-routing technologies that minimize vehicle energy consumption. It is found that eco-driving research focuses mainly on transient performance while eco-routing research concerns steady-state or quasi-steady-state performance to balance model accuracy and real-time performance.

Author Contributions: L.H. contributed to the Investigation and Methodology for developing the vehicle model and conducting case studies for the applications. J.T. was responsible for the Investigation and Methodology related to the powertrain model and its corresponding applications. G.Z. supervised the entire manuscript and handled the Conceptualization, Writing, Review and Editing of the entire manuscript. All authors participated in Writing—Original Draft Preparation. All authors have read and agreed to the published version of the manuscript.

Funding: This research received no external funding.

Data Availability Statement: No new data were created or analyzed in this study. This article is a comprehensive survey of existing literature, and all data referenced are available in the cited publications.

Conflicts of Interest: The authors declare no conflicts of interest.

References

1. Zhang, R.; Yao, E. Electric vehicles' energy consumption estimation with real driving condition data. *Transp. Res. Part D Transp. Environ.* **2015**, *41*, 177–187. [CrossRef]
2. Sciarretta, A.; Guzzella, L. *Vehicle Propulsion Systems: Introduction to Modeling and Optimization*; Springer: Berlin/Heidelberg, Germany, 2005.
3. Kleiner, J.; Stuckenberger, M.; Komsiyska, L.; Endisch, C. Advanced monitoring and prediction of the thermal state of intelligent battery cells in electric vehicles by physics-based and data-driven modeling. *Batteries* **2021**, *7*, 31. [CrossRef]
4. Tang, J.; Dai, W.; Archer, C.; Yi, J.; Zhu, G. Cycle-based LQG knock control using identified exhaust temperature model. *Int. J. Engine Res.* **2023**, *24*, 3047–3060. [CrossRef]
5. Isermann, R.; Schaffnit, J.; Sinsel, S. Hardware-in-the-loop simulation for the design and testing of engine-control systems. *Control Eng. Pract.* **1999**, *7*, 643–653. [CrossRef]
6. Tang, J.; Zhu, G.G.; Men, Y. Review of engine control-oriented combustion models. *Int. J. Engine Res.* **2022**, *23*, 347–368. [CrossRef]
7. Tang, J.; Pal, A.; Dai, W.; Archer, C.; Yi, J.; Zhu, G. Stochastic Bayesian optimization for predicting borderline knock. *Int. J. Engine Res.* **2023**, *24*, 793–807. [CrossRef]
8. Pranckevičius, T.; Marcinkevičius, V. Comparison of naive bayes, random forest, decision tree, support vector machines, and logistic regression classifiers for text reviews classification. *Balt. J. Mod. Comput.* **2017**, *5*, 221. [CrossRef]
9. Huo, D.; Meckl, P. Power management of a plug-in hybrid electric vehicle using neural networks with comparison to other approaches. *Energies* **2022**, *15*, 5735. [CrossRef]
10. Afram, A.; Janabi-Sharifi, F.; Fung, A.S.; Raahemifar, K. Artificial neural network (ANN) based model predictive control (MPC) and optimization of HVAC systems: A state of the art review and case study of a residential HVAC system. *Energy Build.* **2017**, *141*, 96–113. [CrossRef]
11. Agha, R.A.A.R.; Sefer, M.N.; Fattah, P. A comprehensive study on sign languages recognition systems using (SVM, KNN, CNN and ANN). In Proceedings of the First International Conference on Data Science, E-Learning and Information Systems, Madrid, Spain, 1–2 October 2018; pp. 1–6.
12. Zhou, M.; Jin, H.; Wang, W. A review of vehicle fuel consumption models to evaluate eco-driving and eco-routing. *Transp. Res. Part D Transp. Environ.* **2016**, *49*, 203–218. [CrossRef]
13. Di Martino, A.; Miraftabzadeh, S.M.; Longo, M. Strategies for the modelisation of electric vehicle energy consumption: A review. *Energies* **2022**, *15*, 8115. [CrossRef]
14. DieselNet. Emission Test Cycles: ECE 15 + EUDC/NEDC. Available online: https://dieselnet.com/standards/cycles/ece_eudc.php (accessed on 12 September 2024).
15. Heywood, J.B. *Internal Combustion Engine Fundamentals*; McGraw-Hill: New York, NY, USA, 1988.
16. Goering, C.; Cho, H. Engine model for mapping BSFC contours. *Math. Comput. Model.* **1988**, *11*, 514–518. [CrossRef]
17. Zalacko, R.; Zöldy, M.; Simongáti, G. Comparative study of two simple marine engine BSFC estimation methods. *Brodogr. Teor. Praksa Brodogr. Pomor. Teh.* **2020**, *71*, 13–25. [CrossRef]
18. Furze, S.; Barraclough, S.; Liu, D.; Melendi-Espina, S. Model based mapping of a novel prototype spark ignition opposed-piston engine. *Energy Convers. Manag.* **2024**, *309*, 118434. [CrossRef]
19. Friso, D. Brake thermal efficiency and BSFC of diesel engines: Mathematical modeling and comparison between diesel oil and biodiesel fueling. *Appl. Math. Sci.* **2014**, *8*, 6515–6528. [CrossRef]
20. Kreyer, J.; Müller, M.; Esch, T. *A Map-Based Model for the Determination of Fuel Consumption for Internal Combustion Engines as a Function of Flight Altitude*; Deutsche Gesellschaft für Luft-und Raumfahrt-Lilienthal-Oberth eV: Bonn, Germany, 2020.
21. Stuhldreher, M. *Fuel Efficiency Mapping of a 2014 6-Cylinder GM EcoTec 4.3 L Engine with Cylinder Deactivation*; Technical Report; SAE Technical Paper; SAE: Warrendale, PA, USA, 2016.
22. Salameh, G.; Chesse, P.; Chalet, D.; Talon, V. *Experimental Study of Automotive Turbocharger Turbine Performance Maps Extrapolation*; Technical Report; SAE Technical Paper; SAE: Warrendale, PA, USA, 2016.
23. Zhang, N.; Wei, L.; Wang, X.; Long, Y. The Effect of Driving Cycle and Shifting Pattern on Vehicle Emissions. In *Proceedings of the 19th Asia Pacific Automotive Engineering Conference & SAE-China Congress 2017: Selected Papers*, Springer: Singapore, 2019; pp. 293–300.
24. U.S. Environmental Protection Agency. EPA Federal Test Procedure (FTP). 2024. Available online: https://www.epa.gov/emission-standards-reference-guide/epa-federal-test-procedure-ftp (accessed on 14 November 2024).
25. DieselNet. Emission Test Cycles: Worldwide Harmonized Light Vehicles Test Cycle (WLTC). 2024. Available online: https://dieselnet.com/standards/cycles/wltp.php (accessed on 12 September 2024).
26. Teixeira, J.A.G.; Sandoval, O.R.; Caetano, B.C.; Baeta, J.G.C. *Turbocharger Performance Prediction: A Review of Map Modelling*; Technical Report; SAE Technical Paper; SAE: Warrendale, PA, USA, 2020.
27. Watson, N.; Janota, M.S. *Turbocharging the Internal Combustion Engine*; Red Globe Press: London, UK, 1982.

28. Lechner, G.; Naunheimer, H. *Automotive Transmissions: Fundamentals, Selection, Design and Application*; Springer Science & Business Media: Berlin/Heidelberg, Germany, 1999.
29. Al-Ajmi, E.N.R.M.R.; Alajmi, H.A. Modelling and Simulation of Automatic Transmission. *Int. J. Innov. Sci. Res. Technol.* **2021**, *6*, 1671–1681.
30. Wang, B.; Lu, Z.; Chen, X.; Wang, P.; Luo, W.; Xu, H. Modeling and analysis of the dynamic efficiency of manual transmission/reducer. *Int. J. Automot. Technol.* **2015**, *16*, 417–426. [CrossRef]
31. Kim, B.; Choi, S.B.; Kim, Y. Observer-Based Control for a Torque Converter Lock-Up Clutch Utilizing a Steady-State Torque Map. *IEEE Trans. Control Syst. Technol.* **2022**, *31*, 1208–1219. [CrossRef]
32. Eastwick, C.N.; Johnson, G. Gear Windage: A Review. *J. Mech. Des.* **2008**, *130*, 034001. [CrossRef]
33. Zhang, N. Experimental Investigation into the Transient Response of an Automotive Powertrain with an Automatic Transmission during Gear Shifts and Torque Reversals. In Proceedings of the FISITA World Automotive Congress, Yokohama, Japan, 22–27 October 2006; Society of Automotive Engineers, of Japan Inc.: Chiyoda-ku, Japan, 2006.
34. Ryu, W.S.; Kim, H.S. Ratio Control of CVT by Considering the CVT System Efficiency. *Trans. Korean Soc. Automot. Eng.* **2006**, *14*, 157–163.
35. Bhatt, P.; Mehar, H.; Sahajwani, M. Electrical motors for electric vehicle—A comparative study. In Proceedings of the Recent Advances in Interdisciplinary Trends in Engineering & Applications (RAITEA), Indore, Madhya Pradesh, 14–16 February, 2019 .
36. Cao, Z.; Mahmoudi, A.; Kahourzade, S.; Soong, W.L. An overview of electric motors for electric vehicles. In Proceedings of the 2021 31st Australasian Universities Power Engineering Conference (AUPEC), Perth, Australia, 26–30 September 2021; pp. 1–6.
37. Ahssan, M.R.; Ektesabi, M.M.; Gorji, S.A. Electric vehicle with multi-speed transmission: A review on performances and complexities. *SAE Int. J. Altern. Powertrains* **2018**, *7*, 169–182. [CrossRef]
38. Ren, Q.; Crolla, D.; Morris, A. Effect of transmission design on electric vehicle (EV) performance. In Proceedings of the 2009 IEEE Vehicle Power and Propulsion Conference, Dearborn, MI, USA, 7–10 September 2009; pp. 1260–1265.
39. Beser, E.K. Electrical equivalent circuit for modelling permanent magnet synchronous motors. *J. Electr. Eng.* **2021**, *72*, 176–183. [CrossRef]
40. Yildiz, A.B. Electrical equivalent circuit based modeling and analysis of direct current motors. *Int. J. Electr. Power Energy Syst.* **2012**, *43*, 1043–1047. [CrossRef]
41. Du, G.; Cao, W.; Hu, S.; Lin, Z.; Yuan, T. Design and assessment of an electric vehicle powertrain model based on real-world driving and charging cycles. *IEEE Trans. Veh. Technol.* **2018**, *68*, 1178–1187. [CrossRef]
42. Park, G.; Lee, S.; Jin, S.; Kwak, S. Integrated modeling and analysis of dynamics for electric vehicle powertrains. *Expert Syst. Appl.* **2014**, *41*, 2595–2607. [CrossRef]
43. Sheela, A.; Suresh, M.; Shankar, V.G.; Panchal, H.; Priya, V.; Atshaya, M.; Sadasivuni, K.K.; Dharaskar, S. FEA based analysis and design of PMSM for electric vehicle applications using magnet software. *Int. J. Ambient Energy* **2022**, *43*, 2742–2747. [CrossRef]
44. Jape, S.R.; Thosar, A. Comparison of electric motors for electric vehicle application. *Int. J. Res. Eng. Technol.* **2017**, *6*, 12–17. [CrossRef]
45. Larminie, J.; Lowry, J. *Electric Vehicle Technology Explained*; John Wiley & Sons: Hoboken, NJ, USA, 2012.
46. Miri, I.; Fotouhi, A.; Ewin, N. Electric vehicle energy consumption modelling and estimation—A case study. *Int. J. Energy Res.* **2021**, *45*, 501–520. [CrossRef]
47. Huria, T.; Ceraolo, M.; Gazzarri, J.; Jackey, R. High fidelity electrical model with thermal dependence for characterization and simulation of high power lithium battery cells. In Proceedings of the 2012 IEEE International Electric Vehicle Conference, Greenville, SC, USA, 4–8 March 2012; pp. 1–8.
48. Xu, B.; Oudalov, A.; Ulbig, A.; Andersson, G.; Kirschen, D.S. Modeling of lithium-ion battery degradation for cell life assessment. *IEEE Trans. Smart Grid* **2016**, *9*, 1131–1140. [CrossRef]
49. Ramadass, P.; Haran, B.; Gomadam, P.M.; White, R.; Popov, B.N. Development of first principles capacity fade model for Li-ion cells. *J. Electrochem. Soc.* **2004**, *151*, A196. [CrossRef]
50. Nanda, J.; Remillard, J.; O'Neill, A.; Bernardi, D.; Ro, T.; Nietering, K.E.; Go, J.Y.; Miller, T.J. Local state-of-charge mapping of lithium-ion battery electrodes. *Adv. Funct. Mater.* **2011**, *21*, 3282–3290. [CrossRef]
51. Farhad, S.; Nazari, A. Introducing the energy efficiency map of lithium-ion batteries. *Int. J. Energy Res.* **2019**, *43*, 931–944. [CrossRef]
52. Huang, C.; Wang, Z.; Zhao, Z.; Wang, L.; Lai, C.S.; Wang, D. Robustness evaluation of extended and unscented Kalman filter for battery state of charge estimation. *IEEE Access* **2018**, *6*, 27617–27628. [CrossRef]
53. Zhou, Z.; Howey, D. Learning operando impedance function for battery health with aging-aware equivalent circuit model. *arXiv* **2024**, arXiv:2407.06639.
54. Pisu, P.; Rizzoni, G. A comparative study of supervisory control strategies for hybrid electric vehicle. *IEEE Trans. Control Syst. Technol.* **2007**, *15*, 506–518. [CrossRef]

55. Gao, J.; Zhu, G.; Strangas, E.; Sun, F. Equivalent fuel consumption optimal control of a series hybrid electric vehicle. *Proc. Inst. Mech. Eng. Part D J. Automob. Eng.* **2009**, *223*, 1003–1018. [CrossRef]
56. Vu, T.M.; Moezzi, R.; Cyrus, J.; Hlava, J.; Petru, M. Parallel hybrid electric vehicle modelling and model predictive control. *Appl. Sci.* **2021**, *11*, 10668. [CrossRef]
57. Rajamani, R. *Vehicle Dynamics and Control*; Springer Science & Business Media: Berlin/Heidelberg, Germany, 2011.
58. Pauwelussen, J.P. *Essentials of Vehicle Dynamics*; Chapter 2; Butterworth-Heinemann: Oxford, UK, 2014.
59. Pacejka, H. *Tire and Vehicle Dynamics*; Elsevier: Amsterdam, The Netherlands, 2005.
60. Gillespie, T. *Fundamentals of Vehicle Dynamics*; SAE International: Warrendale, PA, USA, 2021.
61. Mechanical Simulation. VehicleSim. 2024. Available online: https://www.carsim.com (accessed on 18 September 2024).
62. IPG Automotive. CarMaker. 2024. Available online: https://www.ipg-automotive.com/en/products-solutions/software/carmaker/ (accessed on 18 September 2024).
63. Balkwill, J. *Performance Vehicle Dynamics: Engineering and Applications*; Elsevier: Amsterdam, The Netherlands, 2018. [CrossRef]
64. Milliken, W.F.; Milliken, D.L.; Metz, L.D. *Race Car Vehicle Dynamics*; SAE International Warrendale: Warrendale, PA, USA, 1995; Volume 400.
65. Sweeting, W.J.; Hutchinson, A.R.; Savage, S.D. Factors affecting electric vehicle energy consumption. *Int. J. Sustain. Eng.* **2011**, *4*, 192–201. [CrossRef]
66. Abousleiman, R.; Rawashdeh, O. Energy consumption model of an electric vehicle. In Proceedings of the 2015 IEEE Transportation Electrification Conference and Expo (ITEC), Dearborn, MI, USA, 14–17 June 2015; pp. 1–5.
67. Powell, B.; Bailey, K.; Cikanek, S. Dynamic modeling and control of hybrid electric vehicle powertrain systems. *IEEE Control Syst. Mag.* **1998**, *18*, 17–33.
68. Nan, S.; Tu, R.; Li, T.; Sun, J.; Chen, H. From driving behavior to energy consumption: A novel method to predict the energy consumption of electric bus. *Energy* **2022**, *261*, 125188. [CrossRef]
69. Chang, D.J.; Morlok, E.K. Vehicle speed profiles to minimize work and fuel consumption. *J. Transp. Eng.* **2005**, *131*, 173–182. [CrossRef]
70. Ejsmont, J.; Taryma, S.; Ronowski, G.; Swieczko-Zurek, B. Influence of temperature on the tyre rolling resistance. *Int. J. Automot. Technol.* **2018**, *19*, 45–54. [CrossRef]
71. Xie, Q.; Claudio Filho, H.; Feng, G.; Clandfield, W.; Kar, N.C. Advanced vehicle dynamic model for EV emulation considering environment conditions. In Proceedings of the 2017 IEEE 30th Canadian Conference on Electrical and Computer Engineering (CCECE), Windsor, ON, Canada, 30 April–3 May 2017; pp. 1–4.
72. Kambly, K.R.; Bradley, T.H. Estimating the HVAC energy consumption of plug-in electric vehicles. *J. Power Sources* **2014**, *259*, 117–124. [CrossRef]
73. Yuksel, T.; Michalek, J.J. Effects of regional temperature on electric vehicle efficiency, range, and emissions in the United States. *Environ. Sci. Technol.* **2015**, *49*, 3974–3980. [CrossRef] [PubMed]
74. Argonne National Laboratory. Autonomie. 2021. Available online: https://vms.taps.anl.gov/tools/autonomie/ (accessed on 14 November 2024).
75. Yao, J.; Moawad, A. Vehicle energy consumption estimation using large scale simulations and machine learning methods. *Transp. Res. Part C Emerg. Technol.* **2019**, *101*, 276–296. [CrossRef]
76. Wipke, K.B.; Cuddy, M.R.; Burch, S.D. ADVISOR 2.1: A user-friendly advanced powertrain simulation using a combined backward/forward approach. *IEEE Trans. Veh. Technol.* **1999**, *48*, 1751–1761. [CrossRef]
77. Piccoli, B.; Han, K.; Friesz, T.L.; Yao, T. Estimating fuel consumption and emissions via traffic data from mobile sensors. In Proceedings of the 2013 51st Annual Allerton Conference on Communication, Control, and Computing (Allerton), Monticello, IL, USA, 2–4 October 2013; pp. 472–477.
78. Saerens, B.; Rakha, H.; Ahn, K.; Van Den Bulck, E. Assessment of alternative polynomial fuel consumption models for use in intelligent transportation systems applications. *J. Intell. Transp. Syst.* **2013**, *17*, 294–303. [CrossRef]
79. Passenberg, B.; Kock, P.; Stursberg, O. Combined time and fuel optimal driving of trucks based on a hybrid model. In Proceedings of the 2009 European Control Conference (ECC), Budapest, Hungary, 23–26 August 2009; pp. 4955–4960.
80. Park, S.; Rakha, H.; Ahn, K.; Moran, K. Virginia tech comprehensive power-based fuel consumption model (VT-CPFM): Model validation and calibration considerations. *Int. J. Transp. Sci. Technol.* **2013**, *2*, 317–336. [CrossRef]
81. Wang, J.; Rakha, H.A. Fuel consumption model for conventional diesel buses. *Appl. Energy* **2016**, *170*, 394–402. [CrossRef]
82. Hung, W.T.; Tong, H.Y.; Cheung, C.S. A modal approach to vehicular emissions and fuel consumption model development. *J. Air Waste Manag. Assoc.* **2005**, *55*, 1431–1440. [CrossRef]
83. Frey, H.C.; Rouphail, N.M.; Zhai, H. Speed-and facility-specific emission estimates for on-road light-duty vehicles on the basis of real-world speed profiles. *Transp. Res. Rec.* **2006**, *1987*, 128–137. [CrossRef]
84. Barth, M.; An, F.; Younglove, T.; Scora, G.; Levine, C.; Ross, M.; Wenzel, T. The development of a comprehensive modal emissions model. *NCHRP Web-Only Doc.* **2000**, *122*, 11–25.

85. Lindgren, M. A transient fuel consumption model for non-road mobile machinery. *Biosyst. Eng.* **2005**, *91*, 139–147. [CrossRef]
86. Hua, L.; Tang, J.; Dourra, H.; Zhu, G. A Gray-Box Surrogate Vehicle Energy Consumption Model Capable of Real-Time Updating. *IEEE/ASME Trans. Mechatron.* **2023**, *28*, 2092–2100. [CrossRef]
87. Hua, L.; Dourra, H.; Zhu, G.G. Route and Speed Co-optimization for Improving Energy Consumption of Connected Vehicles. *IEEE/ASME Trans. Mechatron.* **2024**, *29*, 2830–2838. [CrossRef]
88. Oglieve, C.J.; Mohammadpour, M.; Rahnejat, H. Optimisation of the vehicle transmission and the gear-shifting strategy for the minimum fuel consumption and the minimum nitrogen oxide emissions. *Proc. Inst. Mech. Eng. Part D J. Automob. Eng.* **2017**, *231*, 883–899. [CrossRef]
89. Saerens, B.; Vandersteen, J.; Persoons, T.; Swevers, J.; Diehl, M.; Van den Bulck, E. Minimization of the fuel consumption of a gasoline engine using dynamic optimization. *Appl. Energy* **2009**, *86*, 1582–1588. [CrossRef]
90. Hua, L.; Wei, W.; Dourra, H.; Zhu, G.G. Minimal energy transient motion control of electrical connected vehicles. *IEEE/ASME Trans. Mechatron.* **2022**, *27*, 2116–2124. [CrossRef]
91. Resa, J.; Cortes, D.; Marquez-Rubio, J.F.; Navarro, D. Reduction of induction motor energy consumption via variable velocity and flux references. *Electronics* **2019**, *8*, 740. [CrossRef]
92. Sciarretta, A.; Back, M.; Guzzella, L. Optimal control of parallel hybrid electric vehicles. *IEEE Trans. Control Syst. Technol.* **2004**, *12*, 352–363. [CrossRef]
93. Emadi, A.; Lee, Y.J.; Rajashekara, K. Power electronics and motor drives in electric, hybrid electric, and plug-in hybrid electric vehicles. *IEEE Trans. Ind. Electron.* **2008**, *55*, 2237–2245. [CrossRef]
94. Ahn, K.; Rakha, H. Field evaluation of energy and environmental impacts of driver route choice decisions. In Proceedings of the 2007 IEEE Intelligent Transportation Systems Conference, Bellevue, WA, USA, 30 September–3 October 2007; pp. 730–735.
95. Qi, X.; Wang, P.; Wu, G.; Boriboonsomsin, K.; Barth, M.J. Energy and mobility benefits from connected ecodriving for electric vehicles. In Proceedings of the 2017 IEEE 20th International Conference on Intelligent Transportation Systems (ITSC), Yokohama, Japan, 16–19 October 2017; pp. 1–6.
96. Javaid, A. Understanding Dijkstra's Algorithm. 2013. Available online: https://ssrn.com/abstract=2340905 (accessed on 12 September 2024).
97. Shao, S.; Guan, W.; Bi, J. Electric vehicle-routing problem with charging demands and energy consumption. *IET Intell. Transp. Syst.* **2018**, *12*, 202–212. [CrossRef]
98. Aljohani, T.M.; Ebrahim, A.; Mohammed, O. Real-Time metadata-driven routing optimization for electric vehicle energy consumption minimization using deep reinforcement learning and Markov chain model. *Electr. Power Syst. Res.* **2021**, *192*, 106962. [CrossRef]
99. Basso, R.; Kulcsár, B.; Egardt, B.; Lindroth, P.; Sanchez-Diaz, I. Energy consumption estimation integrated into the electric vehicle routing problem. *Transp. Res. Part D Transp. Environ.* **2019**, *69*, 141–167. [CrossRef]
100. Miao, C.; Liu, H.; Zhu, G.G.; Chen, H. Connectivity-based optimization of vehicle route and speed for improved fuel economy. *Transp. Res. Part C Emerg. Technol.* **2018**, *91*, 353–368. [CrossRef]

MDPI AG
Grosspeteranlage 5
4052 Basel
Switzerland
Tel.: +41 61 683 77 34

Actuators Editorial Office
E-mail: actuators@mdpi.com
www.mdpi.com/journal/actuators

www.ingramcontent.com/pod-product-compliance
Lightning Source LLC
Chambersburg PA
CBHW040829250926
51850CB00007BA/1294
* 9 7 8 3 7 2 5 8 5 0 6 7 9 *